PLATE 1 (*Frontispiece*)

Geology of Western Shetland (*Mem. Geol. Surv.*)

NATURAL ENVIRONMENT RESEARCH COUNCIL
INSTITUTE OF GEOLOGICAL SCIENCES

MEMOIRS OF THE GEOLOGICAL SURVEY OF GREAT BRITAIN
SCOTLAND

The Geology
of
Western Shetland

*(Explanation of One-inch Geological Sheet Western Shetland;
comprising Sheet 127 and parts of 125, 126 and 128)*

BY

W. MYKURA, B.SC., F.R.S.E. AND
J. PHEMISTER, D.SC., F.R.S.E.

with a contribution by
P. A. SABINE, D.SC., F.R.S.E.

EDINBURGH
HER MAJESTY'S STATIONERY OFFICE

The Institute of Geological Sciences
was formed by the
incorporation of the Geological Survey of Great Britain
and the Museum of Practical Geology
with Overseas Geological Surveys
and is a constituent body of the
Natural Environment Research Council

ISBN 0 11 880160 0

PREFACE

THE WESTERN SHETLAND map comprises Sheet 127 of the one-inch-to-one-mile Geological Survey series together with parts of sheets 125, 126 and 128. This means that it has been possible to include the whole of the Walls Peninsula of Shetland and its adjacent islands. The descriptions in the present Memoir, however, exclude that part of the map which lies to the east of the Walls Boundary Fault and which is more properly treated in connexion with the Central Shetland (128) Sheet.

The original survey of the Western Shetland area was carried out between 1931 and 1934 by the late G. V. Wilson, Dr. S. Buchan, the late D. Haldane, Mr. J. Knox and Dr. T. Robertson. While the area was completely covered by field maps on the scale of six-inches-to-one-mile, no one-inch map of the district was published at that time because of the intervention of urgent economic work elsewhere. Following a decision to rectify this lack after a lapse of many years, it was desirable that the original survey should be updated and the re-examination was entrusted to Mr. W. Mykura in 1961. Although his revision was based on the maps of the original surveyors the application of modern techniques of structural mapping and sedimentology necessitated a complete coverage of the Walls area on what was virtually a resurvey. This has resulted in a very considerable re-interpretation of the strata and structures which has an important bearing not only on Shetland geology but on the interpretation of the Scottish Caledonides as a whole.

The memoir was written by Mykura with the exception of Chapters 3, 13 and a large part of 16. These were written by Dr. J. Phemister based on the original survey by Haldane supplemented by his own field studies, particularly in Muckle Roe. Chapter 2 is a summary by Mykura of the relevant part of a paper by Phemister which is in press at the time of writing. The mapping of the ground east of the Walls Boundary Fault is based on work by J. Phemister, Dr. F. May and Dr. P. A. Sabine.

The identification of the Old Red Sandstone fish remains was carried out by Dr. R. S. Miles and the fossil plants were named by Professor W. G. Chaloner. Mr. P. J. Brand collected many of the fossils and compiled the faunal lists. The geological photographs of the area were taken by the late W. Fisher.

The new chemical rock analyses quoted in this memoir are by Messrs. W. H. Evans, J. M. Nunan and C. Park (Laboratory of the Government Chemist) and the isotopic age determinations by Dr. N. J. Snelling. Mr. Mykura was assisted in his petrographical work by advice from Mr. R. W. Elliot and F. May.

The memoir was edited by Mr. G. S. Johnstone with the assistance of Dr. J. D. Peacock.

KINGSLEY DUNHAM
Director

Institute of Geological Sciences
Exhibition Road
London
SW7 2DE

iii

CONTENTS

ILLUSTRATIONS

TEXT-FIGURES

PLATES

TABLES

LIST OF SIX-INCH MAPS

Geological six-inch maps included wholly or in part in the Western Shetland Sheet of the one-inch geological map of Scotland are listed below together with the initials of the original surveyors and the dates of the survey. The surveyors were: S. Buchan, D. Haldane, J. Knox, T. Robertson and G. V. Wilson. Sheets Shetland 34, 35, 36 SW, 40, 41, 42 (western sheets), 46, 47 (western sheets), 50 and 51 (western sheets) were revised between 1961 and 1965 by W. Mykura. Parts of sheets Shetland 36 SE and 42 (eastern sheets) were revised by P. A. Sabine, and parts of 47 (eastern sheets) and 51 NE by F. May. J Phemister has collected much additional information in the areas of sheets Shetland 24, 29 and 36.

The maps are available for public reference at the Institute of Geological Sciences, West Mains Road, Edinburgh.

Shetland							
24 SW	D.H.	1933		42 NW	T.R.	1934
SE	D.H.	1933		NE	D.H.	1931
29 NW	D.H.	1933			T.R.	1931–4
NE	D.H.	1932–3		SW	G.V.W.	1933
SW	D.H.	1931			T.R.	1934
SE	D.H.	1931–2		SE	T.R.	1931–4
34 NW	D.H.	1933			G.V.W.	1933
	S.B.			46 NW	S.B.	1933
SE	T.R.	1934		NE	G.V.W.	1933
35 SW	S.B.	1934		SW	S.B.	1933–4
36 NW	D.H.	1931		SE	G.V.W.	1933
NE	D.H.	1931			S.B.	1934
SW	T.R.	1934		47 NW	G.V.W.	1932–3
SE	D.H.	1931–3		NE	G.V.W.	1931–2
	T.R.	1934		SW	G.V.W.	1931–3
40 NE	T.R.	1934		SE	G.V.W.	1931–3
41 NW	J.K.	1933		50 NE	G.V.W.	1931–3
NE	J.K.	1933		51 NW	G.V.W.	1931–2
	S.B.	1934		NE	G.V.W.	1931
SW	J.K.	1933		SW	G.V.W.	1931
SE	J.K.	1933		SE	G.V.W.	1931
	S.B.	1934				

Chapter 1
INTRODUCTION

LOCATION AND AREA

THE REGION described in the following account comprises most of the area shown on the one-inch-to-one-mile Geological Survey map of Western Shetland. It includes the Walls Peninsula and the south-western part of Northmaven, which are part of the main island of Shetland (Mainland), as well as the adjoining islands, the largest of which are Muckle Roe, Vementry, Papa Little, Papa Stour and Vaila (Fig. 1). A description of the island of Foula, which is 14 miles (22·5 km) SW of the Walls Peninsula, is also included. The total land area involved is approximately 115 square miles (298 km²). In addition a brief account is given of the submarine topography of the sea area adjoining Western Shetland.

The narrow strip of metamorphic and igneous rocks east of the Walls Boundary Fault (Fig. 2) is not described in this memoir, as it will be dealt with in the account of the Geology of Central Shetland.

PHYSICAL FEATURES

The area forms a geological and morphological entity, as it is separated from the eastern part of Shetland Mainland by the north–south trending Walls Boundary Fault. This major dislocation, which is considered by Flinn (1961, p. 589; 1969, pp. 290–1) to be the northern continuation of the Great Glen Fault, separates two areas of strongly contrasting geology and topography. The ground to the east is formed of metasediments, migmatites and partially re-crystallized 'granites', all of which have a consistent north–south trend at this latitude and give rise to smooth, rounded north–south trending ridges and partially drowned valleys. The country west of the fault has a much more complex structural pattern and is composed of a variety of rock types, each of which has imparted its distinctive character to the landscape.

The northern coastal belt of the Walls Peninsula and Vementry Island, which are formed of a varied suite of metamorphic rocks, have a rugged topography with many rock-knolls and escarpments, reminiscent of the Lewisian Gneiss topography of the north-west Highlands (Plate VA). South of this belt the Walls Peninsula is to a large extent made up of the Walls Sandstone, an Old Red Sandstone formation composed predominantly of sandstone with subordinate volcanic rocks. Along the northern part of its outcrop the Walls Sandstone has a consistently steep south-south-easterly dip, is composed of massive sandstones with conglomerates and intercalated lavas, and gives rise to a rugged terrain with prominent west-south-west trending ridges and escarpments and a high proportion of exposed rocks. Farther south, particularly in the eastern and central parts of the peninsula, the Walls Sandstone has a much more complex structure and contains fewer massive sandstones and no conglomerates or lavas. In consequence the topography has no regular directional pattern and consists of scattered areas with abundant rocky outcrops separated by large areas of flat or undulating peat bog. In the west of the peninsula the Walls Sandstone forms

1

Sandness Hill (817 ft [249 m] OD) and Stourbrough Hill (567 ft [173 m] OD)
the highest hills in the area. In the south-west of the peninsula and in Vaila the
more regularly folded Walls Sandstone again produces more abundant rocky
outcrops and some rudimentary scarp and dip topography.

In contrast to the irregular landscape produced by the Walls Sandstone, the
softer and more gently inclined Melby Sandstone, which crops out at Sandness
in the north-west corner of the peninsula, forms low, fertile ground reminiscent
of the landscape of Orkney. The Melby Sandstone is overlain by a series of
basalt and rhyolite lavas which form almost all of the island of Papa Stour.
Because of their near-horizontal disposition the Papa Stour volcanic rocks form
a featureless plateau bounded by magnificent sea-cliffs.

The southern part of the Walls Peninsula is formed of igneous rocks of the
Sandsting Complex, which gives rise to an undulating topography with a number
of prominent rounded hills, and relatively few inland exposures, but with impressive coastal cliffs.

The south-western part of Northmaven is formed principally of diorite with
numerous gabbroic bodies and many granite veins. The diorite is bounded on
the east by a strip of granite, $\frac{1}{2}$ mile (0·8 km) wide, which in turn adjoins a
series of north–south trending metamorphic rocks. Both the diorite and granite
form rugged rocky outcrops, while the metamorphic rocks to the east have
produced a smoother terrain with north–south trending ridges. The third major
igneous body of Western Shetland is the granophyre, which forms the greater
part of Muckle Roe and gives rise to extremely rugged, rocky terrain with
extensive screes that rises to a height of 557 ft (169 m). The granophyre also
forms the most continuous sheer sea-cliffs in Western Shetland.

The drainage of the Walls Peninsula has a radial aspect, with nearly all
streams flowing directly to the sea from a central east–west trending divide. In
the northern part of the peninsula there are a number of drowned valleys, such
as those now occupied by Brindister Voe [290 560][1] and Aith Voe [350 580],
whose orientation is determined by north to north-north-east trending faults
and in the west the Deep Dale [183 548] and the valley of the Burn of Dale are
excavated along bands of relatively soft Old Red Sandstone sediments. Apart
from these examples, structural control of drainage is found on Muckle Roe,
where a number of streams follow the lines of north-north-west trending crush
belts.

The effects of the ice sheets, which overrode the area in the Pleistocene
Period, and the subsequent rise of sea level around Shetland have been major
factors in the shaping of the present topography. The north-west and westward
movement of ice over the greater part of the area has produced a strongly ice-
moulded terrain, particularly in the more rocky ground, and has been responsible for the numerous scoured-out hollows which are now occupied by inland
lakes. Some of the major straits such as the Swarbacks Minn between Vementry
and Muckle Roe may be partially due to glacial overdeepening. The virtually
continuous rise of sea level since the last glacial maximum has produced the
'drowned landscape' topography of Shetland. The most characteristic features
of this are the numerous long, shallow, at times twisting, sea lochs which are
locally termed 'voes'. The continuous subsidence of the land coupled with the

---

[1]National Grid references are given in this form throughout the Memoir. All lie within the
100-km grid H U.

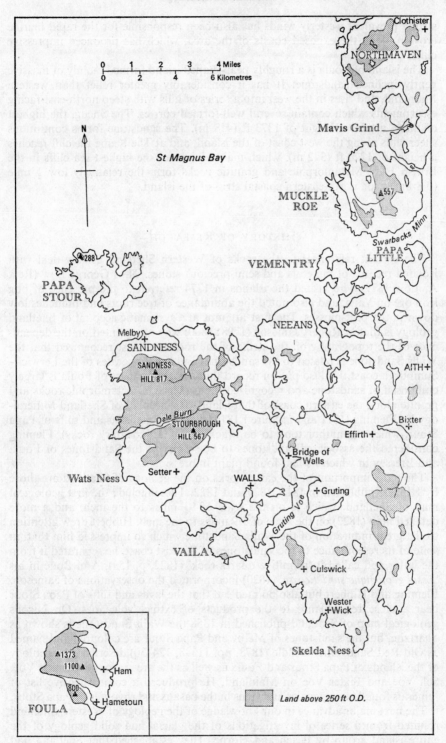

FIG. 1. *Sketch-map showing the area covered by One-inch Geological Sheet Western Shetland*

B

strong prevalent westerly winds has also been responsible for the rapid marine erosion along the exposed coasts of the area, which has produced impressive cliffs.

The island of Foula is a roughly pear-shaped island formed mainly of massive gently inclined sandstone. It has a considerably greater relief than western Mainland, and rises in the west into a series of hills with steep north-east-facing escarpments which contain several well-formed corries. The Sneug, the highest of these, attains a height of 1373 ft (418 m). The sandstone forms continuous sheer cliffs along the west coast of the island, and at The Kame the cliff reaches a height of 1220 ft (372 m), which makes it one of the highest sea cliffs in the British Isles. Metamorphic and granitic rocks form the relatively low ½ mile (0·8 km) wide north-eastern coastal strip of the island.

HISTORY OF RESEARCH

The earliest references to the rocks of Western Shetland usually deal with isolated records of minerals and semi-precious stones. Thus George Low (Rev. A. Low 1879) who visited the islands in 1774 referred to the presence of bog iron ore on Vaila, and also noted the abundance of tree roots and branches low down in the peat-mosses. The first attempt at a systematic account of Shetland geology is that of Robert Jameson (1798; 1800) who commented on the dependence of the topography on the strike of the rocks and first recognized that the Walls Sandstone is related to the sandstones and conglomerates of the Lerwick-Sumburgh coast. He also put on record that the greater part of Foula is largely composed of sandstone and recognized the presence of metamorphic rocks and granite along the eastern coast of that island. The account of Shetland Mineralogy by Fleming (1811 and *in* Shireff 1817) describes the lavas and tuffs of Papa Stour, which the author took to be 'wacken' (i.e. sedimentary rocks). Fleming considered the associated sandstones to be similar to the Sandstones of Foula and Bressay in which he had found plant impressions.

The most important of the early works on the geology of Shetland are those by Hibbert published in 1819, 1820 and 1822. These include the first geological map of Shetland, which is on the scale of 9½ miles to the inch, and a more detailed map (1822) on the scale of 3·9 miles to the inch. Hibbert drew attention to the great induration of the Walls Sandstone which so impressed him that, in spite of its resemblance to the sandstones of the east coast, he separated it from the latter and called it 'primitive quartz rock' (1822, p. 158). Ami Boué in his '*Essai géologique sur l'Écosse*' (1820) incorporated the observations of Jameson, Fleming and Hibbert but also pointed out that the lavas and tuffs of Papa Stour bear a close resemblance to the products of extinct volcanoes. On Nicol's geological map of Scotland published in 1858 the Walls Sandstone is shown as quartzite, but the sandstones of Melby and Papa Stour are coloured and named as Old Red Sandstone. Heddle (1878, pp. 113–8, 124–30) described the geology of the islands of Papa Stour and Foula as well as the areas around Gruting Voe, Seli Voe and Bixter Voe on Mainland. He produced a comprehensive list of minerals found in amygdales and veins in the basalts and rhyolites of Papa Stour.

The next major advance in our knowledge of the geology of Western Shetland resulted from a series of investigations of the glacial and solid geology of the entire island group by Peach and Horne. They established that Shetland had first been overrun by Scandinavian ice and that it later nourished its own local ice cap (Peach and Horne 1879b). Of major importance was their discovery of plant remains in the indurated Walls Sandstone near the village of Walls (Geikie 1870; Peach and Horne 1879b; Peach and Horne *in* Tudor 1883, pp. 384–408), which proved that this sandstone is of Old Red Sandstone age. They also recorded the presence of tuffs and lavas in the Walls Sandstone and produced a short description of the Sandsting Complex. Some of the results of Peach and Horne's work had been incorporated in Geikie's account of the Old Red Sandstone of Western Europe (1879). Geikie had himself examined Papa Stour (1879, pp. 418–21) and demonstrated the contemporaneous character of its tuffs and basic lavas, but concluded that the rhyolite was a large intrusive sill.

At the beginning of this century Lewis (1907; 1911) made a detailed study of the vegetation zones of the peat deposits of the Walls Peninsula and produced the first and, as yet, only account of the post-glacial vegetational and climatic history of Shetland.

The comprehensive account of the Old Red Sandstone of Shetland by Finlay (1926, pp. 553–72; 1930, pp. 671–94) deals with the sediments, contemporaneous volcanic rocks and plutonic complexes of Western Shetland and Foula. Finlay discovered a new plant locality at Watsness in the Walls Sandstone, and established that the rhyolite of Papa Stour is a lava flow. He gave a full description of the petrography of both the contemporaneous igneous rocks and the plutonic complexes of Sandsting, Muckle Roe and Northmaven. Like earlier geologists he concluded that these complexes form part of an intrusive sheet, which probably underlies the Walls Syncline.

The geological mapping of Western Shetland on behalf of the Geological Survey was carried out during the period extending from 1931 to 1934 by G. V. Wilson, D. Haldane, T. Robertson, J. Knox and S. Buchan (Summ. Prog. 1932, p. 63; 1933, pp. 77–9; 1934, pp. 70–3; 1935, pp. 67–9). The results of their work are incorporated in this memoir and on the one-inch Geological Map of Western Shetland. Some of the more significant advances resulting from this survey include the discovery of a fish bed in the Melby Sandstone, whose fauna was shown by Watson (1934, pp. 74–6) to be similar to that of the Achanarras Limestone of Caithness and of the Sandwick Fish Bed of Orkney, the recognition that the northern boundary of the Walls Sandstone is an unconformity over much of its course, and the discovery of a magnetite ore-body of considerable size on Clothister Hill, in the north-eastern corner of the area. A detailed account of the subsequent exploration and the geology of the Clothister magnetite has since been produced by Groves (1952). Charlesworth (1956) briefly summarized the evidence for the presence of corrie glaciers in late–glacial times in Western Shetland.

Flinn (1961, p. 581) has suggested that the Walls Boundary Fault is the northward continuation of the Great Glen Fault and has more recently (Flinn 1969, p. 291) claimed that it may have a post-Old Red Sandstone dextral displacement of 65 km. Flinn (1964, pp. 321–40) has also given an account of the coastal and submarine features around the Shetland Islands and has recognized several submarine shelves. Miller and Flinn (1966, p. 107, table 4) obtained radiometric age dates of Shetland rocks. Flinn and others (1968, pp. 10–19) also carried out an investigation into the radiometric age of sediments and volcanic rocks of Western Shetland.

During 1964 a Swedish party headed by Hoppe studied post-glacial lake deposits in Shetland and carried out an investigation into the directions of ice-movement in the islands (Hoppe and others 1965). Samples of lake deposits collected by them were dated by the C14 method.

Mykura and Young (1969) have described the field relationships, petrology and chemistry of the sodic scapolite which occurs in crush belts and veins within and close to the Sandsting Granite-Diorite Complex.

Summary of Geology

The following statement sets out the geological formations present in Western Shetland:

Superficial Deposits (Drift)

Recent and Pleistocene

Blown sand
Peat
Present beach deposits
Stream- and lake-alluvium
Glacio-fluvial sand and gravel
Boulder clay and morainic drift

Solid Formations

Sedimentary and Bedded Volcanic Rocks

Old Red Sandstone

Foula Formation (probably Middle or Upper)
Melby Formation (probably high Middle): Papa Stour Rhyolites and Acid
Tuffs
Papa Stour Basalts
Melby Sandstone containing
Melby Fish Beds
Walls Formation (probably low Middle)
Sandness Formation (probably low Middle or Lower), including Clousta
Volcanic Rocks with basalts, andesites, ignimbrites and agglomerates

METAMORPHIC ROCKS
(not in stratigraphic order)

Metamorphic Rocks of Foula

Metamorphic Rocks of the Walls Peninsula

SOUTH: Snarra Ness Group (mainly hornblendic)
West Burra Firth Group (calcareous)
Neeans Group (platy, feldspathic)
NORTH: Vementry Group (mainly hornblendic)

Metamorphic Rocks of Lunnister

EAST: Banded Gneiss Group
Calcareous Group
Green Beds Assemblage
WEST: Western Unclassed Group

Enclaves of metamorphic rocks in the Northmaven–Muckle Roe igneous complexes

Ve Skerries Gneiss

IGNEOUS ROCKS
Minor intrusions (probably all late-Caledonian post-tectonic)

Basic and sub-basic dykes and sills: basalt, dolerite, quartz-dolerite, porphyrite, microdiorite, spessartite and unclassed dykes.

Acid dykes, sheets, laccoliths and irregular intrusions: felsite, quartz-porphyry, feldspar-porphyry and microgranite.

Caledonian plutonic complexes (mainly post-orogenic)

Sandsting Complex: granite, microgranite, porphyritic micro-adamellite, granodiorite, syenodiorite, diorite, gabbro. Associated minor intrusions of melanic microdiorite.

Vementry Granite: granite.

Muckle Roe Granophyre: granophyre, leucogranite.

Northmaven Complex: granite, diorite, gabbro-diorite, gabbro and subordinate ultrabasic types.

The superficial deposits are shown on a separate 'Drift' edition of the one-inch map.

METAMORPHIC ROCKS

The metamorphic series enumerated above form completely distinct units which cannot be correlated with each other, and whose relative ages are as yet unknown.

The *Metamorphic Rocks of the Walls Peninsula* crop out in a belt 1 to 2 miles (2–3 km) wide, along the north shore of the peninsula. They comprise four groups of metasediments within all of which the foliation has a predominant east-north-easterly trend and is inclined at 40 to 70 degrees to the south-south-east. Two of the groups (Snarra Ness and Vementry) contain a high proportion of hornblende-schist and the calcareous West Burra Firth Group contains much tremolite-schist and many calc-silicate-rich limestones. The rocks have been involved in at least three phases of folding and four distinct metamorphic episodes, two of which were retrograde. During the main phase of prograde metamorphism the rocks of the succession reached equilibrium under pressure-temperature conditions which characterize the topmost subfacies of the greenschist facies.

The lithology and metamorphic and tectonic development of the *Metamorphic Rocks of Foula* are closely similar to those of the Walls Peninsula metamorphic rocks, and it is believed that the two successions may originally have been part of a single larger unit.

Muckle Roe–Northmaven area. The metamorphic rocks exposed in the Muckle Roe–Northmaven part of the area form two distinct structural units. The western unit consists of large and small masses of contact-altered paragneiss, composed of hornblendic, pelitic and psammitic beds, within the Northmaven–Muckle Roe igneous complex. The eastern unit, which forms the *Metamorphic Rocks of Lunnister*, is composed of four groups of gneiss and schist, termed from east to west the Banded Gneiss Group, the Calcareous Group, the Green Beds Assemblage and the Western Unclassed Group. These have a general north-south trend and a near-vertical inclination. They are cut and bounded by at least four major faults which fan out northwards and form the northward continuation of the Walls Boundary Fault. The zone of dislocation formed by

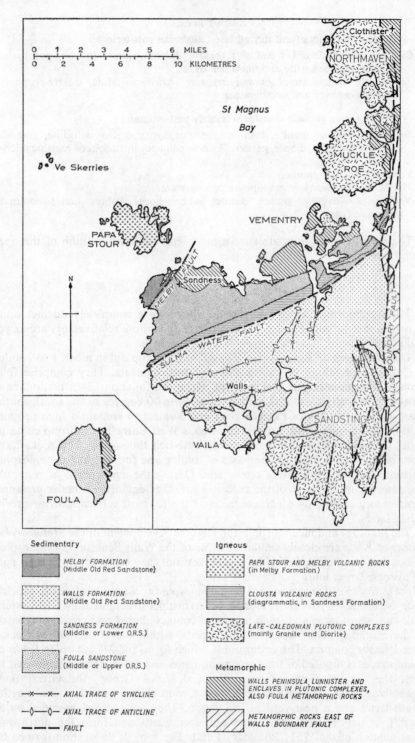

FIG. 2.   *Distribution of the principal rock groups in Western Shetland*

these faults is termed the *Busta–Lunnister Fault Zone*. It has a wedge-shaped outcrop, which widens from 200 yd (180 m) near Busta to 675 yd (617 m) at the Ness of Haggrister and over a mile (1·6 km) at the northern margin of the sheet. The rocks within the zone have been extensively sheared and mylonitized by a series of dislocations which preceded the later brittle movements that produced the faults.

Ve Skerries. The Ve Skerries, a series of low rocky islets situated in the Atlantic Ocean 3½ miles (5·6 km) NW of Papa Stour, are composed predominantly of granite-gneiss with lenticular masses of foliated granite and two bands of hornblende-schist.

SEDIMENTARY AND CONTEMPORANEOUS VOLCANIC ROCKS

The greater part of the Walls Peninsula is composed of the Walls Sandstone (Finlay 1930) which is of possibly Lower and Middle Old Red Sandstone age. This has been affected by two periods of folding, which produced firstly a complex east-north-east trending synclinorium and later a series of north to north-north-east trending folds, which are only locally developed. The Walls Sandstone consists of two major stratigraphic units which are separated from each other by the east-north-east trending Sulma Water Fault (Fig. 2). The lower unit, here called the *Sandness Formation*, is north of the fault. It rests unconformably on the metamorphic rocks of the Walls Peninsula, has a consistent east-north-easterly trend and a steep dip to the south-south-east. It is composed mainly of massive sandstones with some bands of breccia and conglomerate and contains the *Clousta Volcanic Rocks*, a suite of basic lavas, ignimbrites, tuffs, and predominantly acid agglomerates, near its top. Its exposed thickness varies from 4500 to 6000 ft (1370–1800 m) in the east to possibly 12 000 ft (3650 m) in the west, and its lithology indicates that it was deposited largely in a fluvial or basin-marginal environment and that the source of the sediments lay to the north-east.

The upper unit of the Walls Sandstone, here termed the *Walls Formation*, crops out south of the Sulma Water Fault. It comprises up to 30 000 ft (9150 m) of generally hard, indurated sediments, consisting mainly of fine- to medium-grained sandstones with intervening siltstones and shales whose lithology bears some similarity to flysch deposits and turbidites. Its sparse fossil content indicates that it was laid down in fresh water. The outcrop of the Walls Formation contains two intersecting belts of intensely folded sediments, whose axes trend respectively east-north-east and north to north-north-east. The finer-grained sediments within these fold belts have taken on a cleavage and lineation, whose geometry reflects that of the major folds.

Fish-scales and -plates have been found in several localities in the Walls Formation, while poorly preserved plant remains occur both in the Walls and Sandness formations. The fish remains appear to indicate that the Walls Formation is of Middle Old Red Sandstone age, while the plant remains from the Sandness Formation do not rule out a Lower Old Red Sandstone age for the latter.

The *Melby Formation* crops out in the north-west corner of the Walls Peninsula and on Papa Stour, and is separated from the Walls Sandstone and the metamorphic rocks by the north-east trending Melby Fault. It is composed of a lower sedimentary group and an upper volcanic group. The sedimentary group is made up mainly of soft red and grey sandstones and siltstones of fluviatile origin and contains two fish beds, known as the *Melby Fish Beds*, which have yielded an abundant fish fauna of high-Middle Old Red Sandstone age. The volcanic group is composed of a lower series of basalts, which are not present on Mainland, and an upper series of thick rhyolite flows which crop out at Melby on Mainland and form the greater part of Papa Stour.

The sediments which form most of *Foula* consist of soft grey to buff-coloured sandstone with subordinate shales and siltstones. They have an exposed thickness of about 6000 ft (1800 m) and form an open syncline plunging gently southwards. Sedimentary structures indicate that the sandstones were laid down in a mainly fluvial environment and were derived from a west to west-north-westerly source. Though plant fragments have been found at several horizons, they are not sufficiently diagnostic to determine the age of the beds.

LATE-CALEDONIAN IGNEOUS COMPLEXES

Shetland contains a number of late-Caledonian non-foliated granite–diorite complexes which are confined to a north–south trending belt bounded on the east by the Walls Boundary Fault and are probably all to some extent inter-connected. The present area contains from north to south: the southern end of the Northmaven–North Roe Complex, which occupies a large part of the Northmaven and North Roe peninsulas north of the area described, the Muckle Roe Granophyre, the Vementry Granite and the Sandsting Complex.

The southern part of the *Northmaven Complex* consists predominantly of diorite with irregular masses of gabbro-diorite and gabbro and a few very small patches of ultrabasic rock. The diorite is extensively veined by granite, which also forms a continuous outcrop along its eastern side. The Northmaven Complex contains several large irregular masses of contact-altered metamorphic rocks.

The *Muckle Roe Granophyre* is a leucocratic microgranite which has a micrographic texture in its western and central parts but passes into a more normal leucocratic granite further east. It is petrographically distinct from the *Vementry Granite*, which crops out in part of northern Vementry and has an outer rim of coarse-grained leucocratic granite and a small central area of slightly darker porphyritic granite.

The *Sandsting Complex* crops out in the southern part of the Walls Peninsula and is composed of a varied rock-suite, which includes leucocratic quartz-rich granite, microgranite, porphyritic micro-adamellite, granodiorite, diorite and some small masses of gabbro. Like the Northmaven Complex it contains several enclaves of hornfelsed country rock. There are also a number of areas of diorite and sandstone which are intensely veined by granite. Radiometric age determinations of the granite and diorite by N. J. Snelling have given dates of 360 ± 11 million years and 369 ± 10 million years respectively. This suggests that the complex is of basal Upper or late Middle Old Red Sandstone age. It is intruded into the Middle Old Red Sandstone Walls Formation and has a thermal aureole of varying width. As the minor structures, such as cleavage, minor folding and lineation, of the tectonized parts of the Walls Formation do not extend into this thermal aureole it is concluded that granite emplacement preceded or was contemporaneous with the first phase of folding in this formation.

MINOR INTRUSIONS

Basic and acid minor intrusions are abundant in Northmaven and Muckle Roe and in the northern part of the Walls Peninsula.

The *basic intrusions* are predominantly dykes with a north-north-west to northerly trend, and they become progressively more abundant in a northerly direction. The following four petrographic suites are present throughout the area:

1. Dolerite and basalt, usually olivine-free and commonly uralitized.
2. Quartz-dolerite
3. Basic porphyrite
4. Spessartite and microdiorite.

In addition there are a large number of dykes of highly altered rocks which have been shown on the geological map as unclassed basic and sub-basic types. No lamprophyres of the types found in Orkney have been recorded, and, though no radiometric age dates are available, it is believed that all the basic dykes and sills of Western Shetland are of late-Caledonian age.

Acid intrusions are represented by north-east to north-north-west trending dykes as well as by roughly concordant sills, sheets and laccoliths within the northern part of the outcrop of the Walls Sandstone. The acid intrusions consist of banded and spherulitic felsite, quartz- and feldspar-porphyry and, more rarely, microgranite. The dykes generally occur in swarms, which form two distinct suites, one composed of north-east trending dykes centred on the Vementry Granite, and the other, made up of north to north-north-west trending dykes cutting and centred upon the Muckle Roe Granophyre.

PLEISTOCENE AND RECENT

In the Shetland Islands the effects of two ice sheets can be recognized. The earlier ice sheet which approached Shetland from the east is thought by some investigators to have originated in Scandinavia. The second formed a local ice cap which was centred on the middle of Shetland and spread outwards from there in all directions. During the final deglaciation of Shetland there was a period when several areas of high ground nourished small local glaciers. In Western Shetland all evidence from glacial striae, ice-moulded topography and glacially-transported pebbles and blocks indicates that ice moved to the south-west in the southern part of the Walls Peninsula, to the north-west in the northern part of the peninsula and in Muckle Roe and to the west in North-maven. The glacial deposits of the area consist of grey to brownish sandy till with abundant stones. At one locality sand and gravel with an interglacial peat bed, probably Hoxnian, underlies boulder clay. Moraines are rare and deposits and features formed by glacial meltwaters are virtually absent. This appears to be due to the fact that Shetland has been subsiding since the last glacial maximum and that all low ground which may have contained such features in late-glacial times is now submerged. It can be shown that the earliest post-glacial lake deposits of Shetland are approximately 10 000 years old and that there has been a rise of sea level of at least 30 ft (9 m) in the last 5500 years.

REFERENCES

BOUÉ, AMI. 1820. *Essai géologique sur l'Écosse*. Paris.

CHARLESWORTH, J. K. 1956. The Late-glacial History of the Highlands and Islands of Scotland. *Trans. R. Soc. Edinb.*, **62**, 769–928.

FINLAY, T. M. 1926. The Old Red Sandstone of Shetland. Part I: South-eastern Area. *Trans. R. Soc. Edinb.*, **54**, 553–72.

—— 1930. The Old Red Sandstone of Shetland. Part II: North-western Area. *Trans. R. Soc. Edinb.*, **56**, 671–94.

FLEMING, J. 1811. Mineralogical Account of Papa Stour, one of the Zetland Islands. *Mem. Wernerian Nat. Hist. Soc.*, **1**, 162–75.

FLINN, D. 1961. Continuation of the Great Glen Fault beyond the Moray Firth. *Nature, Lond.*, **191**, 589–91.

—— 1964. Coastal and Submarine Features around the Shetland Islands. *Proc. Geol. Ass.*, **75**, 321–39.

—— 1969. A geological Interpretation of the Aeromagnetic Maps of the Continental Shelf around Orkney and Shetland. *Geol. Jnl*, **6**, 279–92.

—— MILLER, J. A., EVANS, A. L. and PRINGLE, I. R. 1968. On the age of the sediments and contemporaneous volcanic rocks of western Shetland. *Scott. Jnl Geol.*, **4**, 10–19.

GEIKIE, A. 1879. On the Old Red Sandstone of Western Europe. *Trans. R. Soc. Edinb.*, **28**, 345–452.

GROVES, A. W. 1952. Wartime Investigations into the Haematite and Manganese Ore Resources of Great Britain and Northern Ireland. *Ministry of Supply, Permanent Records of Research and Development*.

HEDDLE, M. F. 1878. *The County Geognosy and Mineralogy of Scotland, Orkney and Shetland*. Truro.

HIBBERT, S. 1819–20. Sketch of the Distribution of Rocks in Shetland. *Edinb. Phil. Jnl*, **1**, 269–314, **2**, 67–79, 224–42.

—— 1822. *A Description of the Shetland Islands*. Edinburgh.

HOPPE, G., SCHYTT, W. and STRÖMBERG, B. 1965. Från Flät och Forskning Naturgeografi vid Stockholms Universitet, *Särtrych ur Ymer.*, H 3–4, 109–25.

JAMESON, R. 1798. *An Outline of the Mineralogy of the Shetland Islands, and the Island of Arran*. Edinburgh.

—— 1800. *Mineralogy of the Scottish Isles*. 2 vols. Edinburgh (Shetland, vol. 2, 185–224).

LEWIS, F. J. 1907. The Plant Remains in the Scottish Peat Mosses. III. The Scottish Highlands and the Shetland Islands. *Trans. R. Soc. Edinb.*, **46**, 33–70.

—— 1911. The Plant Remains in the Scottish Peat Mosses. IV. The Scottish Highlands and Shetland, with an Appendix on Icelandic Peat Deposits. *Trans. R. Soc. Edinb.*, **47**, 793–833.

LOW, REV. G. 1879. *A Tour through the North Isles and part of the Mainland of Orkney in 1774*. Kirkwall.

MILLER, J. A. and FLINN, D. 1966. A Survey of Age Relations of Shetland Rocks. *Geol. Jnl*, **5**, 95–116.

MYKURA, W. and YOUNG, B. R. 1969. Sodic scapolite (dipyre) in the Shetland Islands. *Rep. No. 69/4, Inst. geol. Sci.*

PEACH, B. N. and HORNE, J. 1879a. The Old Red Sandstone of Shetland. *Proc. R. Phys. Soc. Edinb.*, **5**, 80–7.

—— —— 1879b. The Glaciation of the Shetland Isles. *Q. Jnl geol. Soc. Lond.*, **35** 778–811.

SHIREFF, J. 1817. *General View of the Agriculture of the Shetland Islands*. Edinburgh (Minerals, Section 5: Appendix on the Economical Mineralogy of the Orkney and Zetland Islands, pp. 105–35 by John Fleming, *q.v.*).

SUMM. PROG. 1932. *Mem. geol. Surv. Gt Br. Summ. Prog.* for 1931.

SUM. PROG. 1933. *Mem. geol. Surv. Gt Br. Summ. Prog. for 1932.*

―― 1934. *Mem. geol. Surv. Gt Br. Summ. Prog. for 1933.*

―― 1935. *Mem. geol. Surv. Gt Br. Summ. Prog. for 1933.*

TUDOR, J. R. 1883. *The Orkneys and Shetland; Their Past and Present State.* London.

WATSON, D. M. S. 1934. Report on Fossil Fish from Sandness, Shetland. *Mem. geol. Surv. Gt Br. Summ. Prog. for 1933,* Pt. I, 74–6.

Chapter 2

THE LUNNISTER METAMORPHIC ROCKS

Introduction

THE LUNNISTER Metamorphic Rocks crop out in the north-eastern corner of the present area (Fig. 3). They form a triangular outcrop which has its apex at the head of Busta Voe and widens northwards to $1\frac{1}{4}$ miles (2 km) at the northern margin of the map. This area is bounded by two converging faults. On the west the metamorphic rocks are brought against unfoliated granite of the Muckle Roe–Northmaven Complex by the Haggrister Fault (Fig. 27b) which is seen at the west end of the Bight of Haggrister to be a reversed fault dipping W at 70° and which causes severe shattering of the schists and some shattering of the granite. On the east the Walls Boundary Fault separates the Lunnister rocks from the foliated granite-with-schist of the Delting Injection Complex (see Flinn 1954). The two faults converge southwards into the Walls Boundary Fault-zone which passes under Busta Voe close to its western coast. As the Lunnister Metamorphic Rocks have been described in detail elsewhere (Phemister in press), only a short summary of their lithology and structure is given in the present chapter.

The Lunnister Metamorphic Rocks are believed to include several geological formations. They comprise banded gneisses, siliceous, sericitic and graphitic schists, greenschists and calcareous schists with limestone bands. All of these are greatly deformed, sheared and, in places, mylonitized and phyllonitized. They also contain the magnetite deposits and skarn rocks of Clothister. Towards the southern end of their outcrop the Lunnister rocks occupy only a narrow strip of ground which is about 400 yd (365 m) wide on the north side, 200 yd (180 m) wide on the south side, of the Sullom Voe–Busta Voe isthmus. Within this strip the rocks are so greatly crushed and mylonitized that only exceptionally can an outcrop be referred to a specific unit of the Lunnister group. The eastern and western limits of the strip are ill-defined owing to the severity of the crushing which has occurred along the two lines of fault. Close to the northern margin of the map the Lunnister Metamorphic Rocks consist of four groups (Fig. 3), which have been named from west to east (1) the Western Unclassed Group, (2) the Green Beds Assemblage, (3) the Calcareous Group and (4) the Banded Gneiss Group. It has proved possible to correlate these groups with geological units recognized in the Ollaberry–Gluss district of Northmaven, as shown in Fig. 4 and the table on p. 15.

This table summarizes, on the left, the groups and the lithology of the rocks which enter, or may enter the Lunnister area from the north; and, on the right, the groups of the Lunnister Metamorphic Rocks as subdivided for description in the following pages and their lithology.

Banded Gneiss Group

The rocks of the Banded Gneiss Group are best exposed along the west shore of Sullom Voe south of the Houb of Lunnister (Fig. 3). They are mainly

14

Foliated granite-with-schist of the Delting and Yell Sound Injection Complex
Walls Boundary Fault

Ollaberry–Gluss area	*Lunnister area*
Bardister Gneiss (hornblende-gneiss, hornblende-schist banded with mica-schist)	Banded Gneiss Group (epidotic, horn-blendic and albitic gneisses, granuli-tized and mylonitized)
Calcareous Group (quartzose calc-schists with limestone and muscovite-schist bands)	Calcareous Group (calc-schists, limestone bands, quartz-schist and quartzite)
Greenschist Group (laminated green and white schists, greenschist with 'pebbly' albites, intercalations of muscovite-schist and quartzite)	Green Beds Assemblage (chloritic, sericitic albite-schists with some fibrous amphibole and pyroxene, local amphi-bolite and felted amphibole-schist, quartzose schist graphitic in part, local conglomeratic greenschist, magnetite ore and skarn rocks at Clothister)
Siliceous Group (quartzites, fissile mus-covite-schists, intercalations of green-schist and calcite-epidote-chlorite-schist)	
sheared junctions	
Hornblendic Group (hornblende-schist, banded hornblendic and feldspathic gneiss pyroxenic in part)	Western Unclassed Group (mainly mylon-itic and crushed rocks, hornblendic gneiss at south end of South Ness)
Intrusive contact in the north, faulted junction in the south	Haggrister Fault

Unfoliated granite of the Western Plutonic Complex

striped hornblendic and epidotic gneisses with which more micaceous and albitic components are interbanded. The foliation is vertical and the strike varies from north-east to north-west. The rocks are everywhere shattered, crushed or flasered and, in places, they are converted to banded mylonite. On the south coast of the Ness of Haggrister the group is cut out, since limestone, unknown in the Banded Gneiss Group, adjoins the Walls Boundary Fault. Green chloritic epidotic blastomylonite, which is exposed 70 yd (65 m) west of the fault, might possibly represent a faulted or thrust intercalation of the banded gneiss within the Calcareous Group.

CALCAREOUS GROUP

The Calcareous Group is composed of a series of micaceous calcareous schists, together with some more quartzose and micaceous schists, and thin bands of dull grey microcrystalline limestone. The rocks are well exposed in the Ness of Haggrister and discontinuously farther north. Both the outcrop of the group and the strike of the foliation have a general N–S trend, but there are many local deviations in strike. The softer schists are greatly contorted and often flasered. The harder are shattered, disjointed and noded, while the limestones are deformed into small, complex folds. Near the mouth of the Lunnister Burn the rocks of this group are interflasered with gneiss of the Banded Gneiss Group.

On the low hill [347 721] about 250 yd (230 m) N of the Loch of Lunnister, limestone and quartzitic rock are associated in a contorted mass resembling

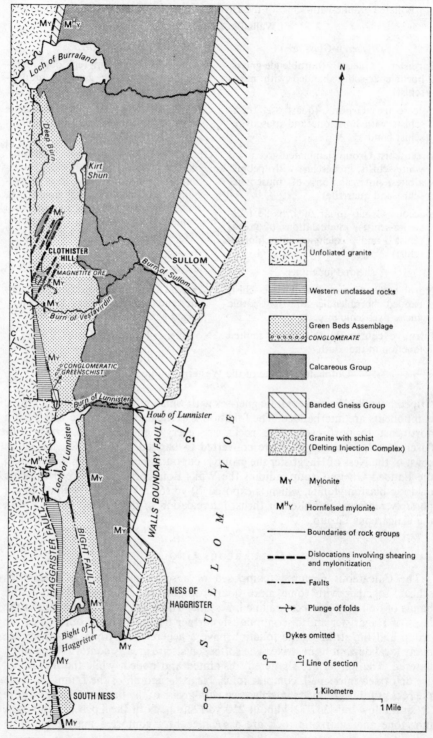

Fig. 3.   *Geological sketch map of the metamorphic rocks of the Lunnister area*
For section C-C₁ see Fig. 4.

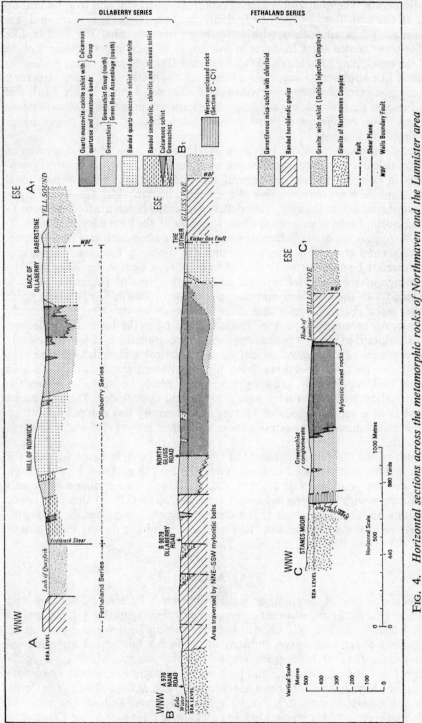

FIG. 4. *Horizontal sections across the metamorphic rocks of Northmaven and the Lunnister area*

Sections A–A₁ and B–B₁ are entirely in the area north of the present sheet. (Locations: A = [349818], A₁ = [377808], B = [335793], B₁ = [778365]). The line of section C–C₁ is shown in Fig 3.

conglomerate. The quartzitic material occurs as spheroidal and oval bodies, up to 1 ft (30 cm) long, in contorted, thinly banded, micaceous, more and less calcareous schist alternating with hard, more siliceous ribs. The boulder-like bodies show healed shear fractures at low angles to their largest principal plane and the enveloping beds are milled to fine grain. Though many of the bodies are isolated like boulders, in other places they are connected in a noded structure. No conglomerate has been observed in the Calcareous Group farther north and it is believed that this rock is a pseudo-conglomerate produced by local intense disruption or boudinage of hard bands and reshaping of the pieces under continued shear stress.

Along the south coast of the Ness of Haggrister grey phyllitic micaceous schists, calcareous schists, and bands of limestone are involved with green-schists in complex folding. In general the folds are tight with vertical axial planes, but there are narrow zones of corrugated beds in which the overall dip is almost horizontal. In the cliffs immediately east of the beach the folds are seen to plunge north. Midway between the beach and the Walls Boundary Fault (Fig. 3) a crush-zone is exposed. Calcareous members are abundant east of this zone, micaceous and green schists are predominant to the west. The close interleaving of calcareous beds with rocks ascribed to the Green Beds Assemblage and the prevalence of shearing and mylonitization in the Ness of Haggrister section, combined with the westward reappearance of the Calcareous Beds inland north of the beach (Fig. 3), imply that their association at the Ness is essentially tectonic. All specimens from the Ness coast section in the Institute's collection are of mylonitized rocks. The area appears to be dominated by a series of closely spaced planes of movement which are now vertical and which diverge in a narrow fan northwards. Along these planes mylonitization without obvious rupture and, more rarely, crushing have taken place. Between the planes the beds are alternately close folded with shearing and open folded. The impression gained is of a series of beds of varying but generally low competence which have been compressed between two more competent groups converging towards the south.

South of the Ness of Haggrister the Calcareous Group is found only on the western shore near the head of Busta Voe (Fig. 5). The rocks in the dislocation zone are here shattered and crushed and are largely indeterminate but a calcareous zone with limestone has been traced for 250 yd (230 m) along the coast. Granite with schist inclusions of the Delting Complex crops out on the north-easterly turn of the coast and the limestone outcrop must lie very close, within 25 yd (23 m), to the Walls Boundary Fault.

GREEN BEDS ASSEMBLAGE

The rocks of this assemblage include siliceous, feldspathic, sericitic and chloritic schists. Feldspathic schist composed of comparatively large grains, 0·5 to 1·00 mm, of albite-oligoclase in a sericite-chlorite base, forms an important component of the assemblage. In many of the feldspathic schists aggregates of fibrous colourless or pale green amphibole are present. More basic rocks composed of green hornblende and plagioclase with some saussurite or epidote are minor components. The group has therefore been distinguished as the Green Beds Assemblage though it is not so typical of Green Beds as the series of laminated green schists and metabasaltic rocks which lies west of the Calcareous

Group in the Ollaberry area of North Mainland. Graphitic schists are present in the Lunnister Green Beds Assemblage of this area, and as these have not been recorded in association with either the Green Beds or the Calcareous and Siliceous groups around Ollaberry, the correlation of the groups in the two areas is imperfect. The assemblage includes also an important component of quartzose schists which may represent the quartzites of the banded quartz- and muscovite-schists of the Ollaberry section. It has not been found possible to make any satisfactory separation in Fig. 4 between the siliceous, graphitic and greenschist components of the assemblage.

Busta isthmus and South Ness. No rock in the narrow part of the dislocation zone between the Haggrister and Walls Boundary faults in the Busta–Sullom isthmus [348 680] is sufficiently recognizable for positive assignment to the Green Beds Assemblage. On South Ness, however, sericitic and chloritic albite-quartz-schists exposed along the eastern coast are appropriately included in the group. They are polymetamorphic schists which show an earlier foliation on which has been superimposed a schistosity which is related to the prominent folding of the rocks. This schistosity is traversed by mylonitic schlieren.

Haggrister–Lunnister area. Along the central part of the south coast of the Ness of Haggrister tightly folded and sheared chloritic and micaceous schists are interlayered with calcareous schists and limestones of the Calcareous Group. The latter are not seen west of the crush-zone [360 700] which crops out near the most southerly point of the Ness, and the crush zone is taken as a convenient position for a mapping line between the Calcareous Group and the Green Beds Assemblage. At the western end of the coastal cliffs, close to the Bight Fault (Fig. 3), the rocks are deformed into tight northward plunging folds and, in places, severely crushed.

Rocks referable to the Green Beds Assemblage are exposed near the east and north-east shores of the Loch of Lunnister where they include epidotic hornblende-albite-phyllonite, albite-hornblende-schist, epidote-hornblende-pyroxene-mylonite and a crushed rock which appears originally to have been a phyllonitized albite-schist. All these rocks contain varying amounts of finely divided graphitic matter, and a black pyritous schist has been recorded near the south-west and west banks of the loch. The group also includes a granulitized schistose grit which is exposed in a knoll 500 ft (150 m) S of the loch.

Deformed conglomerate has been recorded by Dr. F. May in the area [344 722] between the Loch of Lunnister and the Burn of Vestavirdin. This contains a variety of pebbles, including quartzite and granulitized tonalite, set in a dark greenish schistose matrix composed of green hornblende, brown biotite, chlorite, epidote, optically positive sodic plagioclase and local quartz. The pebbles have been deformed into elongate ovoids, the collected quartzites ranging up to 8 in (20 cm) in length and up to $2 \times 1\frac{1}{4}$ in (5×3 cm) in cross section. Though none of the pebbles appear to be volcanic ejectamenta, the matrix of the conglomerate can only have been derived from basaltic material. Both in the nature and shape of the pebbles and in the basic composition of the matrix, this conglomerate is comparable with the Funzie Conglomerate of Fetlar (Summ. Prog. 1930, p. 83; Flinn 1956). No similar conglomerate has been found elsewhere on the Mainland of Shetland.

Vesta Virdin–Clothister area. In the Burn of Vestavirdin, just south of Clothister Hill, pyritous graphitic schists and a massive dark green epidotic amphibolite are exposed. North of the burn a large quarry [342 728] exposes green schists

C

faulted against graphitic schists. The green schists contain a band up to 2 ft (60 cm) thick of garnet-magnetite rocks within a zone of calcsilicate rock, an association which is the same as that of the ore and skarn of Clothister Hill (p. 20). The rocks in the quarry also contain thin white bands composed of mylonitic calcite and rock fragments, which may have been ribs of limestone.

The country rocks of Clothister Hill were seen in trenches dug during the exploration of the magnetite deposit (pp. 285 and 287). They consist of banded or foliated quartz-, quartz-albite- and albite-schists in which sericitic, chloritic and tremolitic laminae are common and biotite is rare. Graphitic phyllonite was recorded in one locality close to the ore margin. Practically all the specimens show either mylonitization or phyllonitization, followed by recrystallization. Many have a second schistosity superimposed on an earlier schistosity or foliation. The inclined bores sunk to investigate the orebody at depth (Fig. 30, pp. 285–7) encountered chloritic and graphitic schists, the foliation of which is inclined at 35° to 60° to the west. Nine borings put down on positions of high magnetic anomaly 388 to 450 yd (350–400 m) N of the orebody proved mainly sericitic, chloritic and graphitic quartz- and albite-schists which have been subject to two foliations and are in places smashed or mylonitized. The most westerly bore and the upper 169 ft (51·5 m) of the most easterly bore traversed calcareous quartz-schists, suggesting that tectonic slices of rocks belonging to the Calcareous Group are present in this area. In some of the bores a rock described as 'granite injection schist' has been recorded. Though no specimen is available this material is unlikely to be injection gneiss in the usually accepted sense of the term.

CLOTHISTER HILL MAGNETITE AND SKARN

The Clothister Hill orebody, the exploration, shape, size and quality of which are described on pp. 285–7, lies in schists of the Green Beds Assemblage. Neither the ore nor the skarn show evidence of the severe deformations which affected the country rocks, but they are affected by minor slipping and faulting. It is believed that the magnetite ore was introduced and the skarnization of the country rocks induced along a narrow pre-existing dislocation belt. The skarn sheath enclosing the magnetite body consists essentially of garnet, hornblende, pyroxene and epidote and its chemical analysis shows that it is rich in lime (Phemister in press, table 3). This suggests either that the ore has replaced a lens of limestone or that there has been an accession of lime and iron from an unknown source. The petrographical evidence tends to favour the latter hypothesis, as in one specimen the magnetite has replaced hornblende-albite-schist and as the calcite associated with the skarn is a late introduction. The source of the iron is uncertain; it may be derived from the pyrite which is abundant in some of the rocks of the Green Beds Assemblage, and which in some schists adjacent to the ore deposit is replaced by magnetite.

The ore and skarn, though close to the margin of the Northmaven Plutonic Complex, are not in contact with the plutonic rock. Both Groves (1952, pp. 294–5) and Phemister consider that the emplacement of the ore was not related to the intrusion of the granite part of the complex. No scapolite has been found in the skarn, nor have any boron or fluorine minerals.

Summary. From the petrography of the complex group of rocks forming the

Green Beds Assemblage and the skarn and magnetite ore of the Clothister–Lunnister area the following conclusions can be drawn:

1. The metasediments comprised carbonaceous silts and basic tuffs with which thin basic igneous sheets were associated. The metasediments are characterized by abundance of a plagioclase with about 10 per cent of the anorthite molecule.
2. This group of rocks was already metamorphosed to greenschist and associated types before a metamorphism of an essentially dynamic type produced a second foliation and mylonitization.
3. The magnetite ore and skarn sheath are only locally affected by shear.
4. There is direct evidence that the ore is replacive in greenschist, but no similarly direct microscopic evidence that it has replaced mylonitized schists. There is some evidence of early oxidizing conditions prior to the ore mineralization.

WESTERN UNCLASSED ROCKS

The rocks occupying the strip along the margin of the Northmaven Granite are probably mainly referable to the Green Beds Assemblage. Where seen near the Haggrister Fault which bounds the granite, they are greatly shattered. The shattered rock includes a blastomylonite which has tight folds with horizontal axial planes and in which there is vague evidence of an earlier foliation. Other exposures show mylonitic and phyllonitic rock ranging in type from schistose grit to graphitic schist. One foliated crush-rock contains grains of blastomylonite. At South Ness (Fig. 3) and at Clothister there are some banded hornblende-feldspar rocks which can be referred with considerable assurance to the banded hornblendic gneiss of the Fethaland Series (Fig. 4). These rocks show intricate folding, plastic disruption and differential mylonitization superimposed on an earlier foliation. Mylonitized feldspathic rocks from the upper part of the most westerly boring on Clothister Hill are thought to represent feldspathic members of this group, while basic rock from a lower level in the same bore could be either a basic member of this group or a metabasalt from the Green Beds Assemblage which has been mechanically intercalated.

All the rocks show some form of post-deformation recrystallization tending towards an unstrained or unorientated condition, but without any mineralogical change indicative of further metamorphic transformation. There is no evidence of thermal alteration by the granite except in one anomalous case. Late veining by potassium-feldspar has been noted. There is also evidence of post-deformation veining by calcite and analcime and post-fault veining by calcite and calcic zeolite.

STRUCTURAL INTERPRETATION

The dislocations affecting the rocks in the Busta–Haggrister fault-zone are of two types. Of the more obvious type are the faults which produce sharp breaks and have only minor shatter and shear belts. Examples are the Haggrister Fault and the Walls Boundary Fault, which are shown in Fig. 3 by the usual fault ornament. Of the second type are the dislocations which affect the rocks between these faults. They are of a type involving shearing, refoliation, mylonitization and phyllonitization. Petrographical evidence (see Phemister in press) shows that in the central area mylonitic and phyllonitic deformation affected a schistose 'green beds' assemblage and a probably older formation of banded hornblendic gneisses and that a later deformation caused a second folding which induced a new cleavage or schistosity and varying degrees of flasering and shearing.

The general distribution of the rocks in the Lunnister area and its northward continuation (Fig. 4) suggests that they occupy a tight northward plunging synform with a near-vertical axial plane. The core of this synform contains variably competent sediments which are bounded by massive banded gneiss on both flanks. In the area which falls within the Western Shetland map the gneisses on the eastern flank are mylonitized and marginally interleaved with the rocks of the core. On the western flank the structural relations of the gneiss with the metasediments are almost completely obliterated by the granite intrusion, but the vestiges of greatly deformed gneiss and of mylonites which are considered to represent gneiss indicate that this flank also formed a zone of essentially mechanical dislocation. The structure is now that of a vertical wedge thinning towards the south into the position of the Aith Voe–Busta Voe Fault. At the south end of the wedge the rocks are entirely mylonite and only rarely, as in the case of the limestone on the coast north of Busta, has even the rock group been determined. As the wedge broadens northwards the rocks of the core become identifiable at Haggrister as representing the Calcareous, Green Beds and, possibly, the Siliceous groups of North Mainland. They are, however, so contorted, interfolded, and locally mylonitized that separation of them into these groups is uncertain and even distinction of the core rocks from the gneiss of the eastern flank is conjectural. From Lunnister northwards separation into the groups becomes possible. The rocks of the core continue to be highly contorted and flasered, and mylonitization appears to be more frequent and increasingly more severe towards the western flank.

The structure and polymetamorphic condition of the rocks in the Lunnister area may have been produced by a sequence of events as adumbrated by one of the following hypotheses.

1. Rock formations already regionally metamorphosed were further deformed by folding in a tight northward-plunging syncline, the axial plane of which was vertical and along the sides of which smaller scale tight folding and shearing took place. Dislocation involving flasering and mylonitization of beds and mechanical interpolation of groups occurred along the smaller folds on the flanks and probably also along more central planes.
2. The metamorphic formations were involved in a local recumbent fold along the upper and lower limbs of which low-angled thrusting and imbrication, with concomitant mechanical metamorphism, took place. The folded rocks were later rotated into verticality on the north-eastern limb of a great fold on a NW–SE axis which affected the existing N–S strike of the Northmaven metamorphic formations and the E–W strike of their counterpart along the northern coast of the Walls Peninsula.
3. Formations already regionally metamorphosed to higher grade gneisses and lower grade schists were re-orientated in a fold of great amplitude with steep NW–SE axial plane as outlined in (2) above. On the north-eastern limb of this fold a succession of subsidiary complex folds and dislocations, accompanied by much mechanical reconstruction, was induced by constriction of the limb towards the culmination of the fold at a position immediately south of the Lunnister–Busta area.

Phemister's detailed account has demonstrated a succession of periods of deformation affecting the already metamorphosed Lunnister rocks and offers some support thereby to the second hypothesis. This hypothesis is supported also by the evidence in the Eela Water area (see the Geological Sheet of Northern Shetland, 1968) of a major displacement of hornblendic paragneisses which appears to have been caused by sharp folding and lateral movement on NNE–SSW lines. On the other hand the evolution of a NW–SE fold of such magnitude as the broad structure of the metamorphic rocks of the Northmaven–Walls region indicates must have been prolonged in time and involved subsidiary, locally intense tangential adjustments at intervals. The third hypothesis may therefore be considered preferable.

REFERENCES

FLINN, D. 1954. On the time relations between regional metamorphism and permeation in Delting, Shetland. *Q. Jnl geol. Soc. Lond.* **110,** 177–201.

—— 1956. On the deformation of the Funzie conglomerate, Fetlar, Shetland. *Jnl Geol.* **64,** 480–505.

GROVES, A. W. 1952. Wartime Investigations into the Haematite and Manganese Ore Resources of Great Britain and Northern Ireland. *Ministry of Supply, Permanent Records of Research and Development.*

PHEMISTER, J. 1975. The Lunnister Metamorphic Rocks, Northmaven, Shetland. *Bull. geol. Surv. Gt Br.,* in press.

PRINGLE, I. R. The structural geology of the North Roe area of Shetland. *Geol. Jnl,* **7,** 147–70.

SUMM. PROG. 1930. *Mem. geol. Surv. Gt Br. Summ. Prog. for* 1929.

Chapter 3

THE METAMORPHIC ROCKS IN THE MUCKLE ROE–NORTHMAVEN COMPLEX

INTRODUCTION

WEST OF the Haggrister Fault (p. 262), three large and several smaller areas of metamorphic rocks appear within the outcrop of the intrusive complex which extends from the south shore of Muckle Roe to the northern margin of the Western Shetland Geological Sheet (Fig. 5). The largest areas lie, (i) around Skipadock, between the head of Mangaster Voe and Sullom Voe, (ii) in the Busta peninsula, and (iii) along the east coast of Muckle Roe. The smaller areas are, (i) at Djubi Dale [337 743], (ii) in the island of Egilsay, (iii) around Houlls Water in the Busta peninsula, and (iv) at Mill Lochs [317 635] and Quhaap Knowe [320 633] in the south of Muckle Roe. The metamorphic rocks form part of a banded group in which black or dark green foliated hornblendic rocks are predominant, mica-schist important, and quartzo-feldspathic schists minor components. They are collectively referred to in the following pages as 'the gneiss' when only a broad reference to the group is required.

LITHOLOGY

DJUBI DALE ENCLAVE

This mass forms a conspicuous bluff on the right bank of the Djubi Dale stream 700 yd (640 m) N by E from the north end of Glussdale Water close to the northern margin of the present Sheet. It is composed of dark foliated hornblendic rock with feldspathic bands. The strike of the foliation varies between NE–SW and NNW–SSE and is in places contorted. The dip is steep. The form of the mass is roughly oval, 150 yd (140 m) long in a NNW–SSE direction and 100 yd (90 m) broad. On the east side it is bordered by granite, though the contact is concealed, and granite crops out within 100 yd (90 m) on the north, south, and south-west of the hornblendic exposures. The mass is therefore a large enclave in the granite. No xenolith of similar rock has been noted in the surrounding granite and there appears no justification for regarding the Djubi Dale enclave as a roof pendant. Thin sections (S 43766[1], 44104–6) show that it is a contact-altered epidotic andesine-hornblende-gneiss with pelitic laminae in which new biotite and pyroxene have been developed and hornblende and feldspar recrystallized. The very fine grain of the rock and the occurrence of larger grains of feldspar which are swathed in amphibole and show varying directions of streaks of included material, indicate that the original foliated rock had been sheared, and possibly milled, prior to the contact-alteration.

---

[1]Numbers preceded by S refer to thin sections in the Petrographical Department Collections of the Institute of Geological Sciences.

24

SKIPADOCK AREA

At the south end of Northmaven the gneiss crops out over an area of about 1 km² between the west coast of Sullom Voe at Skipadock and the high ground of Hurda Field. On its north margin, east of the main road, the contact between gneiss and diorite as mapped is sharp and irregularly stepping southward as the ground rises, as if the diorite lay above the gneiss. On its north-eastern margin the gneiss abuts against granite with which it remains in contact southwards to the coast at Skipadock. The northern half of this eastern boundary is concealed; the southern half is well exposed and is mapped as zigzag, with longer E–W limbs, down to the coast, as if the granite alternately broke across and penetrated along the foliation of the gneiss, but had an overall vertical disposition. A small outcrop of gneiss occurs between the granite and the sea on the coast about 100 yd (90 m) E from the main contact; it is not known whether it represents an enclave or an eastward extension of the gneiss outcrop. Several dyke-like bodies of granite are intrusive in the gneiss at and to the north of Skipadock. On the south the gneiss is bounded by the sea and on the west by the Mangaster Voe Fault. Along the north margin the strike of the gneiss is consistently N–S with high westerly or vertical dip. Near the eastern border the strike is approximately E–W with the dip high in the north and inclined at 20°–30° to the north-west at the coast. In the western part of the outcrop the strike and dip are variable, due presumably to the proximity of the fault. Thus the gneiss appears to have a structure resembling the southern half of an oval bowl, the long axis of which is NNE–SSW.

The nature of the rocks is most readily observed in the coast exposures around Skipadock. They are foliated, in part thinly laminated, and locally contorted. They comprise hornblende-schist, garnetiferous pelitic schist, and banded quartzo-feldspathic granulites with pelitic laminae. All have a hornfels aspect. Petrographical examination shows that some have been sheared prior to contact-alteration (S 53596, 53601). New pyroxene has been formed in the hornblendic rocks (S 29418, 33750, 53596) by contact-alteration, and late impregnation by quartz has reconstructed thermally altered pelitic schist (S 44231).

EGILSAY

About 1½ miles (2·5 km) W of the Skipadock outcrop the gneiss reappears on the island of Egilsay and the islet of Black Skerry, that is on the west side of the Mangaster Voe Fault and west of the 1-mile-(1·8-km)-broad diorite which occupies the Islesburgh peninsula. The gneiss occupies the whole of Black Skerry and the southern extremity of Egilsay where its outcrop is 200 yd (180 m) broad in an E–W direction and extends for 50 to 100 yd (45–90 m) inland to a deeply embayed contact with diorite. Along the north-west margin the gneiss is separated from diorite by a dolerite dyke, which suggests that the pre-dyke junction of gneiss and diorite was vertical here. A small outcrop of gneiss, about 100 yd (90 m) long by 50 yd (45 m) broad, completely surrounded by diorite, lies north of this dyke. The strike of the foliation is contorted but trends mainly NNW–SSE and ENE–WSW; the dip is vertical. If it is assumed that the Egilsay and Black Skerry outcrops are continuous the length of this mass of gneiss would be at least 500 yd (450 m). This considerable extent, the embayed contact with

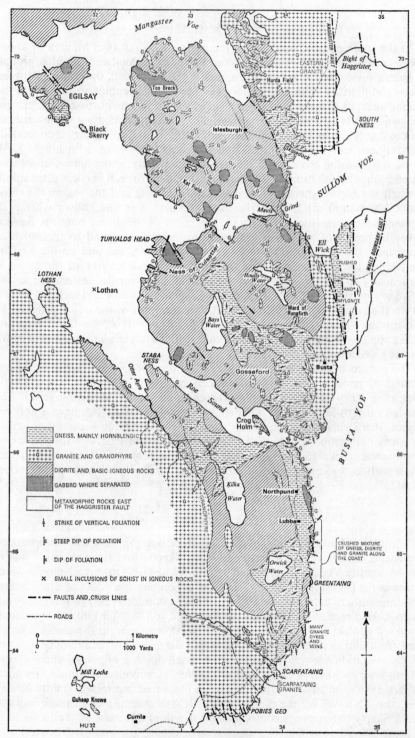

FIG. 5. *Distribution of metamorphic rocks in the Muckle Roe–Mangaster Voe area*

diorite, and the existence of the outlying mass suggest that the gneiss here is pendant from an eroded roof. It is cut by many granitic and pegmatitic veins and larger dykes.

The gneiss in this area appears from the specimens available to be in the main a banded pelitic and quartzo-feldspathic type (S 53600) in which the layers are folded in close isoclines (S 53599). Hornblende-schist bands are also present (S 55247). The rocks are thermally altered with production of granoblastic oligoclase, unorientated idioblastic biotite, and pinite in the pelitic layers, granoblastic andesine and olive-green hornblende in the hornblende-schist. All have experienced a late permeation by quartz which gives them in hand specimen a glazed or sparkling aspect correlating with the replacement and dispersion of the hornfelsed folia by quartz as seen in thin sections. Specimens from the out-lying mass in Egilsay (S 44282A, 55175) show the contact of gneiss and diorite; microscopic aggregates of biotite in the latter may represent small partly assimilated xenoliths and there has clearly been restricted transfer of material between country and igneous rock. There is no record of xenoliths, large or small, in the vicinity of the gneiss–diorite contact.

BUSTA PENINSULA

On the Busta peninsula two main areas are occupied by the gneiss. The smaller, which lies in and east of Houlls Water, includes four small outcrops which may represent one mass intruded by granite. The rock here consists of banded siliceous granulite and pelitic and hornblendic schists (S 29414, 33741, 53594) which have all been hornfelsed, hydrothermally altered, and soaked by quartz. The foliation is vertical and the strike is consistently E–W except in the most south-easterly outcrop where it is N–S. The gneiss of all these outcrops is surrounded by heterogeneous granite-diorite complex and is intruded by massive granite.

The main body of gneiss covers an area of fully one third of a square mile (0·86 km²) and is continuous from the northern end of Bays Water to the eastern end of Roe Sound. On its eastern side it is bounded by the dyke-form body of granite which extends from this position for 10 miles (16 km) northwards to Ronas Voe. Contact relations are concealed by drift, but the interposition of outcrops of gneiss and granite suggests that this contact is intricate in detail. It has, however, a persistent N–S trend on the large scale and therefore is probably vertical or steep. The foliation of the gneiss near the granite margin strikes N–S to NNW–SSE. Between Roe Sound and the Ward of Runafirth the gneiss occupies a belt about ¼ mile (400 m) wide between granite on the east and diorite on the west and outcrops of diorite appear within this belt at 200 to 400 yd (180–360 m) NW and 200 yd (180 m) NE of Roe Bridge. The western margin runs N–S for ½ mile (800 m) but is deeply indented in an E–W direction cor-responding with the E–W to ENE–WSW strike of foliation in the gneiss. At the Ward of Runafirth the gneiss belt changes direction to E–W, continuing for ½ mile (800 m) towards Bays Water and thence north-westward for a further ½ mile (800 m) by the head of Bays Water to the Ness of Coulsetter. The gneiss ends against the rising ground of the Ness of Coulsetter along an undulating but sharp margin which appears to follow the foliation. Beyond this margin no considerable body of gneiss has been noted but inclusions of schist and basic gneiss are found in the diorite on the west coast ⅓ mile (0·5 km) S of Turvald's

Head and on Lothan Skerry; these are represented in the collection by thermally altered pelitic schist (S 53592) and hornblende-schist (S 55168). South of the main belt six considerable masses of gneiss, the largest 200 yd (180 m) long by 60 yd (55 m) broad, have been mapped within the diorite along a $\frac{1}{2}$-mile (800-m) stretch between Bays Water and Roe Sound.

Topographically the gneiss of the main belt rises from sea level to almost 300 ft (91 m) OD south of the Ward of Runafirth, but in general its outcrops occupy rather lower ground than the eminences of diorite and gabbro, for example those of the Ward of Runafirth and the Skeo of Gossaford [337 668], and its course along the hollow running west to Bays Water ends at a level between 150 and 200 ft (46 and 61 m) OD against the Ness of Coulsetter diorite which rises to 300 ft (91 m) OD. In the lower portion of this hollow outcrops of gabbro and of granite appear inside the gneiss belt, and a large body of gabbro, diorite, and granite also is enclosed south-west of the Ward of Runafirth. These relations can be interpreted in two ways: that this extensive area of gneiss forms either the lower part of an uneven roof intruded irregularly from below, or an irregular floor overlain by a sheet of gabbro-diorite. The numerous outcrops of gneiss isolated in the igneous rock between Bays Water and Roe Sound would represent, in the first case, pendants from the roof; in the second case windows in the floor of the intrusion or rafts within its mass. The outcrops of igneous rock within the gneiss area between Bays Water and Busta would represent, in the two cases respectively, plug-like projections into the gneiss roof or outliers of the intrusive sheet. No field evidence of superposition can be adduced in support of either alternative.

The rocks collected from the gneiss outcrops in the Busta peninsula represent a banded group of siliceous and quartzo-feldspathic granulites, pelitic schists, and hornblende-schist and -gneiss. Unusual types are recorded from the most westerly of the isolated masses, 200 yd (180 m) SW of the south end of Bays Water, where in hard hornblende-schist a thin irony band is found. Two specimens from this band include an amphibolite (S 34995) with greenish blue hornblende and a garnet-magnetite-hornblende rock (S 55166) which closely resemble the hornblendic and garnetiferous skarn round the Clothister magnetite deposit. These rocks appear to have preserved their pre-diorite character and if it may be assumed from their unusual mineralogical features that they are geologically the equivalent of the Clothister magnetite then the conclusion that the latter is of pre-granite–diorite complex date (p. 20) is supported. The presence of such rocks so far on the other side of the Haggrister Fault suggests also that the structural association of the Clothister greenschist assemblage with the banded hornblendic group of gneisses is of wide extent across the regional strike; but it remains uncertain whether the juxtaposition of the two groups is tectonic, as supposed for the Clothister area (p. 22), or successional, perhaps unconformable. The specimens of the banded group of the Busta peninsula are thermally altered in greater or lesser degree and many show later impregnation by quartz.

MUCKLE ROE

The gneiss on Muckle Roe occupies two main belts (Fig. 5). The smaller extends from a short distance west of Roe Bridge westwards for about $\frac{3}{4}$ mile (1·2 km). The coastal exposures of gneiss are to a large extent concealed by drift and towards the western end they are interrupted by outcrops of diorite.

The most westerly outcrop is isolated in diorite and lies directly south across Roe Sound from the most westerly of the small gneiss outcrops in the Busta peninsula. Inland from the coast the gneiss belt attains its greatest width of about ½ mile (800 m) to the west of Kilka Water. Its boundary against diorite is irregular and deeply embayed and in the broadest part outcrops of diorite within the gneiss suggest that the surface between the intrusive mass and the gneiss is rising southwards. This east–west belt of gneiss is not seen to join with the larger belt which runs north–south along the east coast of Muckle Roe but there seems little doubt that the two belts are continuous, perhaps with interruptions of diorite, since the gneiss is exposed again in Crog Holm, in Roe Sound and on the rock which supports the bridge linking Muckle Roe to the mainland. These exposures serve also to indicate the continuity of the Busta peninsula gneiss with that of the Muckle Roe belts.

The larger belt of gneiss on Muckle Roe extends from Roe Bridge south along the east coast of the island to Scarfa Taing, a distance of 1½ miles (2·4 km). Its outcrop is narrow except around Orwick Water where the width is nearly one third of a mile (0·5 km). It is not known whether the outcrop of gneiss 400 yd (360 m) NW of Northpund croft is continuous with the coastal exposures. A number of large gneiss enclaves crop out in granite in the Burn of Scarfataing on the coast south-west of Scarfa Taing (Fig. 5).

Along the coast section relations of the country and the intrusive rocks are obscured by the effects of numerous crush-zones. These include north-easterly and north-westerly lines but the main crushes trend N–S, that is they are parallel to the line of the Walls Boundary Fault the major dislocation of which must lie close to the Muckle Roe coast. In the crush-zones the rocks are shattered, locally sheared, and locally brecciated, while the adjacent massive rock has acquired a dull, greenish black amorphous aspect so that along stretches of the more easterly length of the section field distinction of basic gneiss and basic intrusive is at best doubtful. A traverse along the coast southwards from Roe Bridge shows pinkish weathering gneiss and granulite with which numerous layers of dark hornblendic schist and granulite are interbanded; at 200 yd (185 m) NE of Northpund the basic component is a sheet, about 35 ft (11 m) thick, of massive biotite-hornblende-andesine-gneiss in a group of banded grey and green granulites. Southwards along the 300 yd (270 m) of coast between Northpund and Lubba the exposures are all of dull green massive rock showing no foliation but otherwise resembling the massive basic gneiss. Microscopically the rock is dolerite and is therefore considered to form a basic facies of the intrusive complex. Like the metamorphic rocks the dolerite is cut by dykes and veins of granite and by lines of crush. Between the southern termination of the doleritic material and Green Taing the rocks are a mixture of granite, diorite and gneiss, greatly crushed and often indeterminate, but foliated feldspathic and hornblendic gneiss is exposed in the burn from Orwick Water and along the coast north and south of the burn mouth. At and south of Green Taing the main interest of the section lies in the exposure of several parallel N–S crush-lines in rock which is mainly dioritic with granite veins but includes banded gneiss. The crush-lines are vertical but in the reefs 200 yd (182 m) S of Green Taing banded basic and acid gneiss is seen to crop out below a flat dislocation with diorite on top. The melange of igneous rock and gneiss ends at a fault where the coast bends south-westwards 250 yd (230 m) NE of Scarfa Taing. From this fault to the Burn of Scarfataing the rocks are mainly scapolitized basic gneiss with some quartz-soaked garnetiferous schist but speckled rock of dioritic aspect also occurs and in some of the specimens of scapolitized rock igneous textures as well as foliation can be seen (S 45029–31). Close association of scapolitized gneiss and basic igneous rock occurs also on the isolated outcrop of Scarfa Taing. On the coast south-west of Scarfa Taing the main rock is granite which encloses slabby or wedge-shaped masses of gneiss. This type of large inclusion can be seen also in four exposures in the Burn of Scarfataing along the 500 yd (450 m) stretch up from its mouth. These masses appear generally to be elongated along the N to NE strike and down the vertical dip, but in the exposures on the north bank 300 yd (270 m) upstream from the mouth of the burn the granite has lenticular sheet relations to the foliation of the gneiss. Interposition of granite and gneiss is found also in the outcrops on the low ridge north-east of the burn. Some of the massive unfoliated basic rock in the gneiss of the burn and ridge exposures proves to be uralitized dolerite but the relations of gneiss and basic intrusive rock are obscure along this coast owing to the extensive shattering. It is noteworthy, however, that the gneiss nearest to the major body of diorite in the north-east of Muckle Roe is texturally reconstructed (S 28902, 44621, 45036) and it is inferred that the thermal alteration of the gneiss was induced by the basic intrusive. The only clean contact of gneiss and diorite which has been observed in this area lies in the outcrop west of Orwick Water where the medium-grained igneous rock forming most of the outcrop has a fine-grained chilled facies along the contact with dark banded hornblendic gneiss. The contact is parallel to the general strike of the foliation; its hade is not known. Outcrops of foliated hornblendic rocks north, east and south of Orwick Water link the western outcrop with the gneiss of the coastal exposures.

The distribution of gneiss and basic intrusive rock along the strip south of Roe Sound is analogous with that in the centre of the Busta peninsula and it is here equally uncertain whether the gneiss forms a roof or a floor to the intrusion. The continuity of the gneiss from Scarfa Taing northwards for 2 miles (3·2 km) to the Ward of Runafirth and the parallelism of this long outcrop with the trend of the dyke-like body of granite in the Busta peninsula suggest that the gneiss has a generally wall-like relation to the gabbro-diorite. Exposures on the coast and in the Scarfataing Burn, described above, show that locally at least the igneous rock intrudes into this wall as sheets. The form of the gabbro-diorite is discussed more fully in Chapter 13 (p. 178).

Lithologically the gneiss of Muckle Roe is similar to that of the Busta peninsula. Foliated hornblende-rich rocks are predominant in most areas but quartzite and quartzo-feldspathic granulite occur north of Scarfa Taing and banded grey and green, pink-weathering siliceous types occur south-east of the bridge across Roe Sound. Pelitic schist appears to be rare. Thermal metamorphism of the basic rocks led mainly to textural reconstruction but pyroxene crystallized along with the new granular hornblende in one of the basic layers, though in another specimen it is uncertain whether the pyroxene is not a pre-thermal mineral. Later hydrothermal and metasomatic effects are shown in quartz permeation of garnetiferous schist and biotite-oligoclase-gneiss and replacement of granular hornblende-gneiss by microperthite. In addition to the usual foliation pre-thermal structures traceable in several specimens include a second foliation and shear or fracture planes.

Mill Lochs and Quhaap Knowe. In this area, which lies in the interior of the Muckle Roe granophyre, three outcrops of banded gneiss occur on the west and south shores of the southern of the two Mill Lochs and on Quhaap Knowe. Specimens consist of well-foliated gneiss in which folia of hornblende and andesine alternate with more biotitic and apatite-rich laminae and with quartz-oligoclase layers. The gneiss has been thermally recrystallized with development of granoblastic feldspar and hornblende and locally the amphibole is porphyroblastic. Obscure large aggregates peppered with ore granules in the biotitic laminae may represent cordierite (S 44623–4). In hand specimen the rocks have a sheared aspect and in thin section relict phacoidal structure and recrystallized healed fractures are seen.

Lithologically the rocks conform with the gneiss country rock of Muckle Roe and these small masses are regarded as enclaves deeply sunk within the granophyre intrusion. In the field they have a superficial resemblance to the dark varieties of banded felsite dykes and it is possible that more enclaves of this kind remain to be recognized.

THE BUSTA ISTHMUS

The geology of the Busta isthmus is complicated owing to the southward convergence of faults into the line of the Walls Boundary Fault and it is difficult to assign the rocks to their uncrushed originals (p. 22). However the pelitic schists and foliated hornblendic rocks cropping out round Ell Wick and astride the main road west of the Busta road junction are very similar to those of the Skipadock area in lithology and mineral composition and like them lie on the north-east side of the southward continuation of the Mangaster Fault. On the other side of this fault gneiss appears to occur only as enclaves in the igneous rock; the only specimen available comes from the faulted mixture of granite, diorite, and gneiss on the west side of Ell Wick and is a pelitic rock which has been thermally altered, impregnated by quartz, and finally scapolitized (S 53595). East of but close to the assumed line of the Mangaster Fault and 150 yd (140 m) S of the main road an outcrop has provided two specimens, one of which is a thermally altered sheared rock (S 44330), resembling the albite-schists of the Clothister area. The other is a crushed sheared rock composed largely of phacoids of quartz and of alkali-feldspar in chloritic quartzo-feldspathic rock powder. This shows no sign of thermal alteration (S 55248) and may represent an early acid phase of intrusion. The outcrop from which these specimens come is shown in Fig. 5 as caught between two dislocations and as tectonically separated from the main strip of gneiss which is unshaded. Specimens from the east shore of Ell Wick include a corundum-bearing pinitized pelite (S 53586) and a mylonitized and folded epidote-hornblende-plagioclase-gneiss with magnetite porphyroblasts which shows some recrystallization but no thermal reconstruction (S 44278). Thermally altered semi-pelitic schist outcrops by the roadside [346 677] 150 yd (140 m) W of the Busta road junction (S 28914); the thin section gives particularly clear evidence of early folding overprinted by the thermal reconstruction. East of the fault zone which controls the coastline on the north-east of Ell Wick, the only specimen available is a banded hornblende-feldspar rock which has been sheared with folding and disruption and subsequently recrystallized (S 53585); it is uncertain whether the late re-crystallization is due to purely thermal action.

Thus in the Busta isthmus there are from west to east the following faulted strips running NNW–SSE to N–S: (1) a strip where enclaves of gneiss occur in the igneous rock; (2) a strip where granite is in contact with high grade pelitic hornfels, and which is the probable equivalent of the Skipadock gneiss; (3) a strip of gneiss which has been much sheared but little recrystallized; (4) a strip of mainly indeterminate rock (p. 19). The gneiss of strip 3 may equate structurally either with the Skipadock gneiss or with the hornblendic gneiss of South Ness (p. 15). Further field and petrographical work is required for satisfactory elucidation of the structure and geological correlations in this area.

PETROGRAPHY

The schistose or gneissose rocks within or in contact with the granite–diorite complex in Muckle Roe, the Busta peninsula, and southern Northmaven include foliated basic rocks, pelitic schists, and siliceous or semipelitic granulites. Other types include a contact-altered dyke and an amphibolite interlayered with garnet-magnetite-schist. Most of the specimens examined show contact alteration and many provide evidence of later low temperature alteration by siliceous or alkaline solutions. Scapolitization is locally intense but since this affects also the igneous rocks it is described in Chapter 13.

SILICEOUS AND SEMIPELITIC ROCKS

Flaggy laminated grey and pink granulite was collected from one of a number of xenoliths in granite 950 ft (290 m) WSW of Northknowe, Muckle Roe [342 642]. It is the only quartz-microcline-granulite (S 45032) in the collection from the area under description and is composed of granoblastic quartz between which irregular grains of microcline and bent flakes of chlorite and muscovite are moulded. The microcline shows only patchy cross-hatching and contains thin tongues of perthitic plagioclase. Drops and irregular grains of ore are abundant and zircon is common. The rock has irregular laminae of sericitic aggregate which look like pseudomorphs and locally contain tiny scales of green biotite of new development. In these sericitic laminae there are relics of oligoclase in the microcline. This feature and a tendency of late quartz to enclose smaller grains of quartz, drops of microline, and micaceous aggregate suggest slight potash-silica metasomatism. There is, however, no evidence of high temperature transformation other than the pinitic aspect of the micaceous aggregate which suggests that cordierite or andalusite may originally have been present.

Semipelitic granulite from Egilsay is composed of granoblastic quartz and oligoclase with interstitial potash-feldspar and brown biotite in laminae which alternate with dark folia of oligoclase and biotite with accessory apatite (S 55175). In these the biotite is in part coarse and raggedly terminated as in the quartzose laminae but is also commonly developed in small idioblastic laths. This rock is in contact with and is partly assimilated into a granodioritic facies of the diorite which forms the main rock of the island. Banded semipelitic rocks, from exposures in the Ell Wick area of the Busta isthmus, are greatly altered schists in which shear structures earlier than the thermal metamorphism are preserved; one of them (S 28914) appears to have been mylonitized before thermal alteration. Pelitic layers are composed largely of pinite, probably after andalusite, unorientated aggregates of small flakes of brown mica, sericite, muscovite, and obscure material, and contain small round groups of unorientated biotite and chlorite which may have been formed from decomposed garnet. Another is a banded granulitic gneiss (S 44330) containing green biotite in the quartzose layers, while the pelitic laminae are composed of quartz-sieved oligoclase, microcrystalline newly formed brown biotite, pinitic aggregates after cordierite swathing the plagioclase, and spinellid granules. The texture is that of a sheared rock recrystallized so much that only a folded banding is preserved. The microporphyroblastic and inclusion-filled nature of the plagioclase suggests correlation with the albite-schists of Clothister (p. 20) which are, however, not altered thermally. In this rock addition of quartz may have accompanied recry-stallization of the siliceous layers with concurrent formation of chlorite along strings which may represent old shear fractures.

PELITIC SCHISTS

Pelitic members of the gneiss have been noted or collected from all areas from Skipadock to Scarfa Taing, with the exception of the Otter Ayre [326 665] area, and pelitic xenoliths have been found in the far west of the Ness of Houll [327 672] in the south-west of the Busta peninsula.

All the specimens in the sliced rock collection show strong thermal alteration which in most cases has been followed by low temperature change and late soaking by quartz. Those in which the minerals of contact-alteration are best preserved show the normal change of pelitic schist to hornfels characterized by decussate brown biotite and lenticular or granoblastic pinitic pseudomorphs after andalusite or cordierite (S 53589, 53590); in only one specimen are there relics of fresh andalusite and also, doubtfully, cordierite (S 53591). Fresh corundum occurs in the latter rock and also, though mostly replaced by white mica, in a specimen (S 53586) from the east side of Ell Wick, and in this rock local groups of parallel prismatic grains, about 0·01 mm long, in the sericitic aggregates probably are relics of andalusite prisms. The feldspar of the pelitic schist is oligoclase ($\sim$ An$_{20}$) but may approach andesine ($\sim$ An$_{30}$), as in S 53591. Alkali feldspar is present in the more quartzo-feldspathic layers. It is however uncertain whether this is original since there has been considerable invasion by quartz and recrystallization of the pelite (S 53590): certainly in the case of the Ell Wick pelite (S 53586) the alkali-feldspar is introduced since it forms veins which enclose muscovitized and chloritized relics of the biotite-schist as well as forming a matrix to schist laminae. The feldspar is a Na-K species with β 1·525–1·526, γ 1·529–1·530. Accessory minerals include apatite, which may be very abundant, opaque or dark brown spinellid, minor zircon, and sphene which forms crystals which are small, clear and almost colourless in the hornfels, but large, turbid and brown in the feldspathic veins of the Ell Wick specimen.

While none of the hornfelsed pelitic schists fails to show some later permeation by quartz this process has in some cases proceeded so far that only relict lenses or aggre-gates of the hornfelsed schist remain in a granoblastic mosaic of quartz with minor interstitial feldspar or chlorite (S 53568, 53594, 53599, 53600). The relict schist may be well preserved. From the islet in Houlls Water [340 677] a quartzose specimen shows relics of hornfelsed biotite-schist containing garnet (S 29414). But even in the best preserved hornfelsed schist the degree of sericitization and pinitization is no less than in the relics dispersed in quartz and therefore it is concluded that a low temperature hydrous change preceded the quartz invasion. It is possible that during this period a coarser recrystallization of biotite took place (S 53568) and also at a later stage chlori-tization of biotite. The latter process is not necessarily concurrent with the influx of quartz (S 53568, 53599) since the intensity of chloritization and the proportion of quartz are not directly related.

Two quartz-impregnated pelites are very rich in garnet. In the less quartzose rock (S 44231), from the north side of Skipadock, the pelitic folia consist of a multitude of small unorientated biotite flakes, elongate green and colourless pinitic aggregates, xenomorphic garnet, minor oligoclase ($\sim$ An$_{15}$), abundant accessory apatite and magnetite and, locally, blue tourmaline. In the other, from the north side of Scarfataing Bay, most of the pelitic constituents have disappeared and their folia are represented by laminae of closely packed small garnet idioblasts, rarely over 0·05 mm across, inter-stitial turbid oligoclase ($\sim$ An$_{10}$) and chloritized brown biotite, and accessory apatite (S 44622). In both rocks it appears that garnet has been the most stable of the pelitic constituents though it has been recrystallized, while feldspar, biotite, and pinite have been removed. In these rocks as in several of the others cited above the quartz is rich in inclusions, trains of which pass without interruption from grain to grain, and it is clear that the quartz-bearing solutions have contained active fluid components.

The pelitic schists have been so thoroughly transformed by thermal action and invasion of late siliceous solutions that in general no conclusion can be drawn on the possibility of an episode of shearing of the regionally metamorphosed rocks prior to the thermal recrystallization, such as is shown by the basic types of the country rock. Only in one case (S 53590) does the microscopic texture suggest pre-thermal shear in the locally strong schistosity of the new biotite and the elongate structure of highly serici-tized feldspar.

BASIC ROCKS

The basic members of the gneiss vary from almost monomineralic hornblende-schist through well banded hornblende-feldspar types to banded biotite-hornblende-gneiss with quartz. All show some degree of reconstruction and many clearly have been recry-stallized under conditions of high or moderately high temperature and hydrostatic pressure so that the earlier foliation may be greatly obscured. Schistosity within the folia of the rocks has been extensively destroyed (S 28906, 33750) and may be preserved only as a mimetic recrystallization of hornblende, the crystals of which tend to lie with the prism axis parallel to the former schistose structure (S 44280, 44281A). In some of the rocks, however, pre-thermal textures and structures are partly preserved. For example, the early schistosity is shown by orientation, parallel to the foliation of the rock, of pale green prisms of hornblende which is in part converted to unorientated brown hornblende (S 44325). Again, in a specimen from the Busta Voe coast 200 yd (180 m) NE of Northpund, the feldspar grains contain trains of inclusions of new recrystallized hornblende which lie parallel, transverse, or at any intermediate angle to the preserved foliation of the rock and to the mimetic schistosity in the hornblende folia (S 45036). Two periods of directed stress prior to the thermal metamorphism are thus implied. In rocks from the Skipadock area relics of shearing and cataclastic structures are clearly of pre-thermal date (S 53596 53601) and in a rock from the Ness of Coulsetter [333 676] plane-foliated laminae of very fine-grained minerals are interpreted as mylonite recrystallized in varying degree (S 44280). Sharp folding, shearing, and disruption of the foliation in a rock from one of the middle strips of the Busta isthmus are clearly earlier than the latest recrystallization but it is uncertain whether that event is here due to the plutonic intrusion (S 53585). Early fractures filled by epidote (S 44621, 53605) which cross the foliation but are interrupted by or disappear in the thermally recrystal-lized hornblende and feldspar, indicate pre-thermal fracture and low temperature mineral transport. These examples together provide evidence of a period of folding, shearing, and fracture of the regionally metamorphosed gneiss group prior to intrusion of the granite–diorite complex.

In thin section the basic members of the gneiss are seen to be composed essentially of hornblende and plagioclase; fully half of the sliced specimens contain pyroxene also as an essential mineral. Epidote is abundant in some but commonly absent or minor in those which are rich in pyroxene; zoisite or clinozoisite also occurs. Biotite is common in the more feldspathic members, and pelitic layers in these may contain pinite. Apatite, magnetite. and sphene are accessory and each of these minerals tends to be more abundant along thin laminae.

The feldspar of those members which have been well recrystallized lies in the andesine range, An 40 to 50 per cent, and is clear and granoblastic (S 53605). It is however much more calcic, with An about 75 per cent, in a rock composed mainly of equant brown idioblastic hornblende (S 44281 and A). In this rock, which contains no pyroxene, there is transition, from place to place in the section, between pellucid plates of bytownite and microgranular aggregates of clinozoisite; an earlier schistosity is preserved by trains of tiny hornblende prisms enclosed in the new plagioclase and a vein of prehnite cuts the recrystallized rock. In many specimens the feldspar is too turbid or sericitized for determination but seems from its relief to be less calcic (S 44621). Since this feldspar is granoblastic and encloses new crystals of amphibole and sphene it is possible that the sericitization and less calcic character have been acquired after the thermal alteration. In some specimens, however, large sericitized and epidotized grains are clearly relics from the pre-thermal rock (S 53601).

The usual hornblende of the contact-altered rocks forms stumpy prismatic or equant unorientated crystals with strong pleochroism X golden or straw yellow, Y deep olive-green or brown, Z olive-greenish brown; γ: c 16°–20°. Rarely it shows a tendency to late recrystallization in radiating bunches and microporphyroblasts (S 53596). The olive-brown species may pass marginally and on the prism terminations to a greenish blue species which presents idioblastic prism forms to the granoblastic feldspar (S 44621). Usually some proportion of a pre-thermal pale green to colourless amphibole is present as closely cleaved, almost fibrous prisms schistose parallel to the foliation. Transformation of this type of amphibole to the deep brown equant type can be clearly followed (S 44325, 63505). The pyroxene produced in the thermal recrystallization is colourless or faintly green and occurs as single or aggregated prismatic granules, usually less than 0·05 mm, rarely more than 0·1 mm long. These are enclosed in the granoblastic feldspar and packed among the clusters of brown hornblende with which they interfere as concurrent crystallizations and in no way as replacements (S 46602, 53596). Pyroxene does occur also as larger (up to 0·5 mm) interfering crystals which form aggregates parallel to the foliation (S 29418). They are in part recrystallized to clear prismatic pyroxene (S 53601). The relics of large turbid epidotized feldspar grains and irregular elongate clots of pale green coarse amphibole along with the coarse pyroxene aggregates in this rock, which otherwise is composed of clear granoblastic andesine, colourless pyroxene, brown hornblende, and a little red biotite, suggest that it had been a pyroxenic epidiorite. The presence of discontinuous and in places contorted trains of ore granules through the base suggests also that the rock had been sheared prior to the thermal metamorphism.

Biotite is not a common mineral in the altered basic rocks and occurs in essential amount mainly in the more feldspathic (S 45036) and quartz-bearing (S 44280, 55168) members. In the former the biotite occurs as irregular flakes, darkened by sagenite, interlocking with amphibole and plagioclase in a complex foliated base (p. 29). This rock forms a 12-yd (11-mm) wide mass, perhaps a pre-foliation dyke or sheet, in banded grey and green granulites on the coast of Busta Voe, 200 yd (185 m) NE of Northpund and seems only in minor degree affected by thermal recrystallization. One of the quartz-bearing rocks, an inclusion in diorite forming the Lothan islet [311 676], is a well-banded gneiss in which folia of granoblastic quartz and andesine, on which prismatic biotite is moulded, alternate with others rich in granular monoclinic colourless pyroxene (S 55168); the rock contains also sporadic grains of uralitized optically negative orthopyroxene and numerous granular pseudomorphs in chlorite and birefringent pinite. Biotite, however, is an important essential constituent of a basic schist which forms a minor outcrop in diorite on the northern slopes of Roe Sound [335 664], 200 yd (180 m) NW of Houll. The rock is composed mainly of schistose hornblende and minor granoblastic andesine (S 44325). The schistose hornblende forms pale green, closely cleaved prisms with frayed terminations and is in course of conversion to a clear yellowish-brown hornblende which appears centrally and marginally to the green variety as well as forming stout idioblastic unorientated prisms. In other layers the hornblende though idioblastically recrystallized remains green and is cemented by shapeless unorientated fox-red biotite which has a small optic axial angle (2E $\sim$ 20°). In yet other layers this biotite is the main constituent. Since in this, and in the other rocks cited above, the biotite is an important mineral in certain bands or laminae only, its formation is due primarily to suitable chemical composition of these layers.

Potassium feldspar is present in two specimens of the basic country rock and appears to be derived from granitic material which at both localities veins the diorite in contact
D

with the basic gneiss. One of these (S 45132) is unusual in containing turbid oligoclase in irregularly sutured grains as the main mineral and in being exceptionally rich in sphene and iron ore along well-defined directions. This rock may have been a tuff. It is very variably recrystallized and the veins of potassium feldspar within it enclose fragments of the more crystallized minerals which are in optical continuity with the minerals marginal on the vein. In the other rock, which is highly contact-altered and contains pyroxene and biotite, microcline occurs in small pools among the andesine folia and in large crystals which appear to have a replacing relation to the fine-grained granoblastic plagioclase (S 53601).

OUTLYING XENOLITHIC BODIES OF BASIC GNEISS

The small mass of basic foliated rock enclosed in the granite at Djubi Dale [337 743], on the northern margin of the sheet, is similar in thin section (S 43766, 44104–6) to the andesine-hornblende rocks containing much microgranular pyroxene and local red biotite described above. Granoblastic recrystallization of feldspar is far from complete and turbid microaugen, perhaps augiclasts, are relict.

The small masses enclosed in the granophyre at Quhaap Knowe and Mill Lochs, in the south of Muckle Roe, are foliated andesine-hornblende rocks (S 44624) with folia of biotite and ore-sprinkled pinitic aggregates which may have been cordierite. The hornblende of this rock is shapeless, porphyroblastic, and sieved by plagioclase. This habit, which has not been seen in other thermally altered rocks of the area, and the sieving of plagioclase by optically continuous droplets of quartz indicate a penetrative activity of fluids during immersion of the gneiss in the consolidating granophyre.

Folia of quartz occur in the more feldspathic layers (S 44623) and may represent original quartzo-feldspathic laminae. Potassium feldspar is locally abundant and its varying relation to the plagioclase, cementing, sieving, or enclosing, suggest that it is derived from the granophyre.

From a small body of country rock enclosed in diorite 200 yd (180 m) SW of Bays Water [331 668] two rocks very different in character from all others from Muckle Roe and the Busta peninsula area require mention. One is an amphibolite (S 34995) composed essentially of an unorientated felt of amphibole blades and fibres, pleochroism X straw yellow, Y deep green with a tinge of blue, Z deep greenish-blue and small varying optic axial angle, negative sign. The colour varies from faint in the fibrous forms through pale in the core to deep at the margins of larger prisms. Acting as a sparse cement to the amphibole are aggregates of clear overlapping grains which are demonstrably quartz when sufficiently large for determination but may include feldspar. Crystals and aggregates of magnetite are numerous and fairly evenly distributed through the rock while apatite grains, also numerous, tend to occur in short, roughly parallel chains. Epidote also is abundant forming small aggregates within the amphibole felt and occupying large spaces in which amphibole is scanty. The rock shows no indication of thermal recrystallization and is cut by thin veins of clear mineral which is probably albite. The other rock (S 55166) is composed of an open sponge of magnetite crystals, the small pores of which are filled by bunches of colourless to pale bluish amphibole prisms, up to about 0·05 mm long, while the larger spaces are occupied by aggregates of interfering, straw yellow to bluish-green amphibole, which in sections showing strong Y green, Z greenish blue dichroism gives an off-centre, almost uniaxial figure in convergent light. Much less abundant than amphibole, pale pink, slightly anisotropic, garnet aggregate also occupies large spaces in the magnetite and encloses both magnetite and amphibole; it appears also as idioblastic crystals in the coarser amphibole aggregates. Apatite is a very abundant accessory constituent occurring as stout prismatic grains, up to 0·3 mm long, enclosed both in magnetite and in amphibole and tending to form chains nearly 1 cm long which are roughly parallel to one another and to thin monomineralic amphibole streaks. The only other constituents are quartz and locally associated epidote occupying small spaces between garnet and magnetite or amphibole.

In their association, their unusual mineral composition, and the species of amphibole which is their main constituent, the rocks of this outcrop compare closely with those peripheral to the Clothister magnetite deposit (p. 20); the only difference is in the abundance of apatite in the Bays Water specimen.

Before concluding this section on the petrography of the Muckle Roe–Busta country rocks attention may be drawn to the occurrence of dyke rocks which cut the gneisses but are thermally altered by the diorite, and to the scapolitization which affects some of the gneisses as well as the diorite. These phenomena are described in Chapter 13.

SUMMARY

The petrography of the gneiss group shows: (i) that it is a banded series of hornblende-rich schist and hornblende-plagioclase-gneiss in which pelitic schists, locally garnetiferous, are important and quartzo-feldspathic and quartzose granulites are minor components; (ii) that this series, regionally metamorphosed in the amphibolite facies, experienced later shearing and fracturing at low temperature prior to (iii) thermal metamorphism, locally to the pyroxene-hornfels grade; (iv) that low temperature hydrothermal changes affected the thermally altered rocks with pinitization of cordierite and andalusite and metasomatism by areally active siliceous and local potassium-bearing solutions.

The combined field and petrographical data indicate that:

1. the gneiss group is the equivalent of the striped hornblende-gneiss series, the Burravoe Gneiss, which in the area of One-inch Geological Sheet Northern Shetland extends from Fethaland at the extreme north end of the Shetland mainland to Colla Firth and Eela Water;
2. the gneiss of the Skipadock area may underlie the basic igneous rock and appears to be intruded vertically, but with sheet apophyses along the foliation, by the Eastern Granite;
3. the gneiss outcrops on Egilsay and Black Skerry may represent roof pendants;
4. the gneiss of the Busta–Muckle Roe area represents either a roof or a floor to the basic igneous intrusion. There is no evidence of superposition of either on the other. Topographical distribution suggests the gneiss is the lower rock;
5. the surface between gneiss and basic intrusion is very irregular and separation takes place along generally vertical planes controlled by directions of foliation and cross-fracture in the gneiss;
6. the junction between the gneiss and Eastern Granite is near vertical with granite apophyses perhaps guided by the foliation of the gneiss;
7. no mechanical disintegration of the gneiss is associated with the intrusion;
8. chemical reaction between gneiss and igneous rock has occurred only very locally, on Egilsay;
9. thermal alteration of the gneiss is due to the basic igneous rock and was high locally but not sufficiently prolonged at high temperature to cause complete reconstruction of its more basic hornblende-feldspar components;
10. thermal alteration was followed by hydrothermal transformation and later silicification; locally potash-metasomatism was associated with granitic intrusion;
11. country and intrusive rocks were shattered by movements along the Walls Boundary Fault-zone and both were scapolitized during this period of movement;
12. the occurrence, south of Bays Water, of garnet-magnetite rock and amphibolite resembling the Clothister association (p. 20) implies that skarns of the

Clothister type are, or were, more numerous than suspected. The contrast of the abundance of apatite in the Bays Water rock and its poverty at Clothister suggests that the skarnizing fluids were of different composition in separate channels. If the two occurrences were originally related in space tectonic displacement is implied.

Chapter 4

THE METAMORPHIC ROCKS OF THE WALLS PENINSULA

INTRODUCTION

METAMORPHIC ROCKS form an east–west trending outcrop, 1 to 2 miles (1·5–3 km) wide, which extends along the northern coast of the Walls Peninsula and the adjoining islands from Melby in the west eastward to Papa Little, where it is truncated by the Walls Boundary Fault (Plate II). The succession is made up of four major lithological units and gives rise to relatively rugged terrain with a high proportion of rocky outcrops. The foliation of the metamorphic rocks has an overall east to north-easterly trend and an inclination which ranges from 40° to over 70° to the south or south-east. As the inclination of the foliation agrees roughly, both in direction and amount, with that of the bedding of the overlying Old Red Sandstone sediments, it is assumed that prior to the deposition of these sediments the foliation of the metamorphic rocks was approximately horizontal.

The succession is composed of the following groups, from north to south:

Vementry Group. Hornblende-schists and amphibolites interbedded with quartzo-feldspathic semi-pelites and some quartz-granulites.

Neeans Group. Platy feldspathic muscovite-biotite-schists with large lenticular masses of coarse hornblende-schist and a few thin bands of limestone and calc-silicate rock.

West Burra Firth Group. Tremolite- and mica-schists with numerous limestones.

Snarra Ness Group. Hornblende- and mica-schists with bands and lenses of amphibolite. Subordinate tremolite-schist and quartz-granulite.

The outcrops of these groups do not form continuous bands.

The Vementry Group is exposed only in the north-western peninsula of Vementry and the outcrop of the West Burra Firth Group thins out westwards between Bousta [223 576] and Norby, apparently interdigitating with the Snarra Ness Group. The latter is present only in the area west of Brindister Voe. In the east the West Burra Firth Group is interdigitated with rocks similar to those of the Neeans Group. The present foliation is not everywhere concordant with the lithological boundaries, and detailed evidence (pp. 49–51) supports the view that it does not always coincide with the original bedding planes and other pre-foliation surfaces. The metamorphic rocks are cut by numerous faults and shear belts and the present boundaries between groups are, at least in part, tectonically modified.

A high proportion of the metamorphic rocks of the Walls Peninsula were originally sediments, probably siltstones and sandy siltstones with subordinate sandstones and limestones, together with one thick group of calcareous mud-stones and siltstones with fairly thick bands of limestone. The hornblende-schists and epidiorites of the hornblendic groups may have originated as basic lavas or pyroclastics, but the thick lenticular masses of coarse-grained hornblende-schist

39

and amphibolite, which are present in all groups except the West Burra Firth Group, may represent thick basic intrusions.

The metamorphic rocks contain many quartzo-feldspathic lits and porphyroblasts as well as granite or pegmatite veins. The period of feldspar 'permeation' and granite veining commenced before the onset of the Main Phase of Folding (p. 49) and continued until after its completion. It probably spanned three metamorphic episodes (p. 54). Thermal metamorphism associated with 'permeation' may have been responsible for the local development of high temperature mineral associations (p. 57). Feldspar lits and granite veins are most abundant along the southern margin of the metamorphic belt, particularly within the Snarra Ness Group, which is now structurally the highest member of the series. Here granitic material makes up nearly 20 per cent of the total volume of the rock. Granite veining is intense in the West Burra Firth Group, but is much less pronounced in the Neeans Group, where it forms thinner and less closely spaced veinlets. In the Vementry Group, however, granite and pegmatite veining approaches the intensity of that in the Snarra Ness Group.

The mineral associations within the greater part of the metamorphic belt indicate that during the Main Phase of regional metamorphism pressure–temperature conditions were those associated with the greenschist-amphibolite transition facies (Turner 1968, pp. 303–7). Areas containing minerals indicative of higher temperatures are present only in the most intensely permeated and veined belt (p. 57). Over a large part of the outcrop the metamorphic rocks have suffered retrograde metamorphism which is associated with two periods of movement.

The probable metamorphic and tectonic history of the series can be summarized as follows:

	Deposition of sediments, extrusion of pyroclastics or lavas, emplacement of basic intrusions.
$D_1{}^* = M1S$	**?First folding.** Evidence for this phase is inconclusive and based on the presence of planar and folded trains of inclusions in possible M1P porphyroblasts. Beginning of period of feldspar 'permeation' and granite-pegmatite veining.
$D_2 = M2S$	**Main phase folding,** producing the regional linear and planar tectonite fabric now seen in the rocks, by rotation and new growth of platy and acicular minerals. Texture fine-grained. New minerals include hornblende, biotite, muscovite. In calc-silicate rocks; epidote, clinozoisite, zoisite, tremolite, phlogopite.
M2P	**Mimetic coarsening of fabric and late porphyroblast metamorphism.** Annealing and growth of hornblendes, micas and epidote family minerals. Growth of porphyroblasts of garnet and albite-oligoclase.

---

*The individual tectonic and metamorphic episodes described in this chapter have each been labelled with a combination of letters and numbers, so that a repetition of lengthy titles and descriptions could be avoided in the text.

The first letter of each combination indicates the type of deformation or alteration affecting the rocks. D thus stands for deformation and M for metamorphism.

The numbers 1 to 5 when preceded by D refer to the first to fifth period of deformation, or the first to fifth period of metamorphism if they follow on M.

The letter following the number in the combinations starting with M indicates that the metamorphism was either synchronous (S) with the deformation or has post-dated (P) it. The letter P also indicates that metamorphism took place in a passive environment in which new minerals, whose orientation was not affected by any stress pattern, could grow. Similarly the letter S indicates that the new minerals were formed in an environment subjected to directional stress, and that their alignment follows a planar or linear pattern.

Near end of phase: Possible renewed movement causing rotation of garnets and new growth of mica.

D_3 = M3S **Intense shearing and granulitization,** associated with east- to north-east trending faults. Local belts of mylonite and flinty crush. Retrograde metamorphism: garnet, biotite and, in part, hornblende to chlorite. Granulitization of quartz-feldspar aggregates.

D_4 = M4S **Dextral kink folds and belts of conjugate folds.** Granulitization and 'mortar texture'. Bending of micas. Local biotite to chlorite. New chlorite parallel to axial planes of microfolds.

M5P **Thermal metamorphism,** north-west Vementry. ?Associated with intrusion of Vementry Granite or Muckle Roe Granophyre. New minerals biotite, actinolite.

D_5 **Regional tilting to south-south-east.** (= F1 affecting Sandness and Walls formations, see p. 126). Associated with formation of east to north-east trending faults and crushes close to and along junction between basement and Old Red Sandstone sediments. No mineral reconstruction.

D_6 **NNE to NNW trending major faults** with associated narrow belts of conjugate folding. These large-scale block movements of the basement may be contemporaneous with the NE to NNW trending folds in the Sandsting and Walls Formations, i.e. F2 affecting Walls Sandstone. (p. 134).

Radiometric ages of samples from the metamorphic rocks, determined by the potassium-argon method, have been obtained by Dr. N. J. Snelling. These are as follows:

IGS 67.1   Muscovite from platy muscovite-biotite-gneiss, Neeans Group, south-west slope of Muckle Hoo Field [267 585], 330 yd (300 m) E14°N of south end of Maw Loch. Percentage potash 8·22. Radiogenic argon $1·458.10^{-4}$ scc/gm. *Age* 399·5 ± 15 *m.y.*

IGS 67.2   Muscovite from platy micaceous granulite, Neeans Group, shore between Ayre of Whalwick and Turl Stack [258 586], 220 yd (200 m) N of Whalwick. Percentage potash 8·37. Radiogenic argon $1·558.10^{-4}$ scc/gm. *Age* 416 ± 16 *m.y.*

IGS 67.3   Muscovite from platy micaceous granulite, same locality as IGS 67.2. Percentage potash 8·55. Radiogenic argon $1·576.10^{-4}$ scc/gm. *Age* 413 ± 16 *m.y.*

IGS 67.48   Hornblende from garnet-amphibolite, Snarra Ness Group, south shore of West Burrafirth [244 570], 1250 yd (1150 m) W10°N of Burraview. Percentage potash 0·816. Radiogenic argon $1·593.10^{-5}$ scc/gm. *Age* 435 ± 20 *m.y.*

IGS 67.49   Hornblende from garnet-amphibolite. Same locality as IGS 67.48. Radiogenic argon $1·323.10^{-5}$. *Age* 415 ± 20 *m.y.*

Snelling states that the ages obtained do not differ significantly within the limits of experimental error. The average of these ages is 415·5 m.y., which is virtually the same as that of the ages obtained by Miller and Flinn (1966, pp. 100–3) from specimens from the Shetland East Mainland succession which is 422 m.y., and from the zone of Read's second metamorphism in the Valla Field Block of Unst (Read 1934) which is tentatively dated at 418 m.y. The ages from Western Shetland are not thought to date any of the metamorphic or tectonic events listed above. They merely set a younger limit to the time of metamorphism, possibly pinpointing the last time at which the temperature of the rocks fell below 200°C (cf. Harper 1967). They may thus date the period of uplift that preceded the deposition of the Old Red Sandstone sediments.

LITHOLOGY

VEMENTRY GROUP

The outcrop of the Vementry Group is confined to the north-western peninsula of the Island of Vementry and possibly to a narrow strip along the southern margin of the Vementry Granite in central Vementry. The group consists of feldspathic and quartzose schist and granulite together with hornblende-schist and garnet- or epidote-amphibolite. The acid and basic rock types are inter-banded and occur in roughly equal proportions. Two distinct varieties of horn-blendic rocks are present. The massive poorly foliated, generally coarse-grained hornblende-rich types are here termed amphibolite. They form distinct lenses which give rise to marked topographic features, and probably originated as sills of melanic dolerite or gabbro. The second type are strongly foliated finer-grained hornblende-schists which form an integral part of the metasedimentary succession and may originally have been basic lavas or tuffs.

Amphibolite and *epidotic amphibolite* form bodies, which range in size from massive, coarse-grained lenses over 110 yd (100 m) wide, like that forming Heill Head, to narrow finer-grained black bands only inches wide. Coarse amphibolites (S 30780, 49310) consist of poorly aligned moderately acicular crystals of dark grey hornblende, which in thin section is pale yellowish green and only faintly pleochroic, together with porphyroblasts of pink garnet, which are up to 0·4 in (10 mm) in diameter and generally contain aligned or helicitically arranged inclusions of quartz or epidote. Though epidote forms inclusions within the garnets of some rocks neither epidote nor clinozoisite normally occur together with garnet in the same rock. These calc-silicate minerals form up to 40 per cent by volume of some amphibolites and occur as either slightly elongated crystals of subhedral to euhedral shape, or as crystal networks. The feldspar is albite-oligoclase, which in some cases forms porphyroblasts enclosing bent hornblende crystals. Accessory minerals are sphene, which commonly forms parallel trains of inclusions in hornblende, apatite, zircon, ilmenite, and pyrite. The composition of *hornblende-schists* (S 30720, 49315) is similar to that of amphibolites, though they are finer-grained, have a stronger fabric and contain a higher proportion of albite-oligoclase and subordinate quartz.

The *feldspathic schists* (S 30730, 30728) consist of albite-oligoclase, large garnet porphyroblasts, and epidote, which in some instances (S 49311) forms a high proportion of the rock. The plagioclase is either partially altered to clay minerals and sericitized or has aligned patches of clinozoisite. Micas are sub-ordinate in amount and the biotites are commonly retrogressively altered to chlorite.

The schists are traversed by veins of granite and contain many feldspathic lits. There are also a number of pale non-foliated bands, tens of metres thick, which are composed entirely of sodic plagioclase, quartz and epidote group minerals. Many feldspathic granulites contain small ovoid pods of amphibolite and at one locality quartzo-feldspathic schist contains irregular streaks and patches of pyrite.

The outcrop of the Vementry Group is cut by a number of major north-east to east-north-east trending crush belts (Plate II), and the entire area is traversed by innumerable sub-parallel and intersecting minor crush planes, which are present in almost every thin section from the area (Phase D3 of p. 41). The intensity of the crushing increases northwards, and in the northern half of the peninsula intersecting shear planes have in part obliterated the D2 foliation and lineation. Crush-rock approaching mylonite in texture (S 49316, 49345, 49313) is developed in many crush belts. On the east coast of Swarbacks Head the hornblende-schists are largely reduced to fine-grained greenish chlorite-schist. Over the greater part of the outcrop oligoclase and quartz are intensely granulitized and streaked out into lenticles (S 30721), with minute grains of new epidote (S 49345) and patches of chlorite developed in the feldspar. The lenticles are thinly sheathed in chlorite or in irregular bands of mylonite. In the less sheared specimens (S 49308) quartz is granulitized, but the plagioclase is bent and fractured. Garnets are shattered, rotated and in many cases streaked out (S 47806). They are altered to chlorite, either completely or to varying extents along cracks, and generally enclosed in a sheath of chlorite flakes. Amphibole is in most cases only bent and shattered. Only in the north-east of the peninsula is there appreciable retrogression of hornblende to chlorite (S 49315). Epidote and clinozoisite show little evidence of mechanical break-up and none of secondary alteration, suggesting that much of it may have recrystallised during this phase of movement.

NEEANS GROUP

Platy feldspathic muscovite-biotite-schist forms two belts separated, in the eastern part of the area, by the West Burra Firth Group (Plate II). In both belts the predominant rock type is pink to buff platy feldspathic and garnetiferous mica-schist or granulite. The texture is locally almost gneissose, but regular feldspar lits, characteristic of true gneiss, are nowhere present. The group contains a number of thick belts of more massive quartz-epidote-feldspar-granulite which form prominent escarpments in the Neeans peninsula (Plate VA). In the eastern and southern parts of Vementry Island the group contains a number of silvery mica-schists which are almost phyllitic.

Though the ratio of hornblende-schist and amphibolite to mica-schist is considerably lower than in the two hornblendic groups, amphibole-rich bands and lenses, which commonly contain a high proportion of epidote and clinozoisite, are present throughout the group. Lenticular masses of poorly foliated amphibolite or amphibolite-epidosite, which may originally have been basic sills, are up to 75 yd (70 m) wide and 380 yd (350 m) long and form prominent topographic features. Rather better foliated hornblende-schist forms many smaller lenses and bands. The latter show intense internal folding and regular alignment of hornblende needles. Limestone and epidosite bands, some over 10 ft (3 m) thick, are present throughout the group, but form less than 1 per cent of the total volume of rock. The limestone and epidosite ribs are associated with both mica-schists and hornblende-schists. Though thin granite and pegmatite veins are present throughout the group, they are generally much thinner than in the other groups and form a lower proportion of the total rock volume.

A small, almost round, outcrop of serpentinite, 55 yd (50 m) in diameter, occurs within the group on Vementry Island, close to the north-west corner of Maa Loch [297 604]. This is composed of 70 per cent antigorite, and 20 per cent talc which forms irregular fibrous patches, together with small grains of opaque ore minerals (S 30733). The serpentinite has no planar or linear fabric, but its field relationships suggest that it was involved in at least some of the deformation undergone by the metamorphic rocks.

The *feldspathic schists* and *granulites* (S 30715, 47760, 47745, 47753) are composed of quartz, albite-oligoclase, muscovite, biotite and garnet. Muscovite forms large plates whereas biotite, which is either strongly pleochroic, deep brown to greenish brown or reddish brown, forms smaller plates, generally bounding the muscovite. Chlorite normally occurs as the retrograde alteration product of biotite and garnet. It is abundant in areas adjoining the major N–S and ENE trending faults and along belts of strong D4 folding (S 49324). Chlorite plates are also developed along and parallel to the axial planes of some D4 folds (S 49319). The thinly foliated phyllitic muscovite-schist exposed in the eastern and southern peninsulas of Vementry Island (S 49319–20, 49326) contains small needles of tourmaline (pp. 52, 58). Some of the pale granulites within the group contain varying amounts of epidote and clinozoisite and, more rarely, tremolite. Some contain interstitial calcite which has partially replaced quartz and plagioclase. Abundant accessory minerals are sphene, apatite, zircon, rutile and pyrite.

The limestones (S 50135, 30746) consist of calcite, quartz, clinozoisite and epidote together with accessory cloudy feldspar and muscovite. Both tremolite and phlogopite (S 49321, 57754) have been recorded in subordinate amounts in the northern part of the group's outcrop, but are more abundant near the junction with the West Burra Firth Group. Limestones commonly grade into, or are interbanded with almost calcite-free calc-silicate-rock, which in many outcrops is closely associated with amphibolite. The most common constituents of calc-silicate-rocks within this group (S 47758, 47759) are epidote and clinozoisite, which forms euhedral to subhedral, frequently zoned, crystals, set in an interstitial matrix of quartz. Some epidotes have central zones of allanite. Sphenes forming both large euhedral and small aligned crystals are abundant. In several specimens (e.g. S 47749), the sphene crystals contain rutile, which forms symplectic intergrowths with sphene. Many calc-silicate rocks contain interstitial calcite, others contain varying amounts of bluish green hornblende or, more rarely, needles of tremolite.

The essential constituents of the amphibolites of this group (S 31005, 50137) are hornblende, which in most areas is pleochroic from pale greenish yellow to bluish green, but in Vementry Island, close to the junction with the Vementry Group (S 30729), is pale green and garnet, which forms large porphyroblasts, in many cases with inclusions of albite-oligoclase or oligoclase, quartz, and sphene. Epidote and clinozoisite form bands within many non-garnetiferous amphibolites. These bands are of slightly finer grain than the adjoining hornblendic bands.

Retrograde metamorphism (M3S) associated with the D3 phase of shearing (pp. 57–58) is much less pronounced than in the Vementry Group and most zones of intense crushing, granulitization and conversion of garnet and biotite to chlorite are associated with the major north–south trending faults. These form belts up to 110 yd (100 m) wide within which the schist is intensely crushed and chloritized and partially mylonitized.

WEST BURRA FIRTH GROUP

The West Burra Firth Group is characterized by the presence of abundant ribs of limestone and bands of tremolite-phlogopite-schist. The limestones are commonly 2 to 12 yd (2–10 m) thick, but in the area between Norby and Bousta one persistent limestone locally reaches a thickness of 65 yd (60 m). Owing to the intense folding undergone by most limestones in the group during the D2 phase it is not possible to estimate their original thickness. Many of the thinner limestones are closely associated with amphibolite and hornblende-schist, which are more common than in the Neeans Group. Feldspathic muscovite-biotite-schits and quartz-granulite form approximately 40 per cent of the total volume of the group.

The rocks of the group are intensely veined by microgranite and pegmatite and 'permeated' by feldspar porphyroblasts, particularly in the eastern part of the outcrop where the group directly underlies the faulted or locally uncon-formable junction with the Old Red Sandstone sediments.

The characteristic mineral assemblage of the *limestones* (S 49299, 30595, 31013) is calcite, clinozoisite or zoisite, tremolite, phlogopite, feldspar (oligoclase in S 31012) and quartz. The latter is partially replaced by calcite. Epidote is subordinate to clinozoisite in this group. In some of the limestones exposed in the area between the Bay of Garth and Bousta (S 33779) the interstitial feldspar is partially altered to a symplectic intergrowth of zoisite and quartz or a more irregular intergrowth of zoisite and sodic plagioclase. In contrast to the limestones of the Neeans Group, most of these limestones contain a much higher proportion of tremolite than zoisite, clinozoisite or epidote. Diopside had been recorded in limestones in two restricted areas within the group: on the south-east shore of the Bay of Garth (S 30599) and close to the east shore of West Burra Firth (S 31012). In the first locality diopside is largely altered to tremolite and clino-zoisite, in the second the diopside is associated with clinozoisite and is partly altered to carbonate and rimmed by pale acicular amphibole (tremolite). No forsterite or pseudomorphs after forsterite have been detected. Sphene is present in nearly all limestones.

Calc-silicate rocks composed largely of epidote and clinozoisite together with quartz, feldspar and muscovite are present within the central part of the group's outcrop (i.e. around Brindister Voe, S 47737), but in the more western (S 47786–7, 50805) and eastern (S 47792) parts of the outcrop, the original impure cal-careous sediments are now represented by strongly lineated *tremolite-phlogopite-schists*, which in most cases contain subordinate calcite, epidote, clinozoisite, quartz, sodic plagioclase, sphene, apatite, ilmenite, pyrite and, in places, chlorite. Near the eastern end of West Burra Firth diopside forms an important component of some tremolite-schists (S 50132, No. 4 of Table I). In these the diopside is shattered and streaked out parallel to the lineation, patchily altered to calcite and partly replaced by tremolite (p. 57).

A characteristic feature both within this group and in the other groups is the euhedral to subhedral shape of most minerals of the epidote family, even in areas affected by later retrograde metamorphism. Many epidotes are zoned (S 47734), and some have a central zone of allanite (S 47738). The zoning suggests that growth took place in a number of stages during the phase of static meta-morphism. In some instances the alignment of the epidote crystals is at a slight angle to the regional foliation, indicating that crystallisation was not always mimetic. Minerals of this family display no compositional and relatively little mechanical alteration in areas strongly affected by later shearing and retrograde metamorphism. Some crystals in specimens from these areas are slightly broken and re-annealed, indicating that these minerals continued to grow under condi-tions of much lower temperature and pressure. The growth of members of the

TABLE 1

Chemical analyses of Walls Peninsula metamorphic rocks

	1	2	3	4	5	6
SiO_2	47·85	48·69	32·85	50·62	72·08	61·62
Al_2O_3	13·74	14·89	8·72	5·31	12·61	13·80
Fe_2O_3	2·81	3·29	0·53	0·17	0·99	4·31
FeO	10·14	6·47	2·71	1·94	3·28	1·10
MgO	7·65	8·12	4·70	19·10	1·50	2·21
CaO	9·85	12·01	27·40	16·21	1·15	13·74
Na_2O	1·71	2·16	1·04	0·55	2·52	0·10
K_2O	1·09	0·53	2·18	0·86	2·58	0·15
$H_2O > 105°$	2·11	1·96	1·17	1·79	1·68	1·22
$H_2O < 105°$	0·21	0·13	0·17	0·22	0·18	0·12
TiO_2	1·94	1·17	0·38	0·22	0·66	0·77
P_2O_5	0·15	0·11	0·13	0·07	0·12	0·03
MnO	0·21	0·19	0·07	0·06	0·10	0·14
CO_2	0·24	0·04	17·66	2·80	0·14	0·35
FeS_2	0·13	n.d.	n.d.	n.d.	n.d.	n.d.
Organic C	—	—	—	0·04	—	—
Allowances for minor constituents	0·16	0·14	0·22	0·29	0·20	0·19
	99·99	99·90	99·93	100·25	99·79	99·85
*Ba	82 ppm	20 ppm	470 ppm	125 ppm	580 ppm	50 ppm
*Co	35	20	<10	<10	<10	<10
*Cr	92	220	80	30	45	56
*Cu	74	22	<10	<10	10	<10
*Ga	<10	10	<10	<10	22	30
Li	12	7	15	25	12	10
*Ni	65	86	22	<10	40	54
*Sr	100	150	740	230	190	480
*V	390	270	60	40	80	120
*Zr	140	85	70	20	260	490
B	6	9	5	3	10	4
F	450	300	800	4000	500	380
S	—	—	—	200	250	—

*Spectrographic determination   n.d. not determined

1. Garnetiferous hornblende-schist, Snarra Ness Group. South coast of West Burra Firth, 220 yd (200 m) N7°E of Hogan [247 569]. S 50131, Lab. No. 1990. Anal. W. H. Evans and J. M. Nunan, spectrographic work by C. Park, (*A. Rep. Inst. geol. Sci.*, 1967, p. 96).
2. Hornblende-schist, Neeans Group. 370 yd (340 m) E7°S of south-west corner of Maa Loch, Neeans peninsula. [267 584]. S 50137, Lab. No. 1984. Anal. W. H. Evans, spectrographic work by C. Park, (*A. Rep. Inst. geol. Sci.*, 1967, p. 95).
3. Limestone with calc-silicate minerals, West Burra Firth Group. 470 yd (430 m) N38°W of Engamoor, Neeans peninsula [257 576]. S 50133, Lab. No. 1988. Anal. W. H. Evans, spectrographic work by C. Park, (*A. Rep. Inst. geol. Sci.*, 1967, p. 95).
4. Tremolite-schist with calcite, diopside and phlogopite, West Burra Firth Group. 240 yd (220 m) N46°W of Engamoor [259 575]. S 50132, Lab. No. 1989. Anal. W. H. Evans, spectrographic work by C. Park, (*A. Rep. Inst. geol. Sci.*, 1967, p. 95).
5. Feldspathized garnetiferous muscovite-biotite schist, Neeans Group. West side of Geo of Djubaberry, 710 yd (650 m) N16°W of south-western end of Maw Loch [263 590]. S 50134, Lab. No. 1987. Anal. W. H. Evans, spectrographic work by C. Park, (*A. Rep. Inst. geol. Sci.*, 1967, p. 95).
6. Calc-silicate rock with epidote, calcite and hornblende. Neeans Group. East side of Geo of Djubaberry, 630 yd (560 m) N12°W of south-western end of Maw Loch [264 590]. S 50135, Lab. No. 1982. Anal. W. H. Evans, spectrographic work by C. Park, (*A. Rep. Inst. geol. Sci.*, 1966, pp. 76–7).

epidote and zoisite groups appears thus to have spanned several major tectonic episodes.

In the vicinity of Bousta (S 33774) zoisite or clinozoisite forms symplectic intergrowths with quartz and acid plagioclase. These intergrowths appear to result from the breakdown of plagioclase porphyroblasts.

The mineral content of *amphibolites* and *hornblende-schists* in the West Burra Firth Group does not significantly differ from that of the amphibolite of other groups. Both pleochroic bluish-green varieties and pale varieties of hornblende have been recorded. Garnet poikiloblasts up to 10 mm in diameter with inclusions of quartz, hornblende and sphene are present in the massive amphibolites. Both the interstitial and porphyroblastic feldspar is albite-oligoclase. Greenish brown biotite is present in many hornblende-schists, and in some cases the orientation of the mica cleavage-plates is at an angle to the long axes of the hornblende (S 47761).

SNARRA NESS GROUP

The Snarra Ness Group is characterized by a preponderance of hornblende-schists over quartz-feldspar-granulite and feldspathic muscovite-biotite-schist. Large and small lenses and pods of poorly foliated amphibolite are abundant. Limestones and epidote-rich rocks are virtually absent over the greater part of the outcrop, but on the south-west shore of the Bay of Brenwell (Plate II) there is a small area in which the lithology is that characteristic of the West Burra Firth Group. It is probable that this outcrop is not an integral part of the Snarra Ness Group, but an infolded or infaulted portion of the West Burra Firth Group.

The *amphibolites* of the Snarra Ness Group (S 47763, 47784, 49298, 50131; Table I) consist of large vaguely aligned plates of hornblende which are pleochroic from yellowish green to bluish green (Plate VI, fig. 4), garnet porphyroblasts sieved with quartz, feldspar and hornblende, together with interstitial quartz and sodic plagioclase. Some amphibolites contain porphyroblasts of deep brown biotite (S 47779) and many contain abundant sphenes which are aligned in trails that pass through the coarse crystals of hornblende and biotite without displacement (S 47762) (p. 55). Both epidote and clinozoisite are present in variable amounts within garnet-free amphibolites, where they are commonly concentrated in bands.

In the area west of the Bay of Brenwell the distinction between the West Burra Firth Group and the Snarra Ness Group is ill defined. Limestones less than $3\frac{1}{2}$ ft (1 m) thick are present in the cliffs on the Neap of Norby, which consist of hornblende-schist and muscovite-biotite-schist with lenses of amphibolite. These rocks are intensely veined by granite and pegmatite and permeated by feldspathic lits and porphyroblasts. They are cut by a number of sub-parallel north-east trending faults and are intensely shattered throughout. Most of these faults, which are parallel to the Melby Fault (p. 265), are fractures apparently formed at a high crustal level as they have produced brittle fracturing and have soft clay in the fault planes.

GRANITE AND PEGMATITE VEINS AND INTRUSIONS

The Snarra Ness, Vementry and West Burra Firth groups are intensely veined by both trondhjemitic microgranite, which is locally almost felsitic in texture and trondhjemitic granite and pegmatite. These trondhjemitic rocks (S 49926, 50804) are composed of sodic plagioclase (oligoclase or albite-oligoclase), quartz and subordinate muscovite and/or biotite. Potash feldspar is present in minute interstitial patches in a number of pegmatites. Many of the veins are pre- or syn-tectonic as they have been involved in the D2 tectonic episode, being themselves weakly foliated and folded either with the foliation or by differential slip along the foliation (Plate VB), rodded or mullioned parallel to the lineation or lensed out into discrete pods. The foliation is not always obvious owing to the virtual absence of platy minerals.

Non-foliated post-D2 trondhjemitic intrusions are equally common. These form either cross-cutting veins, or veins which mimetically follow the folded foliation of the country rock (Plate IVB). Some amphibolites are patchily net-veined by microgranite to produce areas of agmatite. In such intensely veined areas parts of the amphibolite contain feldspar porphyroblasts up to $\frac{3}{4}$ in (2 cm) in diameter, which are round or oval-shaped in cross-section.

In the areas just north of Longa Water and Djuba Water within the Neeans peninsula, and in the ground south of Loch of Collaster [208 571], near Norby, there are some larger masses of trondhjemitic granite. These are not obviously foliated, but some are elongated parallel to the foliation. The largest is 170 yd (150 m) long and up to 55 yd (50 m) wide. They are similar in composition to the granite veins, being composed of oligoclase, slightly granulitized quartz, subordinate greenish brown biotite partially altered to chlorite, muscovite and apatite. Potash feldspar forms only small interstitial patches.

STRUCTURE

INTRODUCTION

The probable stages in the structural development of the metamorphic rocks are set out in tabular form on pp. 40–41. The original bedding planes of the metasediments and contacts of igneous extrusive and/or intrusive bodies are now largely obliterated and the evidence for the deformation or deformations which produced the planar fabric of the rock prior to the Main Phase cannot be demonstrated in the field (p. 55). The foliation, lineation and related minor folds now visible in the rocks were formed during the Main Phase (D2) of folding and during the subsequent mimetic coarsening of the fabric (p. 56). In parts of the area, especially in the north-west corner of Vementry Island, both lineation and foliation have been partly obliterated by intense shearing and shattering along innumerable east to north-east trending crush belts (D3). At a later period (D4) dextral kink folds and belts of conjugate folds with consistently east-south-east trending axes were formed, probably at a high crustal level. These did not greatly modify the D2 fabric, but in the belts of intense D4 folding the D2 lineation is, in places, slightly obscured. In north-west Vementry shear planes attributed to the D3 phase of crushing are folded by D4 conjugate folds.

The period of folding which produced the complex east-north-east trending synclinorium within the Walls Sandstone (p. 126) appears to be responsible for the present east to north-easterly trend and the south-south-easterly inclination of the foliation of the metamorphic rocks (D5). Apart from many sub-parallel fault planes close to the junction between the two formations, this period of folding does not appear to have produced minor structures within the meta-

morphic basement. The second phase of folding recognized within the Walls
Sandstone (p. 134) may here have produced open north–south trending synforms
and antiforms (D6) which express themselves as bulges in the outcrop of the
junction between the metamorphic rocks and Old Red Sandstone (Plate II).
Major faults with trends in the north-north-west to north-east sector, which
traverse the Neeans peninsula may also be associated with this phase of earth-
movement. Some of these faults are accompanied by belts of box- and conjugate-
folds, whose axial trends are unrelated to those of the D4 folds.

GEOMETRY OF MAIN STRUCTURAL ELEMENTS

Foliation. The strike and dip of the foliation in the metamorphic succession are
roughly parallel to those of its upper, partially sheared, junction. The strike
varies from S20°W to W40°N and its mean is W10°S (Plate IIIA). The inclination
of the foliation close to the junction is almost everywhere roughly parallel to the
dip of the adjoining Old Red Sandstone, though discordances of up to 30° are
locally recorded. As the junction between the two formations does not have the
appearance of a major dislocation (pp. 73–74 and p. 53) it is assumed that prior
to the D5 phase of folding the foliation was horizontal or gently undulating with
local 'dips' not greatly exceeding 30°.

Lineation. Linear structures are very pronounced in all areas except the northern
peninsula of Vementry. Their mean plunge is 30° to S20°W and the variations
in inclination and trend are shown in Plate IIIA. The local reversals in inclination to the north-east close to the east shore of the Neeans peninsula and along
the west shore of The Rona may be due to late large-scale (?D6) flexuring
associated with major NNW-trending faults. If the succession is 'rotated' and
unfolded back to its pre-D5 attitude, it is seen that the pre-Old Red Sandstone
plunge of the linear structures ranged from horizontal to possibly 30° to NW.

MAIN PHASE OF FOLDING (D2)

Fabric. The D2 tectonite fabric is strongly developed in most rock types except
some quartzites and some epidote-clinozoisite rocks, which superficially have no
preferred mineral orientation. The fabric pattern which ranges from planar
(s-tectonite) to linear (l-tectonite) is, in this area, determined by the rock type
rather than the local or regional stress pattern. Throughout the area the micaceous schists, gneisses and granulites have a predominantly planar fabric owing
to the parallel orientation of platy minerals. The linear fabric in these rocks is
largely due to a poorly developed small-scale rodding of the feldspathic lits,
and the elongation of feldspar augen and the 'bending' of platy minerals around
these structures. Micaceous schists with a strong linear fabric (l-s tectonites)
(Plate IVA) contain a fairly high percentage of an acicular mineral, such as
tremolite, or, less commonly, hornblende. Some lenses of epidotic amphibolite
have a pronounced s-fabric, which is brought out by the banding of hornblende-
and epidote-rich layers, but this fabric is in many cases disrupted by irregular
internal folding (p. 50) along the margins of such basic lenses. Strong l-fabrics
are present in tremolite-schists, tremolite-bearing limestones and the outer
zones of amphibolite pods. Limestones and calc-silicate-rich limestones which
are devoid of nematoblastic minerals have an internal banding which is intensely
folded around axes parallel to the lineation (Plate IVc). As these minor folds

weather out into parallel rods or tubes they impart a strong macroscopic linear fabric to the outcrop. In areas where the foliation is not folded, the limestone has a strong layering due to the differential weathering of limestone and calc-silicate or feldspar-rich ribs.

Though there are no zones which have a predominantly l- or s-tectonite fabric which is independent of lithology, there are areas in which the linear elements are partially obliterated by later folding or shearing parallel to the foliation. Thus D2 lineations are very faint in belts of intense D4 folding, in the intensely shattered parts of north-west Vementry, and in the narrow belt of sheared and faulted rocks close to the junction with the Old Red Sandstone.

Apart from minor folds the most commonly developed megascopic linear structures within the series are the rodding of feldspar lits and, less commonly, of microgranite veins. Along the coast of the Bay of Garth [216 581] bands of quartzite and quartz-granulite adjoining a thick limestone have a well-developed mullion structure. In cross-section the mullioned quartzite ribs have a characteristic lobate form with narrow embayments.

Early D2 minor folds. Small similar folds, which range in wavelength from several millimetres to 5 ft (1·5 m) are present in all lithological units throughout the series. In rock types with a predominantly planar fabric the axial planes of the folds are parallel to the regional foliation and the tightness of the folds ranges from isoclinal to tight (cf. Fleuty 1964, p. 47). The folds generally have rounded hinge zones and long, straight limbs (Plate IVB, c). The folds in the Vementry Group appear to be tightest being isoclinal with intensely corrugated hinge zones. Isoclinal folds with less corrugated hinge zones occur in the northern part of the Neeans peninsula, but farther south the folds are more open and less complex.

In the lithological units characterized by a linear fabric, such as limestones, tremolite-schists, and some amphibolites, the D2 minor folds have axial planes whose inclination bears no obvious geometrical relationship to the attitude of the regional foliation. Many axial planes are curved and adjacent planes are not necessarily parallel. Flow folds in which the thickness of individual folia varies widely are developed in amphibolites and, more rarely, in platy mica-schists. In limestones the style of folding is in places extremely plastic, with individual laminae disrupted and feldspar lits or granite veins reduced to small irregular fragments. The plunge of fold axes in most limestones is, however, parallel to the regional lineation.

Flinn (1967, p. 283) has shown that, in Eastern Shetland, the most common folds in his East Mainland Succession are 'internal folds' in the flaggy semi-pelites. These folds are found in 'beds' whose bounding planes are not deformed, but which exhibit intricate flowage folds within them. Flinn suggests that these folds are associated with pinch- and swell-structures and result from inhomo-

---

EXPLANATION OF PLATE IV

A. South shore of Bay of Garth. [214 580]. D_2 lineation in mica-schist. (D 977).
B. North shore of West Burra Firth at south-west corner of Crockna Vord peninsula. [251 573]. D_2 fold in hornblende-schist, with mimetic granite veins along folded foliation planes. (D 953).
C. South-west slope of Crockna Vord, just north of West Burra Firth. [251 575]. D_2 fold in limestone. (D 952).

geneous flow within them. In West Shetland the flow folds in the limestones and
the disharmonic folds in the pods of banded epidotic amphibolite are of similar
origin.

Granite veins cutting the schist commonly exhibit ptygmatic folding and, in
some instances, stepping of the vein along the planes of schistosity is seen (Plate
VB). Many granite veins follow the folded foliation of the country rock. These
veins are usually thicker in the axial zones of the folds than in their limbs. As
many of these veins have branches which cut the regional foliation without
displacement, it is likely that most are 'mimetic' in the sense that they follow
originally deformed planes and were intruded just after or during a late stage
of the folding (Plate IVB).

Local post-D2 pre-D4 minor folds. All the folds described above are thought to
have been formed during the main phase of folding, when the rock acquired its
present schistosity. These folds have axial planes which are either parallel to
the schistosity or are curved or disrupted by flowage. In addition to these folds
there are a number of small isolated areas with folds whose axes are parallel to
the regional lineation but whose axial planes are almost normal to the foliation.
Minor folds of this type are seen on the south shore of West Burra
Firth [244 571]. Many have a narrow or acutely angled hinge area, an interlimb
angle of 100°–120° and a rudimentary axial plane cleavage which is almost
normal to the main schistosity. Small interconnected veins of microgranite
follow both the original schistosity and the later cleavage, indicating that this
folding took place before the phase of granite and pegmatite vein emplacement
had ceased. It thus appears to be the result of a weak localized phase of movement which took place after the main phase folding but before the D4 phase of
folding, which everywhere post-dates the granite and pegmatite veining.

PHASE OF INTENSE SHEARING AND GRANULITIZATION (D3)

The north-western peninsula of Vementry Island is traversed by a number
of major east to north-east trending shear-belts as well as innumerable intersecting minor faults of variable trends and small throws (Plate VB). In the
northern part of this peninsula the faulting is in places sufficiently intense to
have obliterated both the lineation and foliation of the rocks. In the less intensely
sheared parts of the area the faults make an angle of 15°–40° with the foliation.
Most of these are normal faults with throws of less than 8 in (20 cm). The faults
invariably cut the microgranite and pegmatite veins but are themselves folded
by D4 conjugate folds. All fault planes are sharply defined and contain no soft
crush-rock, suggesting that they were formed under relatively high confining
pressures. Similar faults intersecting the foliation at a low angle are present on
the northern coast of the Neeans peninsula.

KINK BANDS AND CONJUGATE KINK-FOLDS (D4)

Small straight-hinged angular monoclines of the type which have been termed
kink bands, kink zones and knee-folds (Ramsay 1967, p. 436) are developed in
the thinly foliated rocks throughout the series. The axes of these folds have a
consistent east-south-easterly trend (Plate IIIB), which in many areas is normal
to the trend of the D2 linear structures. The steep limb of the folds, which varies
in length from 2 in (5 cm) to just over 1 ft 8 in (50 cm) is almost always on the

E

south side of the fold-crest when viewed from the west-north-west, producing a consistent dextral shift of the foliation planes. In rare cases the axial plane of the fold has become a slip plane along which small displacement has taken place. Individual folds can be traced for several yards along the crest, and adjacent folds are arranged *en echelon*. These folds have folded but not obliterated the D2 linear fabric. In areas with closely and regularly spaced kink folds joint planes with slight or negligible displacement are developed roughly parallel to the axial planes of the folds.

Along a number of fairly well-defined east-south-east trending belts the intensity and amplitude of the D4 folds is greatly increased. In these zones kink bands are accompanied or replaced by conjugate folds, which are straight-hinged angular folds with two intersecting axial planes (Johnson 1956). Conjugate folds are best developed in flaggy or platy granulites or schists, and many fine examples are exposed on the north coast of the Neeans peninsula, 300 yd (270 m) N of Whalwick [258 585]. In the most intensely deformed parts of these zones the style of the folds becomes less brittle. Thus along the coast just north-east of the Bay of Garth [217 582] a series of tight E 10°S trending flexural-slip folds with rounded hinge zones are seen. In both zones of intense D4 folding which cross Vementry Island (Plate IIIB), the folds locally lose their conjugate symmetry and box-shaped outline and become flexural-slip folds with one axial plane. In the most intensely deformed area of the northerly zone fairly open east–west trending similar folds with an axial-plane cleavage are developed. In the areas of intense D4 folding bedding-plane slip has partially obliterated the macroscopic D2 linear fabric. There is also some correlation between the intensity of D4 folding and the platiness of the mica-schists and granulites involved. As the axes of the kink-folds and conjugate folds throughout the series lie within the plane of the D2 schistosity, their plunge varies according to the trend and inclination of this schistosity, and in some areas such as 250 yd (230 m) N of Houlma Water [268 579], where there is a marked local variation in the attitude of the foliation along the hinge of a D5 fold, the plunge of the D4 kink folds changes abruptly. Evidence such as this suggests that the D4 folding preceded and was unconnected with the D5 phase of folding and regional tilting.

REGIONAL TILTING TO THE SOUTH-SOUTH-EAST (D5)

As the regional trend and inclination of the foliation within the metamorphic rocks is more or less parallel to the strike and dip of the adjoining sedimentary rocks of Old Red Sandstone age it can be assumed that the tilting of the two formations took place during the first phase of folding that affected the Walls

EXPLANATION OF PLATE V

A. General view of Neeans, looking east from north-east slope of Crockna Vord. [257 583]. Characteristic topography formed by the metamorphic rocks of the Neeans Group. (D 950).

B. West coast of Vementry Island, 33 yd (30 m) N of Whal Geo. [283 612]. Platy feldspathic gneiss with hornblendic bands, with tight D_2 folds, small normal faults and thin ptygmatically folded granite veins. (D 913).

C. Summit of Muckle Hoo Field, Neeans peninsula. [268 586]. Platy feldspathic gneiss with D_4 kink folds. (D 943).

Sandstone (F1 of p. 126). The mean trend of the F1 fold axes within the latter is E15°N. This phase of folding has produced no minor folds or other small-scale penetrative structures within the metamorphic succession.

In the southern part of the Neeans peninsula there is a marked change in the regional strike of the foliation along a line trending approximately E10°S. This line appears to mark the outcrop of a fold hinge which has developed into a rotational fault plane. Another example of late rotation of entire blocks within the metamorphic succession is found on the north-west coast of the Neeans peninsula, south-west of Turl Stack [257 586] where, in an area bounded by faults, the lineation has been rotated clockwise by 60° relative to the regional lineation (Plate III).

LATE NORTH-NORTH-WEST TRENDING FOLDS (D6)

Both the metamorphic rocks and the overlying Old Red Sandstone are affected by two gentle NNW-trending folds. In the Norby–Bousta area a large open N15°W trending syncline extends from the Bay of Garth southwards to the junction with the Old Red Sandstone rocks and can be traced for a short distance into the latter. Along the eastern margin of the Neeans peninsula the lineation of the metamorphic rocks is folded into a NNW-trending anticline and probably, to the east, a complementary syncline. These folds are reflected by the sinuous outcrop of the junction between the metamorphic rocks and the Old Red Sandstone and also, to some extent, by the dips within the Old Red Sandstone. It is possible that these folds belong to the same period of deformation as the F2 folds within the Walls Sandstone (p. 134).

NORTH-EAST TO NORTH-NORTH-WEST TRENDING FAULTS AND ASSOCIATED CONJUGATE FOLDS

The major north-east to north-west trending faults cutting the metamorphic rocks are shown on Plate II. In addition to the exposed faults, major faults are probably present in Brindister Voe, close to the east coast of the Neeans peninsula, and in The Rona, near the east coast of Braga Ness. There is no definite evidence as to the age relationships of these faults to the NNW-trending D6 folds.

These faults have wide crush belts, and where they traverse platy schist or granulite they are associated with zones up to 110 yd (100 m) wide, of conjugate or chevron-folding. The fault crossing the shore at the Ayre of Starastet [254 584], 770 yd (700 m) SW of Turl Stack has a number of branches which converge southwards. Along the coast the zone of sharp-crested folds is 88 yd (80 m) wide and the fold axes plunge at 10°–30° to between S15°E and S40°E, which is at a slight angle to the trend of the main fault. Most folds in this zone are conjugate, but have slightly larger wavelengths and amplitudes than the typical D4 folds. Many folds are cut by small contemporaneous faults which have soft clay in their shear planes.

Conjugate and chevron-folds which plunge to S10°E and are associated with branch-faults of the N–S trending fault, which crosses the coast [260 587] just south of Turl Stack, are seen to traverse and locally refold a belt of east-south-east trending D4 conjugate folds. It can thus be demonstrated that the folds associated with the major faults are younger than the regionally developed D4 folds.

METAMORPHIC HISTORY

The relationships of the various phases of mineral growth in the metamorphic rocks to the major tectonic episodes can sometimes be determined by an inter-pretation of the textural relationships of the constituent minerals. Some evidence is provided by porphyroblasts of garnet and feldspar, which contain relict fabrics of aligned or helicitic inclusions of other minerals (cf. Rast 1958; Sturt and Harris 1961; Zwart 1960a, b). Provided that the mode of origin and the history of crystallisation of the inclusions within the porphyroblasts can be satisfactorily established, the relationship of the mineral fabric within the crystal (Si) to the surrounding fabric (Se) provides evidence for the ages of the two fabrics relative to the growth of the porphyroblast. Thus, if the internal fabric is aligned throughout the crystal and set at an angle to the external fabric it may mean that the two fabrics are of different generations, and that the porphyroblast was formed under static conditions prior to the onset of the period of stress

---

EXPLANATION OF PLATE VI

PHOTOMICROGRAPHS OF METAMORPHIC ROCKS OF THE WALLS PENINSULA

FIG. 1.   Slice No. S 47741. Magnification × 16. Crossed polarisers. Garnet in horn-blende-schist with small aligned inclusions, of which all but the outermost are aligned at a high angle to the external fabric. The internal (Si) fabric may be of M1P age, the outer (Se) fabric was developed during the M2P phase. Note the limited deflection of the Se fabric by the garnet. East coast of Neeans, 1235 yd (1130 m) N15°W of Brindister Broch [283 584].

FIG. 2.   Slice No. S 49322. Magnification × 31. Plane polarized light. Garnet in micaceous gneiss with concentrically arranged penetration inclusions. These were probably formed by the preferential resorption of unstable zones in the garnet. Vementry Island, north-west shore of Uyea Sound, 700 yd (640 m) NNW of Vementry House [304 603].

FIG. 3.   Slice No. S 47761. Magnification × 16. Plane polarized light. Hornblendic gneiss, with large hornblendes enclosing garnets. The growth of the large amphiboles appears to have taken place during a late stage of static growth (probably M2P) and its orientation may be mimetic after D2 fabric. Neeans peninsula, 280 yd (255 m) N42°W of NE corner of Lang Loch [270 592].

FIG. 4.   Slice No. S 49298. Magnification × 16. Plane polarized light. Amphibolite with garnets which show no trace of rotation and have not significantly eyed the external (Se) fabric. Sphenes are aligned parallel to the hornblende and mica. The coarsening of amphiboles and garnets is of M2P age. West coast of Geo of Bousta, 230 yd (210 m) NW of Muckle Bousta [223 577].

FIG. 5.   Slice No. S 30728. Magnification × 42. Crossed polarisers. Sheared garnet-oligoclase-gneiss, with thin lenses and bands of mylonite. The garnet is rotated and peripherally altered to chlorite. Vementry Island, 250 yd (230 m) SSE of Head of Corbie Geo [286 613].

FIG. 6.   Slice No. S 47806. Magnification × 38. Plane polarized light. Granulitized quartz-oligoclase-gneiss. Garnet rotated and largely altered to chlorite by post-M2P shear movements (S3). Vementry Island, near west coast, 800 yd (720 m) N18°E of SW tip of Heill Head [283 612].

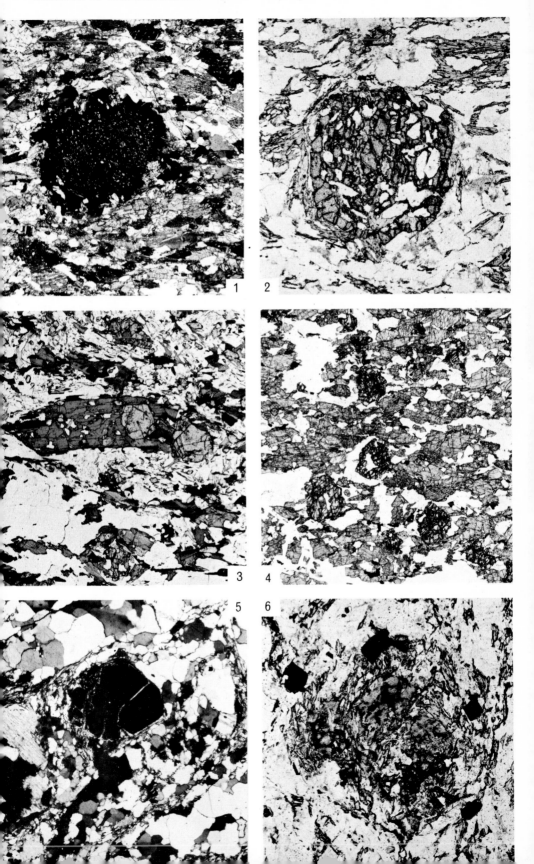

which shaped the external fabric. Rotation of the porphyroblast during this kinematic phase would lead to a bending of platy minerals producing an 'eyed' structure, around the crystal. Porphyroblasts which continued to grow during the kinematic phase would have an outer zone of spirally curved trails of possibly larger inclusions (cf. Powell and Treagus 1970). Post-kinematic porphyroblasts, which grew under static conditions after the tectonite fabric of the rock had formed, do not distort the surrounding fabric and may contain inclusion trails continuous with that fabric (Sturt and Harris 1961, fig. 1).

Unfortunately, the evidence from porphyroblasts in this area is less conclusive than that in the ideal case mentioned above. Garnet porphyroblasts with an internal fabric set at an angle to the plane of the schistosity and around which the latter is deflected are common (S 47741, 47784) (Plate VI, fig. 1). In most of these, however, there is some evidence of movement and rotation after the Se fabric had formed (S 47795). Many garnet porphyroblasts in this series show marked changes in the character and distribution of inclusions from the centre outwards. Several have a core of small aligned or helicitic inclusions and an outer part in which rather larger quartz, biotite and hornblende inclusions are either concentrically arranged or show no definite pattern. Others have inner inclusion-free zones and outer zones with concentrically arranged inclusions (S 30729). The shape and alignment of the inclusions in some garnets in this area suggest that these may, in fact, be zones of secondary dissolution, formed by preferential resorption of relatively unstable zones in the garnet (S 49322).

The size of the inclusions in the outer zones of many garnets, however, suggests that they were formed during the phase of post-kinematic static coarsening and annealing of the fabric which coincided with the final growth of the porphyroblasts. In addition to the porphyroblasts of garnet and feldspar, elongate crystals of amphibole, epidote and clinozoisite and their inclusions provide evidence for post-kinematic (M2P) mimetic coarsening and annealing of the tectonic fabric (S 47761, 47762) (Plate VI, fig. 3).

New fabrics and diaphthoretic minerals developed during the D3 and D4 episodes are more easily recognized. Apart from the extensive development of chlorite at the expense of biotite and garnet, nearly all minerals of the affected rocks show signs of shearing and distortion or fracturing which took place at low temperatures and low confining pressures. Textures and alignments of new minerals are directly related to axial planes or other geometric features of newly developed microfolds or shear planes which can, in turn, be related to the large-scale structural features.

POSSIBLE DEFORMATION (D1) AND MINERAL GROWTH (M1S AND M1P) PRIOR TO MAIN TECTONIZING METAMORPHISM

Evidence for the existence of a tectonite fabric (D1) prior to the formation of the main penetrative fabric (D2) is inconclusive. There are many garnet porphyroblasts with an aligned or folded fine-grained Si fabric which is discontinuous with the 'eyed' Se fabric (S 47741, 47784, 47795). The rotation of the garnets could, however, have occurred at some stage after D2. Feldspar 'permeation' and granite veining commenced either before the D2 folding or during an early stage in the M2S phase, as a high proportion of feldspar lits and porphyroblasts and microgranite veins were deformed both on a megascopic and microscopic scale during the D2 folding.

MAIN PHASE FOLDING (D2) AND ASSOCIATED METAMORPHIC EPISODES
(M2S AND M2P)

The preferred orientation of the platy (phylloblastic) and elongate (nemato-blastic) minerals, which imparts the foliation and lineation to the metamorphic rocks, was developed by recrystallization during and just after the Main (D2) Phase of folding. The following two phases of mineral growth were involved in this process:

1. *Syn-kinematic phase (M2S)*. The evidence for the formation of a fine-grained fabric during D2 is provided by the presence of aligned trails of small inclusions, notably in amphibole. These consist of quartz, epidote, hornblende, biotite and, particularly, sphene and lie parallel to the main fabric. The trails of inclusions pass from one large amphibole crystal to another without deflection.

EXPLANATION OF PLATE VII

PHOTOMICROGRAPHS OF METAMORPHIC ROCKS OF THE WALLS PENINSULA

FIG. 1.   Slice No. S 47786. Magnification × 42. Plane polarized light. Tremolite-schist, composed largely of euhedral laths of tremolite, flakes of phlogopite and subrounded crystals of epidote. East shore of West Burra Firth, 200 yd (180 m) S47°E of Broch at head of West Burra Firth [257 571].

FIG. 2.   Slice No. S 33779. Magnification × 42. Plane polarized light. Crystalline limestone with calc-silicate bands, with calcite (bottom), clinozoisite, horn-blende and quartz with zoisite inclusions. Skinhoga peninsula, 380 yd (350 m) SSE of Skerry of Stools [223 579].

FIG. 3.   Slice No. S 31005. Magnification × 16·8. Plane polarized light. Epidote-amphibolite. Bluish green hornblende (with cleavage), subrounded crystals of epidote, interstitial calcite and rare small sphenes. Concentrations of epidote and hornblende occur in roughly alternate bands. Neeans Peninsula, 70 yd (64 m) N of Lang Loch [272 589].

FIG. 4.   Slice No. S 30589. Magnification × 42. Plane polarized light. Hornblende-schist with symplectic intergrowth of zoisite and quartz. Norby, 550 yd (502 m) N10°E of west end of Loch of Collaster [207 578].

FIG. 5.   Slice No. S 50811. Magnification × 42. Plane polarized light. Epidote-quartz rock composed of large euhedral epidote laths set in base of quartz full of tremolite needles. Vementry Island, near south-east shore of Suthra Voe, 1000 yd (910 m) S7°W of summit of Muckle Ward [295 602].

FIG. 6.   Slice No. S 49316. Magnification × 42. Plane polarized light. Flinty crush rock with new growth of radiating needles of fibrous amphibole due to late thermal metamorphism. Vementry Island, east shore of Swarbacks Head, 900 yd (820 m) NNW of summit of Muckle Ward [293 618].

FIG. 7.   Slice No. S 49319. Magnification × 16. Plane polarized light. Muscovite-chlorite-schist with phylloblastic minerals folded by F3 movements. New chlorite plates developed sub-parallel to axial planes of folds and tourma-line prisms (seen in section) parallel to fold axes. Vementry Island, west shore of Cribba Sound, 1520 yd (1400 m) N of SW corner of Muckle Head [296 595].

FIG. 8.   Slice No. S 49294. Magnification × 17·6. Crossed polarisers. Mylonitized quartz-mica-schist composed of alternate streaks of quartz with mortar texture and mylonite containing newly-grown phlogopite and muscovite. South-east shore of Bay of Garth, 1020 yd (930 m) NE of Muckle Bousta. [215 580].

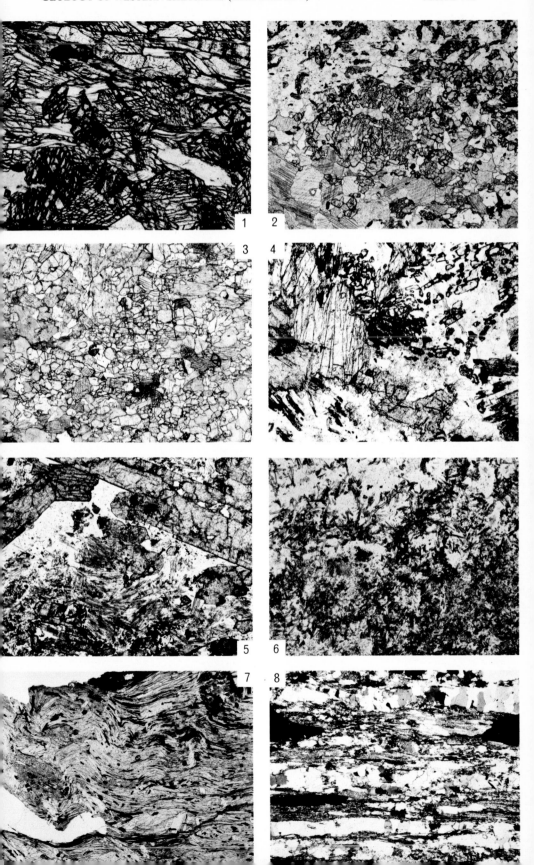

2. *Porphyroblast growth and mimetic coarsening (M2P)*. It is thought that the growth of some garnets commenced before there was any marked mimetic coarsening of aligned minerals, though both processes were to a large extent contemporaneous. Thus there are some garnets which contain inclusions that are finer grained in the centre than near the periphery and others whose Si fabric remains consistent in both size and orientation throughout the crystal. Minerals included in garnets are usually only quartz and calcite (minute patches) in the central zones, but in the outer parts both biotite and amphibole are present. Many of the smaller garnets are enclosed in amphibole crystals (S 47761, Plate VI, fig. 3) and it is suggested that both developed during this period of mineral growth.

The evidence for mimetic coarsening of the entire fabric is best preserved in the amphibolites and in some calc-silicate rocks in which large amphiboles, biotites and epidotes contain aligned inclusions the fabric of which coincides in direction with that of the new fabric. In micaceous schists and gneisses the evidence for static coarsening of the fabric is less obvious, but many plagioclase porphyroblasts contain aligned inclusion trails which are parallel to the external fabric (S 47799).

Metamorphic Facies. The peak of regional metamorphism appears to have been attained during the static phase of mimetic coarsening and porphyroblast development (M2P). Over a large part of the area the mineral assemblages are characteristic of the greenschist-amphibolite transition facies (pp. 42–47). There are, however, two areas, at the eastern end of West Burra Firth and on the shore of the Bay of Garth, where diopside is present in the calc-silicate rocks and limestones, and this is indicative of the amphibolite facies (p. 45). The position of the pockets of diopside-rich limestone and calc-silicate rock coincides with areas of intense feldspathization and granite veining, and may be due to the local effects of thermal metamorphism associated with feldspathization. It is possible that the original extent of the diopside-bearing rock was considerably greater than at present, and that the diopside has been retrograded to tremolite during a late- or post-mimetic phase of deformation (p. 51) described above. The early stages in the break-down of diopside are seen on the eastern shore of West Burra Firth where diopside is rimmed by fibrous amphibole (p. 45).

RETROGRADE METAMORPHISM ASSOCIATED WITH D3 PHASE OF SHEARING AND FAULTING (M3S)

Retrograde metamorphism resulting from intensive shearing and shattering has affected the entire north-western peninsula of Vementry Island, parts of the remainder of Vementry Island, and the northern part of the Neeans peninsula. The principal textural and mineralogical changes are as follows:

1. Intense granulitization and partial mylonitization along narrow shear planes (S 49294, Plate VII, fig. 8) which are most abundant close to faults. These are associated with the alteration of garnet and biotite to chlorite and extensive fracturing and partial chloritization of amphibole (S 49315, 49345).

2. In less intensely sheared areas quartz is invariably streaked out into bands of fine-grained mosaics, and individual crystals have polygonal or amoeboid outlines (S 30728, 49307, 49322, Plate VI, fig. 5). Feldspars are fractured with stepped and strained twin-planes, show turbid alteration and are patchily sericitized (S 49308). Biotite is in most instances completely altered to green chlorite, but muscovite, though fractured and bent, is unaltered. Garnets are broken up by rotation, streaked out and partially altered to chlorite (S 47806, Plate VI, fig. 6). The least

altered garnets are enclosed in a sheath of chlorite (S 30728). Amphibole and minerals of the epidote family are slightly cracked or broken but the latter show no sign of alteration. New epidote has formed in some sheared feldspars during this phase. Iron ores are irregularly streaked out. Sphene, apatite and rutile are unaltered.

RETROGRADE METAMORPHISM ASSOCIATED WITH D4 FOLDING (M4S)

Diaphthoretic changes resulting from D4 folding are confined to the belts of intense folding. Mineral changes in these belts are similar to those in the areas less intensely affected by the earlier D3 shearing but here the platy and elongate minerals are intensely folded, bent and fractured, and in the southern peninsula of Vementry Island new chlorite plates are formed along and parallel to the axial planes of the microfolds (S 49319, 49320, Plate VII, fig. 7). In the D4 fold belts traversing the southern and eastern peninsulas of Vementry Island there are abundant small needles of tourmaline, usually closely associated with chlorite. These commonly have their c-axes parallel to the axes of the D4 microfolds, but this alignment may be mimetic, as the formation of tourmaline may have been due to a subsequent influx of boron-bearing solutions, possibly during the emplacement of porphyry dykes.

LATE THERMAL METAMORPHISM (M5P)

The effects of thermal metamorphism, possibly by the Muckle Roe Granophyre or an unexposed plutonic body, are strongly marked along the east shore of the Swarbacks Head peninsula of Vementry Island. Here the sheared and partly mylonitized schists are hornfelsed, with original chlorite altered to deep green biotite, and radiating needles of pale bluish green amphibole cutting all the earlier fabrics (S 49315, 49316, Plate VII, fig. 6). In sheared amphibolites needles of bluish amphibole commonly grow out from the original fractured hornblendes. Feldspar porphyroblasts are usually turbid or have a newly developed twinning pattern. A specimen of intensely hornfelsed metamorphic rock from Swarbacks Skerry [290 622], just off Swarbacks Head, contains subhedral porphyroblasts of andalusite (S 30726).

Within the metamorphic rocks adjoining the Vementry Granite the effects of thermal metamorphism are considerably less pronounced. Needles of bluish green amphibole cutting the earlier fabric are present but are less abundant than in specimens from Swarbacks Head. In some instances the only sign of thermal metamorphism is the turbid, patchily saussuritized character of feldspars. At the eastern end of Suthra Voe a rib of epidote-quartz rock, within an area cut by a north-east trending dyke swarm (Plate XXIV) consists of near-euhedral epidotes set in a matrix of clear quartz, which contains innumerable bundles of fine parallel fibres of amphibole (S 50811, Plate VII, fig. 5). These bundles are randomly orientated and commonly cross one another. This rock is probably a metamorphosed calc-silicate rock, but may be a recrystallized vein. The presence of tourmaline within chlorite-rich schists (Plate VII, fig. 7) in areas cut by porphyry and felsite dyke swarms may be due to boron-metasomatism during or after the emplacement of the granite magma.

References

A. Rep. Inst. geol. Sci. for 1965, 1966, 76–8.

A. Rep. Inst. geol. Sci. for 1966, 1967, 95–6.

FLEUTY, M. J. 1964. The description of folds. *Proc. Geol. Ass.*, **75**, 461–92.

FLINN, D. 1967. The metamorphic rocks of the southern part of the Mainland of Shetland. *Geol. Jnl*, **5**, 251–90.

HARPER, C. T. 1967. The Geological Interpretation of Potassium-Argon ages of Metamorphic Rocks from the Scottish Caledonides. *Scott. Jnl Geol.*, **3**, 46–66.

JOHNSON, M. R. W. 1956. Conjugate Fold Systems in the Moine Thrust Zone in the Lochcarron and Coulin Forest areas of Wester Ross. *Geol. Mag.*, **93**, 345–50.

MILLER, J. A. and FLINN, D. 1966. A Survey of Age Relations of Shetland Rocks. *Geol. Jnl*, **5**, 95–116.

POWELL, D. and TREAGUS, J. E. 1970. Rotational fabrics in metamorphic minerals. *Mineralog. Mag.*, **38**, 801–14.

RAMSAY, J. G. 1967. *Folding and Fracturing of Rocks*. New York: McGraw-Hill.

RAST, N. 1958. Metamorphic History of the Schichallion Complex. *Trans. R. Soc. Edinb.*, **63**, 413–31.

READ, H. H. 1934. The metamorphic geology of Unst in the Shetland Islands. *Q.Jnl. geol. Soc. Lond.*, **90**, 637–88.

STURT, B. A. and HARRIS, A. L. 1961. The Metamorphic History of the Loch Tummel Area, Central Perthshire, Scotland. *Lpool Manchr geol. Jnl*, **2**, 689–711.

TURNER, F. J. 1968. *Metamorphic Petrology*. New York: McGraw-Hill.

ZWART, H. J. 1960a. Relations between Folding and Metamorphism in the Central Pyrenees, and their Chronological Succession. *Geologie Mijnb.*, **39**, 163–180.

—— 1960b. The Chronological Succession of folding and metamorphism in the Central Pyrenees. *Geol. Rdsch.*, **50**, 203–18.

Chapter 5

VE SKERRIES

THE VE SKERRIES are a series of low rocky islets situated $3\frac{1}{2}$ miles (5·6 km) NW of Papa Stour. They are devoid of soil and vegetation and do not rise more than 20 ft (6 m) above Ordnance Datum. About half the exposed rock is below high water mark. The Ve Skerries extend for about $\frac{3}{4}$ mile (1·2 km) from north-east to south-west and consist of three main islands named from north to south: North Skerry, Ormal and The Clubb, as well as a number of smaller skerries (Fig. 6). The author has not visited these islets and the present account is based on the geological field maps of Messrs. D. Haldane and S. Buchan who visited the Ve Skerries in 1933 and collected rock specimens.

FIELD RELATIONSHIPS

The islands are composed predominantly of pale grey to pink granite-gneiss with varying proportions of mica and hornblende, together with lenticular masses of foliated granite and two thick bands of partially 'permeated' horn-blende-schist. The gneisses are quite unlike any of the metamorphic rocks of the Walls Peninsula or Foula and it is possible that they may eventually be correlated with the acidic and hornblendic orthogneiss forming the north-western part of North Roe which has tentatively been considered to be of Lewisian type (Pringle 1970, fig. 1; Miller and Flinn 1966, p. 98). No radiometric age dates have so far been obtained from the Ve Skerries gneisses.

Two dykes of fine-grained basalt, which is not foliated or metamorphosed, cut the gneisses of North Skerry and Ormal.

North Skerry. North Skerry and the adjacent rocky islets are formed of strongly banded gneiss and foliated granite, whose foliation is near vertical and has a trend which ranges from north-east to east–west. In the eastern part of the island group the more thinly foliated micaceous gneisses exhibit small-scale chevron-folding. Individual folds have straight limbs, narrow rounded hinge zones and inter-limb angle ranging from 90° to 120°. Their wavelength ranges from 1 in (2·5 cm) to $1\frac{1}{2}$ in (3·8 cm) and their amplitude averages $\frac{1}{2}$ in (1·3 cm).

The following lithological units have been recognized in North Skerry and the adjacent islets:

1. Hornblende-gneiss, which forms the two islets about 100 yd (90 m) NNW of North Skerry, but whose relationships to the adjoining group have not been recorded.
2. Pale pink strongly banded, but poorly lineated granite-gneiss, which contains irregular plates rather than continuous laminae of the dark greenish phylloblastic minerals biotite and chlorite. This forms the northern half of North Skerry and in the east is interbanded with:

   (a) A highly distinctive white foliated medium- to fine-grained granite with small black micas and small deep red patches.
   (b) Bands of thinly foliated epidotic mica- and hornblende-gneiss, which are vaguely lineated and are, in places, affected by chevron folding.

60

3. Just within its southern half, the island is traversed by a band of pink granite-gneiss or foliated granite which consists of pink granitic folia up to ½ in (13 mm) thick, separated by continuous folia of muscovite.

4. The southern part of the island consists of a similar pink micaceous granite-gneiss, with bands of mica- and hornblende-schist, and veins of fine- to medium-grained pink granite, some of which are sheared and have a sugary texture.

North Skerry is traversed by a N30°E trending basalt dyke of unstated width, which locally splits into two in the southern part of the island.

Ormal. The regional trend of the foliation on Ormal is north-westerly, but changes locally to east-north-east and north-north-east. The following three more or less distinct north-west trending lithological zones are recognized:

1. A north-eastern belt of pale pink to greenish granite-gneiss or foliated granite with thin bands of pink granite. This belt is similar to, though less well foliated than, the granite-gneiss forming the southern half of North Skerry.

2. A central zone of pale greenish chlorite-rich granite-gneiss, and

3. A south-western zone composed partly of fine-grained hornblende-schist and partly of hornblende-gneiss with irregular, slightly aligned feldspar blebs and streaks.

Ormal contains a narrow north-west trending basalt dyke.

The Clubb. The Clubb is the most south-westerly island of the group and consists of foliated gneiss which trends from E10°N to E45°N and whose inclination ranges from vertical to 60 degrees to the south-east. The predominant rock type is pale pinkish grey thinly foliated biotite-gneiss with some bands of coarse granite-gneiss and veins and lits of pinkish granite. There are also belts of siliceous gneiss interbanded with chlorite-biotite-gneiss. Thin bands of horn-blende-schist have been recorded in the north-eastern and south-western parts of the island.

No data or specimens are available from the Skerries of Reaverack and Helligoblo which are situated between The Clubb and Ormal.

PETROGRAPHY

Granite-Gneiss. The granite-gneiss of the Ve Skerries is not easily distinguished from foliated granite as both have a very similar composition. The feldspathic lits of the granite-gneiss in all islands are composed of a poorly foliated aggregate of feldspar and slightly strained quartz crystals, which are present in the ratio of 7:3. In some sections only potash-feldspar, principally microcline, is present (S 29983), but in most sections albite is also developed in considerable quantity. In many sections a large proportion of the feldspars are sieved with small flakes of sericite (orientated in two directions) and abundant small grains of epidote (Plate VIII, fig. 2). Epidote also forms thin discontinuous veins and isolated smaller crystals.

The micaceous laminae are composed either of large plates of muscovite (S 29983) or of smaller plates of khaki-brown biotite which is partially altered to chlorite.

Foliated Granite. Foliated granites cannot always be distinguished from granite-gneisses. The latter have a lithological banding and the former have a schistosity but no banding, but the two types tend to grade into each other. The foliated granites are generally coarse-grained to very coarse-grained with feldspar crystals, in one instance (S 29982), up to several centimetres in size. The granites

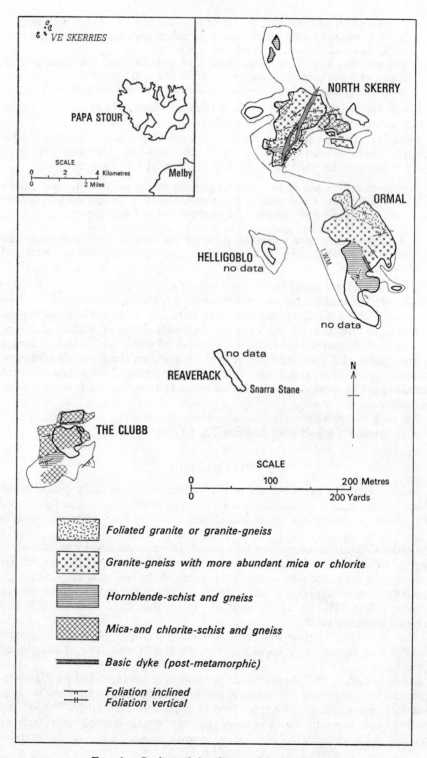

FIG. 6.   *Geological sketch-map of the Ve Skerries*

contain 80 to 90 per cent of feldspar and interstitial quartz, which in many instances is strained or even mylonitized. In the white granite of North Skerry (p. 61) both regular (symplectic) and irregular intergrowths of quartz and feldspar are found. Both potash-feldspar and sodic plagioclase are present, and the relative proportions vary considerably. The white granites of North Skerry, for instance, are formed predominantly of albite-oligoclase which in one thin section (S 29982, Plate VIII, fig. 1) is full of small muscovite flakes, aligned in two directions and cut by a network of epidote veins as well as a series of thin bands of quartz-mylonite. In other white granites (S 29986, 30869), however, the feldspars are free from inclusions.

Platy minerals either between grains or forming folia are uncommon. Rare muscovite has been recorded in some specimens, and all samples contain irregular patches of chlorite. A granite from The Clubb contains small interstitial crystals of hornblende and biotite. Accessory minerals are sphene, allanite and apatite.

Schistose rocks. Mica-, hornblende-, and chlorite-schists with either discrete quartzo-feldspathic lits or irregular blebs are present in both North Skerry and The Clubb. The mica-schists (S 29987, 30865) in most cases have discrete folia of muscovite, khaki-brown biotite and varying proportions of chlorite. The hornblendic rock (S 30866) contains large crystals of hornblende, which is strongly pleochroic from straw-coloured to deep bluish green, and patchily or completely altered to or enclosed in chlorite, together with subordinate small plates of muscovite. The dark portion of one specimen (S 30870) from The Clubb contains only large plates of green chlorite associated with sphene. Epidote is abundant in nearly all specimens, and in one section (S 30868) it forms elongate euhedral crystals up to 1·8 mm long set in a base of green biotite. The pale folia or blebs within the schists consist of elongated, commonly interlocking grains of quartz, microcline, and, in some instances, larger near-euhedral porphyroblasts of sodic plagioclase. Accessory minerals include sphene, apatite, zircon, rutile and, in one specimen (S 30868), small scattered grains of pyrite.

REFERENCES

MILLER, J. A. and FLINN, D. 1966. A Survey of Age Relations of Shetland Rocks. *Geol. Jnl,* 5, 95–116.

PRINGLE, I. R. 1970. The structural geology of the North Roe area of Shetland. *Geol. Jnl,* 7, 147–70.

WALKER, F. 1932. An albitite from Ve Skerries, Shetland Isles. *Mineralog. Mag.,* 23, 239–42.

Chapter 6

METAMORPHIC AND INTRUSIVE ROCKS
OF FOULA

INTRODUCTION

THE ISLAND of Foula is situated in the Atlantic Ocean, 14 miles (22 km) WSW of Wats Ness. It consists mainly of sedimentary rocks of Old Red Sandstone age (Chapter 12) but contains a ½-mile (800-m) wide strip of metamorphic rocks along its east coast (Fig. 21, p. 173). The metamorphic rocks are separated from the sedimentary rocks to the west by a fault and they give rise to relatively flat ground at roughly 100 ft (30 m) OD bounded by low sea cliffs. They consist mainly of metasediments and are cut by a network of microgranite dykes and sills.

The metasediments are made up of alternating bands of garnetiferous quartz' feldspar-granulite (psammite) and garnetiferous mica-schist (semi-pelite and pelite). They contain lenses and layers of amphibolite and hornblende-schist with bands and patches of epidosite. The amphibolites probably originated as basic and ultrabasic sills and dykes. Several bands of crystalline limestone with calc-silicate minerals have been recorded. All the metamorphic rocks contain feldspathic lits and porphyroblasts; they are also cut by two sets of syntectonic granite and pegmatite intrusions. There is a marked similarity between the Foula metamorphic rocks and parts of the Neeans Feldspathic Group of the Walls peninsula (pp. 43–44).

The characteristic mineral associations of the Foula metamorphic rocks are given on pp. 66–68 and these show that the rocks belong to the transition between the greenschist and amphibolite facies of metamorphism (Turner 1968, pp. 302–20). The only aluminium-silicate mineral in Foula which is diagnostic of the amphibolite facies is staurolite and this has been recognized in only one band (p. 67).

The sequence of events in the structural and metamorphic development of these rocks appears to be similar to that of the metamorphic rocks of the Walls Peninsula (pp. 48–57), though not all stages recorded there can be recognized. The following events can be recognized in Foula:

1. Deposition of sediments, emplacement of basic intrusions. (No evidence for a pre-Main Phase period of folding.)
2. Beginning of emplacement of 'pre-tectonic' granite and pegmatite veins, and development of feldspar lits and porphyroblasts. This phase probably continued throughout period (3).
3. 'Main phase' folding, producing tectonite fabric, followed by mimetic coarsening and porphyroblast metamorphism.
4. Late 'brittle' folding with production of kink bands and conjugate folds.
5. Emplacement of network of sills and dykes of porphyritic microgranite (pp. 67–70).
6. Development of N10°–20°E and N10°–20°W–trending faults and some near-horizontal thrusts.

64

METAMORPHIC ROCKS

FIELD RELATIONSHIPS

Metasediments. Though the metamorphic rocks are composed of roughly alternating bands of feldspathized garnetiferous mica-schist and flaggy quartz-feldspar-granulite, major groups in which psammite or pelite bands predominate can be recognized. Many of the mica-schists have been converted into banded gneisses or augen-gneisses by feldspar 'permeation'. In the banded psammite the most common products of feldspathization are lits of feldspar 1 to 8 in (2·5–20 cm) thick, which are concordant with the foliation and are generally rodded. These rods are oval in cross section, up to 1 ft (30 cm) wide and 6 in (15 cm) high. Discrete feldspar porphyroblasts are more common in the amphibolites and mica-schists. The intensity of feldspar permeation is variable within the sequence, and certain parts of the coast section, such as that between Swaa Head and Kinglia [977 397], are virtually unaffected.

Two beds of crystalline limestone, respectively 15 ft (4·6 m) and 4½ ft (1·4 m) thick, are present on the north shore of Little Ham, 300 yd (275 m) S of Ham Voe and at Strem Ness]972 413], in the north-east corner of the island. Other bands appear to crop out on the inaccessible cliffs of Durga Ness [976 383].

Amphibolites. Concordant bands, lenses and pods of coarse-grained amphibolite, amphibole-epidote-rock and epidosite are present throughout the metamorphic rocks. They range in thickness from less than 1 ft (30 cm) to 20 ft (6 m). They are most commonly 4 to 6 ft (1·2–1·8 m) thick and can be traced laterally for several metres. Though a number of ultrabasic bodies appear in the field to be virtually monomineralic, many consist of irregularly interbanded streaks and lenticles of amphibolite and epidosite. Within the larger lenticular masses these bands, which range in thickness from ½ to ¾ in (12–19 mm), are folded, with the individual fold axes parallel to the lineation, but with axial planes randomly orientated and curved when seen in sections normal to the lineation. Close to the margins of the larger bodies the amphibole-epidote-rock is in places interbanded with quartz-feldspar-granulite and both rock types are folded together. The folds in these marginal zones are more regular, and their axial planes are roughly parallel to the regional foliation (p. 49). Many of the smaller amphibolite masses contain roughly equidimensional porphyroblasts of feldspar.

Early (pre- and syn-kinematic) granites and pegmatites. The metamorphic rocks of Foula are cut by both concordant and cross-cutting veins of:

1. Yellowish pegmatite devoid of dark minerals with large white quartz crystals.
2. Pink medium-grained granite with some schistose partings.

The concordant intrusions are folded with the country rock and the cross-cutting veins show typical ptygmatic folds. It is possible that many of the veins were intruded along foliation planes which were already folded and therefore the period of vein formation may have extended to the post-kinematic phase.

Fold styles in metamorphic rocks. Minor folds of two periods are recognized (p. 40). The early folds ascribed to the main phase of deformation are isoclinal or near isoclinal with axes parallel to the rodding of the feldspar lits and the mineral lineation of the gneiss. The axial planes of the folds are sub-parallel to the regional foliation within the metasediments but irregularly curved in the amphibolites and amphibole-epidote-rocks (p. 68). All these folds are small with wave lengths generally less than 2 ft (0·6 m). The area is too small for major structures, which could be ascribed to the folding of this period, to be recognized.

The second set of folds consists of brittle-style kink folds (pp. 51–52) with a regional plunge of 30 degrees to S5°–10°E. These folds are best seen at Baa Head and between Skarf Skerry [978 396] and The Taing where they affect the platy mica-schists. Locally the axial planes of these brittle folds are fractured and pass into small faults.

PETROGRAPHY

Granulites. The characteristic mineral assemblage of the Foula granulites is: quartz, albite-oligoclase with subordinate interstitial potash feldspar, subordinate muscovite and biotite. Epidote and almandine-garnet are present in some specimens. Accessory minerals are zircon, apatite and sphene.

Quartz forms up to 70 per cent of the total volume and occurs in discrete, somewhat lenticular bands, 0·3 mm to several millimetres thick. These are composed of interlocking grains elongated in a ratio of 3 to 1 parallel to the foliation, together with subordinate patches of albite-oligoclase and interstitial microcline

---

EXPLANATION OF PLATE VIII

PHOTOMICROGRAPHS OF METAMORPHIC ROCKS, MICROGRANITE AND SANDSTONE OF VE SKERRIES AND FOULA

FIG. 1. Slice No. S 29982. Magnification × 16·8. Crossed polarisers. Granulitized granite with large crystals of albite-oligoclase, sieved with muscovite. Adjacent feldspar crystals are in optical continuity and separated by streaked out mozaic-quartz. Small near-euhedral crystals of epidote are abundant in the quartz network. Ve Skerries, North Skerry, west coast [103 658].

FIG. 2. Slice No. S 29989. Magnification × 31. Crossed polarisers. Coarse poorly-foliated granite-gneiss composed of quartz, large clear plates of potash-feldspar and albite-oligoclase full of inclusions of white mica, and small grains of epidote. Ve Skerries, Ormal, north coast [105 656].

FIG. 3. Slice No. S 29898. Magnification × 8. Plane polarized light. Garnet-kyanite-staurolite-gneiss, with muscovite and quartz. Large stumpy plates of kyanite with close parallel cleavage (bottom and top centre), smaller plates of golden-yellow staurolite, and subrounded garnets are set in a base of biotite, muscovite, quartz and andesine. Foula, Swaa Head, 860 yd (790 m) NNE of Sloag. [976 401].

FIG. 4. Slice No. S 50823. Magnification × 20. Plane polarized light. Strongly foliated and sheared quartz-biotite-schist composed of lenses of quartz with mortar texture alternating with streaks composed of feldspar, muscovite and reddish brown biotite. Scattered porphyroblasts of oligoclase (left-centre). Foula, south shore of Ham Voe, 110 yd (100 m) E5°N of Brae [974 387].

FIG. 5. Slice No. S 29900. Magnification × 16. Crossed polarisers. Dyke of porphyritic microgranite, with granulitized matrix between phenocrysts of albite-oligoclase. Foula, Swaa Head, 880 yd (800 m) NNE of Sloag [976 401].

FIG. 6. Slice No. S 50829. Magnification × 16. Crossed polarisers. Coarse-grained arkose with subrounded to subangular grains. Ratio of quartz to feldspar grains is 50:50. Some interstitial flakes of muscovite. Matrix forms 15 per cent of total volume, composed mainly of carbonate. Foula, shore of Whiora Wick, 520 yd (470 m) E20°S of Freyars [966 412].

which appears in some cases to be replaced by myrmekite (S 29889). Albite-oligoclase, usually kaolinized in the centre or patchily throughout, also forms irregular blebs or porphyroblasts. In the coarsely banded granulites khaki-brown biotite and muscovite form scattered small curved flakes, but in the more finely banded rocks they form folia, up to 0·3 mm thick, which contain thin lenticles of cloudy plagioclase.

Epidote, which in some cases forms 20 per cent of the total volume, appears as:

1. Large grains (up to 0·8 mm in diameter) which are usually slightly elongated parallel to the foliation, but in some cases (S 50823) cut across it. Elsewhere they are bent or fractured.
2. Indistinct aggregates within or associated with feldspars and probably alteration products of feldspar.

Zircon is the most consistently present accessory mineral. It forms euhedral crystals scattered throughout the quartzose and feldspathic bands. Apatite and sphene are abundant in certain bands.

Garnetiferous mica-schist and gneiss. The following mineral assemblage is present in most of the Foula mica-schists: muscovite, biotite, almandine, epidote, quartz, calcic albite or oligoclase. Accessory minerals are apatite, zircon and sphene. Only one highly aluminous band which contains kyanite, staurolite, biotite, muscovite, quartz and oligoclase has been recorded.

Most of the mica-schists are strongly foliated and consist of:

1. Micaceous folia composed of thick plates of muscovite, which are in many instances mantled by smaller, thinner flakes of biotite which is pleochroic from yellowish-green to khaki-brown.
2. Folia made up of an aggregate of irregular interlocking granulitized crystals of albite-oligoclase and quartz. The quartz grains are generally elongated parallel to the foliation.
3. Feldspar lits consisting of very large (up to 1·5 mm diameter) irregular plates of oligoclase.

The garnets are generally slightly elongated parallel to the foliation, and the larger porphyroblasts contain inclusions of quartz and muscovite. Epidote, as in the granulites (p. 66), occurs both as large crystals and as diffuse aggregates in saussuritized feldspars.

The band of garnet-kyanite-staurolite-schist on Swaa Head (S 29898, Plate VIII, fig. 3) contains garnet poikiloblasts up to 5 mm in diameter, large subhedral crystals of kyanite which form 20 per cent of the total volume, and smaller subhedral crystals of staurolite. Muscovite, in thick plates, forms distinct folia, while biotite is scattered in small plates throughout. Quartz and feldspar (oligoclase) form irregular aggregates of small grains which make up less than 50 per cent of the total volume.

There are also bands of poorly foliated calc-schist which consist of very irregular lenses and folia of:

1. Fine-grained interlocking feldspar-quartz aggregates.
2. Epidote-calcite aggregates with abundant large sphenes, and
3. Poikilitic feldspar full of small grains of amphibole.

Small laths of amphibole and flakes of chlorite are scattered throughout.

Crystalline limestones. The limestones (S 50843) are composed of calcite with

F

varying amounts of epidote, clinozoisite and hornblende. In the Strem Ness peninsula tremolite is associated with calcite.

Amphibolites and Epidosites. The metamorphosed basic rocks are composed of hornblende, epidote, clinozoisite and almandine with subordinate oligoclase or albite-oligoclase. Accessory minerals are sphene, apatite and biotite.

The amphibolites (S 29883, 29886, 29893) are composed of large anhedral stumpy crystals of hornblende which range in diameter up to 2 mm and make up between 70 per cent and 90 per cent of the volume of the rock. In the hornblende-rich bands the crystals show little or no orientation but in the more feldspathic layers and, particularly in the banded hornblende-epidote-rocks (S 50838), hornblende crystals are strongly elongated and aligned. Apart from almandine the remainder consists largely of interstitial anhedral plates of extensively altered oligoclase together with very small interstitial patches of quartz. Sphenes are abundant, both as large euhedral crystals and as clusters of small rounded grains. Apatite normally forms clusters of subrounded crystals. The amphibolites are commonly cut by thin veinlets of epidote and potash-feldspar, locally with penninite.

The epidote-rich bands are weakly foliated and consist of up to 90 per cent epidote which forms large subhedral to euhedral crystals, averaging 0·8 to 1·2 mm in length but reaching 6 mm, with less than 10 per cent interstitial oligoclase together with rare interstitial quartz. Sphene and apatite are both abundant accessory minerals.

Banded epidote-amphibole-quartz rocks (S 29891) are common. These consist of alternate laminae of:

1. Calc-silicate, composed of up to 60 per cent of epidote and interstitial patches of amphibole (?actinolite), which is pleochroic from straw-yellow to bluish green, and
2. a coarse-grained quartz aggregate with individual crystals up to 2 mm across.

Syn-tectonic pegmatites and granites. The syn-tectonic granites have a banded texture both on the macroscopic and microscopic scale. Though in the porphyritic granites (S 29892) the feldspar phenocrysts and the feldspar (untwinned potash feldspar and albite-oligoclase) of the groundmass are relatively undeformed, the quartz occurs as elongated interlocking grains. Biotite and muscovite form small crystals which are in most cases orientated parallel to the foliation, but which in one thin section (S 50825) give rise to radiating aggregates.

MICROGRANITE MINOR INTRUSIONS

The metamorphic rocks of Foula are cut by a network of sills and dykes of pink porphyritic microgranite, which, in the field, show no evidence of deformation by either of the two sets of folds described above (p. 65), but which in thin section contain irregular planes and patches of sheared and granulitized granite.

FIELD RELATIONSHIPS

Ruscar Head Microgranite. By far the most extensive outcrop of pink microgranite is that exposed along a 700-yd (640-m) stretch of the shore at Ruscar Head, near the north-east corner of Foula. This outcrop appears to be part of a large sill which splits both northwards and southwards into a network of sills and dykes. Though in the Ruscar Head exposure neither top nor base of the sill is seen, metamorphic rocks apparently underlying this sill appear to be present in the faulted exposures at the head of the geo just south of Ruscar Head. In the northern half of the Ruscar Head peninsula a number of small masses of intensely granite-veined granulite and gneiss project through the granite. Though it is possible that these are detached enclaves within the granite it is more likely that they are projecting portions of the country rock underlying the sill.

The Ruscar Head microgranite is intensely jointed and in the southern half of its outcrop the following sets of joints are most prominent:

(a) Closely set joints trending S10–20°E and inclined at 70° to WSW.
(b) Two sets of joints, both trending N10–20°E inclined at 70° to WNW and 50–60° to ESE. The trend of the latter joints swings to NNW in northern part of outcrop, and
(c) Near horizontal, gently curving joints inclined eastward at a low angle.

The microgranite is cut by swarms of thin chlorite-bearing quartz veins emplaced mainly, but not invariably, along the S10–20°E trending joints. The swarms are up to 6 yd (6 m) wide and individual veins are 1 to 2 in (2·5–5 cm) (exceptionally up to 5 in. (12·5 cm)) thick and 6 in to 1 ft (15–30 cm) apart. The microgranite also contains a number of lenticular kaolinitized zones of up to 2½ ft (0·8 m) wide and 15 yd (14 m) long, in which the granite is completely white. The orientation of these zones is parallel to the S10°E joints.

Sills and dykes. Both north and south of Ruscar Head, porphyritic microgranite forms a network of sills and dykes cutting the metamorphic rocks. These are most closely spaced in the immediate vicinity of Ruyhedlar Head [975 402] and Swaa Head where individual sills range in thickness from 2 to 5 ft (0·6–1·5 m) and are 6 to 10 ft (1·8–3 m) apart. Though roughly concordant, they tend to transgress the foliation by a series of small steps. The dykes are more widely spaced and generally thinner than the sills. They have a north-easterly trend and a steep inclination to the south-east. There are instances of dykes cutting sills and *vice versa*, but in most cases the microgranite of the intersecting dykes and sills is continuous in all directions, indicating that the network of microgranite was formed in a single intrusive pulse.

South of Swaa Head microgranite sills are less closely spaced, and dykes are very rare. Sills range in thickness from 1 to 7 ft (30 cm–2·1 m) and easily accessible examples occur on both shores of Ham Voe.

Intrusion-breccia. An irregular sill of microgranite with three roughly lenticular masses of intrusion-breccia close to its lower (north-eastern) margin, is exposed on the shore just south-west of The Taing, 670 yd (610 m) NE of Ham Pier. Two of the breccia masses contain subangular clasts of all the types of metamorphic rocks encountered in Foula. The clasts range in size from 4 in to 2 ft (10–60 cm) and are set in a matrix of microgranite. Though the orientation of many clasts is parallel to that of the foliation of the adjoining county rock, a certain proportion of clasts, particularly those composed of amphibolite, are randomly orientated. The most southerly lens of breccia which is up to 5 ft (1·5 m) thick and is closest to the sill consists of subrounded fragments of amphibolite and gneiss, up to 2 in (5 cm) long, normally orientated parallel to the sill margins, set in granulitized granite. The junctions between the lenses of breccia and between breccia and granite are gradational. It is thought that the brecciation of the country rock is the result of gas-streaming ahead of the rising granitic magma. This produced a breccia approximately *in situ*, leaving the orientation of the country rock, which forms most of the clasts, unchanged. Subsequently the intruding magma engulfed the clasts and carried with it and rotated many of the less platy fragments, particularly the amphibolites. It is presumed that within a lens of breccia containing many small ovoid clasts the velocity of the gas-charged magma was sufficient to move and round the clasts.

PETROGRAPHY

The Ruscar Head microgranite is composed of plagioclase phenocrysts set in a granular groundmass rich in quartz, and poor in mafic minerals. The phenocrysts form from 20 to 70 per cent of the total volume of the rock, range in diameter from 3 to 0·6 mm and consist entirely of oligoclase (An > 20) which is occasionally zoned, and is in many cases turbid particularly in the centre. The phenocrysts in places form clusters and they always have irregular or serrate margins with embayments suggesting partial resorption by the matrix. The latter consists of a fine-grained mosaic of equidimensional grains of quartz with intersertal patches of potash-feldspar. The ratio of quartz to feldspar in the groundmass ranges from 40:60 to 75:25 and in some areas within the Ruscar Head Sill there are large irregular patches composed entirely of a coarse quartz mosaic. In some sections (S 29900) there is an increase in the grain size of the matrix in the immediate vicinity of the oligoclase phenocrysts. In a number of sections ferromagnesian minerals are completely absent, in others the only dark minerals are small plates of muscovite and biotite, or, more commonly, irregular patches of penninite or diffuse chloritic material.

Though some hand specimens of the microgranite show foliation, this is less obvious in thin section. All thin sections, however, show varying degrees of granulation of the groundmass (Plate VIII, fig. 5), which may be due to shearing before the granite was completely consolidated. The granular texture which is reminiscent of the texture of many coarse-grained felsites may, however, be an original feature due to the high percentage of quartz in the matrix.

FAULTS AND CRUSH BELTS

Both the metamorphic rocks and microgranite intrusions are cut by two suites of closely-spaced sub-parallel, somewhat curving faults with displacements not exceeding 10 ft (3 m) as well as a number of low-angle thrusts. In the area extending from Ruscar Head to Shoabill the trend of the majority of the faults ranges from due north–south to N20°W, and their inclination ranges from 60° to the west to 50° to the east. At Baa Head, just south of Ham Voe, the country rock is intensely shattered close to these faults and the foliation is locally buckled into open conjugate folds similar to those associated with faults in West Mainland (p. 53).

At Strem Ness and Ruscar Head the following two intersecting sets of faults are developed:

(a) Trend N10–20°E with steep inclination to both east and west.
(b) Trend N20–40°W with inclinations ranging from 55° to the south-west to 65° to the north-east.

Thrust planes accompanied by considerable zones of shattered rock crop out on the west side of Strem Ness (trend north–south, inclination 25° to the west), on the shore between Ruscar Head and Ruyhedlar Head [974 403] (trend south-south-east, inclination to the east-north-east), at Swaa Head (trend east to east-south-east, inclination to the south), and between The Taing and Sloag Burn [977 391] (trend north–south, near horizontal). The extreme variation in the direction of inclination of the exposed sections of thrusts over the whole area would suggest that the thrusts are highly undulating planes, whose regional disposition is near horizontal.

REFERENCE

TURNER, F. J. 1968. *Metamorphic Petrology*. New York: McGraw-Hill.

Chapter 7

OLD RED SANDSTONE:
SANDNESS FORMATION

INTRODUCTION

THE EXPOSED thickness of the sediments and contemporaneous volcanic rocks forming the Sandness Formation (Plates IX, X) ranges from 4500 ft (1370 m) to nearly 7000 ft (2130 m) in the central and eastern parts of their outcrop, but thickens considerably to possibly 12 000 ft (3650 m) farther west. Over the greater part of their outcrop the strata dip consistently to south-south-east at angles ranging from 30° to vertical and they are locally inverted in the Clousta area. In the ground between the Voe of Snarraness and Sandness Hill the consistent south-south-easterly dip is interrupted by two east-north-east trending folds, the Djuba Water Syncline and the almost isoclinal Mousavord Loch Anticline (Plate XII).

A complete sequence through the Sandness Formation is available only in the area east of West Burra Firth, where the following sub-divisions, shown diagramatically in Plate X, have been recognized:

6. Alternating beds of fine-grained sandstone, siltstone and calcareous mudstone, with impersistent beds of argillaceous limestone.

4. Sandstone, fine- to medium-grained, cross-bedded with pebbly beds near base. Some argillaceous intercalations.

3. Conglomerate, forming several lenses, possibly at slightly varying horizons (e.g. Clousta Conglomerate).

These sediments are intercalated with the Clousta Volcanic Rocks (5), composed of basalt and basic andesite lavas, ignimbrites, thick beds of basic tuff and tuffaceous sandstone, and cones of acid agglomerate. Also concordant intrusions of felsite.

2. Sandstone, cross-bedded with scattered pebbles and subordinate argillaceous bands.

1. Basal beds of pebbly sandstone and grit with lenses of breccia.

Plant remains in sediments interbedded with the Clousta Volcanic Rocks appear to be of Lower or Middle Old Red Sandstone age. Whole rock potassium-argon age determinations of basalts by N. J. Snelling are as follows: Basalt lava, Ness of Clousta [299 577] (KA ref 71/73), p.p.m. radiogenic ^{40}Ar = 0·075, Age 323 ± 9 m.y.; basalt lava, Ness of Clousta [304 579] (KA ref 71/74), p.p.m. radiogenic ^{40}Ar = 0·063, Age 336 ± 9 m.y. Both these ages are too low when compared with the age based on the plant evidence. It is likely that the basalts suffered argon loss, possibly during the phase of uralitization which affected most basic lavas in the Sandness Formation (p. 94). Flinn and others (1968, p. 14) obtained a potassium-argon age of 255 ± 4 m.y. from rhyolite, just west of Burga Water [225 543]. This date indicates that the rhyolite suffered argon loss at some later period.

JUNCTION BETWEEN SEDIMENTS AND METAMORPHIC ROCKS

The junction between the Walls Sandstone and the metamorphic rocks which occupy the northern fringe of the Walls Peninsula was stated to be a major fault by both Peach and Horne (1879 p. 786, 1884 p. 365; *in* Tudor 1883 p. 395) and Finlay (1930 p. 673) but the officers of the Geological Survey who mapped the area between 1931 and 1934 (*in* Summ Prog. 1934; p. 73; 1935; p. 68) concluded that along the greater part of its course it is an unconformity locally obscured by small shear planes.

Along its western outcrop, between Sandness and the Bay of Brenwell (Plate IX), there are no good exposures of the contact of the two formations, but the curved nature of its outcrop, the presence of lenticular masses of sedimentary breccia at or near the junction, and the occurrence of thin but persistent beds of conglomerate at a constant vertical distance above it (Plate X) suggest that it approximates closely to the original unconformity. In the ground [208 568] 1 mile (1·6 km) W of the Bay of Brenwell there is a small outlier of breccia and sandstone which occupies an area of 110 yd (100 m) by 40 yd (36 m) and rests with a near-horizontal base on highly inclined metamorphic rocks (Plate IX). The outlier is approximately 250 yd (230 m) N of the outcrop of the main junction whose inclination in this area is approximately 50° to the south. The presence of this outlier suggests that the junction north of its present outcrop (i.e. above ground level) was undulating and in places very nearly horizontal and that it may have been stepped down northwards by faulting. Faults of this type are not easily recognized in the metamorphic terrain.

Another very small fault-bounded mass of indurated greenish sandstone is exposed on the east shore of the Geo of Bousta [223 577] 800 yd (730 m) N of the nearest outcrop of Old Red Sandstone sediments (Fig. 7). The presence of this faulted outlier supports the supposition that above the present western outcrops of the metamorphic rocks the original junction was displaced by a number of faults the throw of which may have been of the order of several hundred feet.

On the headland [229 570] between the Bay of Brenwell and the Voe of Snarraness (Fig. 7), where conglomerate with pebbles of metamorphic rock is underlain by at least 90 ft (27 m) of predominantly mudstone and siltstone (p. 76), the only well-exposed junctions between Old Red Sandstone and metamorphic rocks are two faults trending respectively W10°N and N30°W, which have produced a north-west trending graben of Old Red Sandstone sediments bounded by metamorphic rocks. On the shores of the Voe of Snarraness the junction is formed by a large fault inclined at 50°–55° to the south.

The junction is well exposed on the south shore of West Burra Firth [242 569] between 300 and 420 yd (275 and 385 m) E of Snarra Ness. In this area it is formed by one of a series of sub-parallel faults trending roughly west-north-west and inclined at 50°–80° to the south. These faults cut both the metamorphic rocks and the sediments. The fault plane separating the two rock types is inclined at 68°–80° to S10°–15°W, which is steeper than the foliation of the gneiss (35°–40° to SSW) and the dip of the sandstone (55° to SSW). It forms a belt of fault clay only 1 to 6 in (2·5–15 cm) thick and has produced very little shattering in the adjoining rocks. The adjacent sub-parallel faults are, on average, 8 to 10 yd (7·3–9 m) apart, and form a series of step faults. The west-north-west trending faults are themselves cut by a suite of south-south-west trending faults, which in this area displace the former in a sinistral sense by distances ranging from 10 to 27 ft (3–8 m) thus giving the junction a stepped outcrop.

Farther east the boundary between the sandstone and the metamorphic rock is again exposed in the upper reaches of West Burra Firth, where it is a fault plane inclined at 62° to S10°W. On the east coast of West Burra Firth there is a 10-ft (3-m)-wide fault slice of sedimentary breccia along the faulted junction. This contains angular pebbles of metamorphic rock and appears to have been deposited near the base of the group, suggesting that here the displacement along the fault has not been large. Between 100 and 250 yd (90 and 230 m) E of West Burra Firth the boundary between the metamorphic rocks and the Old Red Sandstone basal conglomerate is undulating and may be a normal junction, but exposures are not sufficiently good to verify this. Farther east, between Longa Water [265 570] and Brindister Voe, the junction is marked by a straight depression, suggesting that it is a fairly large fault (*in* Summ. Prog. 1935, p. 68). One exposure along this junction shows a band of hard mylonite 1 ft (30 cm) thick separating sandstone from gneiss.

In the area east of Brindister Voe the outcrop of the junction between sandstone and metamorphic rock is fairly undulating, suggesting that it may be an unconformity, but as in the areas farther west all exposures of the junction are to some extent sheared. Close to Vementry House the junction is near-vertical and locally inverted, and the overlying sediments contain a basal breccia several inches thick, passing upwards into a pebbly sandstone. On the west shore of the Stead of Aithness the junction is inclined at 30°–45° to the south-south-west, and shows no obvious sign of shearing. This exposure is the only one in the whole peninsula which can be classed as an unconformity not markedly affected by subsequent faulting, though in a number of the cases described above there is little doubt that faulting along the contact is of a minor nature and that the true base of the Sandness Group is not far below the lowest exposed beds.

On the island of Papa Little all junctions between metamorphic rocks and sediments appear to be faulted.

SEDIMENTS BELOW THE CLOUSTA VOLCANIC ROCKS

LITHOLOGY

Basal Sediments (1 of Plate X)

In the eastern part of the area the basal sediment is a grey, medium-grained, poorly graded sandstone with small scattered pebbles of metamorphic rocks and some well-rounded quartz grains. The sequence contains rhythmic units composed of thick cosets of cross-bedded sandstone grading up into greenish siltstone or, less commonly, reddish mudstone. The fine-grained phases of the units are only up to 2 ft (60 cm) thick. Clasts of mudstone, now practically converted to slate, are present both at the bases and within the sandstone cosets.

The characteristic feature of the basal beds is the presence of thin lenticular bands of breccia. In the vicinity of South Loch of Hostigates [313 593] breccia lenses occur 700 to 800 ft (215–245 m) above the lowest exposed sediments. Traced westwards towards Sonso Ness [302 589] the breccias gradually approach the base of the formation. They contain angular to subangular fragments of gneiss and schist up to 1 in (2·5 cm) long and subangular to rounded clasts of vein quartz and their thickness nowhere exceeds 50 ft (15 m). Thin lenticular

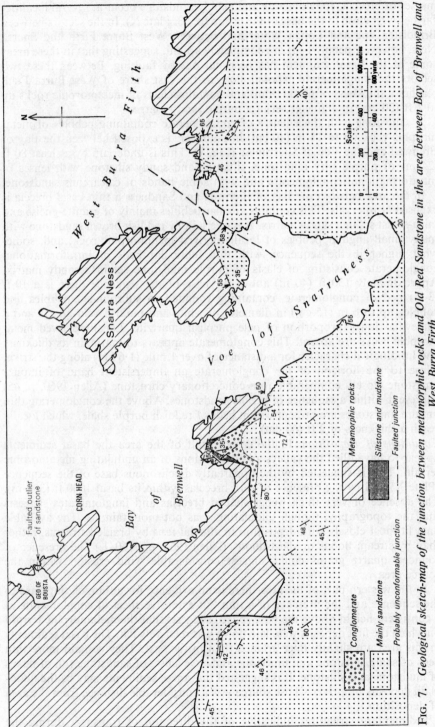

FIG. 7. *Geological sketch-map of the junction between metamorphic rocks and Old Red Sandstone in the area between Bay of Brenwell and West Burra Firth*

masses of breccia, conglomerate and pebbly sandstone occur at several localities farther west close to the base of the group (p. 73, Plate X). In the sectors between Brindister Voe and Longa Water and between West Burra Firth and Snarra Ness, however, pebbly beds are completely absent, suggesting that in these areas some of the basal sediment has been cut out by faulting. Between these two sectors the conglomerate exposed close to the east shore of West Burra Firth reaches a thickness of 60 ft (18 m) and contains clasts of metamorphic rocks up to 1 ft (30 cm) in diameter set in a purple sandy matrix.

A sandstone with thick lenses of conglomerate containing pebbles of metamorphic rock up to 18 in (40 cm) in diameter, is exposed between the Bay of Brenwell and the Voe of Snarraness (Fig. 7). This is underlain by at least 80 ft (24 m) of purple silty mudstone, siltstone and sandy siltstone with lenses of intraformational conglomerate and with some bands of calcareous sandstone.

In the area between the Bay of Brenwell and Sandness a thin basal breccia is exposed at intervals. This contains angular pebbles mainly of granite-gneiss and mica-schist set in a sandy matrix. It is overlain by reddish brown sandstone with many small angular pebbles of locally-derived metamorphic rock, and, somewhat higher in the sequence, with thin lenticular beds of intraformational conglomerate consisting of clasts of purple mudstone set in a sandy matrix. Approximately 130 ft (40 m) above the base of the sequence there is a 10-ft (3-m) bed of conglomerate, containing rounded to subrounded pebbles and cobbles up to 6 in (15 cm) in diameter of white and pinkish vein-quartz and a somewhat lower proportion of pale purplish quartzite. Locally derived metamorphic rocks are absent. This conglomerate appears to maintain its thickness and lithological character for a distance of over 1 mile (1·6 km) along the strike. Close to the horizon of the conglomerate an impersistent band of impure limestone closely comparable to a concretionary cornstone (Allen 1960, p. 45) is present within a sequence of pebble sandstones. Above the conglomerate thin impure limestones are present within beds of reddish purple shale, which locally reach a thickness of 2 ft (60 cm).

Conditions of Deposition. In the eastern half of the area the basal sediments appear to have filled the valleys and depressions of an undulating metamorphic terrain. This is suggested by the apparently diachronous base of the sequence (Plate X) and by the presence of thin breccias within its basal 1000 ft (300 m). The absence of very thick wedge-shaped breccias and fanglomerates suggests that the topography, though undulating, was not mountainous. The available data fit most closely with deposition in alluvial fans by braided streams, rather than by stream floods or sheet-floods. (Allen 1965, pp. 160–1.) The presence of rounded quartz grains could suggest considerable wind erosion within the source area.

In the western part of the area there is no evidence for an undulating base to the sequence and the presence of a thin but extensive sheet of conglomerate at a consistent height above the base (p. 73) suggests that the metamorphic terrain on which the first deposits were laid down was relatively flat. The basal sediments in this area are slightly finer grained than in the east. They contain more thin beds of shale, siltstone and limestone and have lenses of sandstone full of flakes of mudstone and siltstone. It is suggested that these sediments may be the flood-plain deposits of a river whose source was a considerable distance from the area of deposition.

Sedimentary structures indicative of the direction of transport during the deposition of the basal sediments have not been investigated, but the evidence from the overlying beds and the over-all decrease of grain size of the beds in a west-south-westerly direction suggests that these beds may have been derived from a north-easterly source.

Sandstones and Conglomerates below Clousta Volcanic Rocks (2 to 4 of Plate X)

In the area east of Brindister Voe the sediments between the basal beds and the Clousta Volcanic Rocks consist mainly of massive sandstone, with a lenticular conglomerate, up to 280 ft (85 m) thick, near the middle. West of Brindister Voe (Plate IX), however, the conglomerate is not present. Between Brindister Voe and Mousavord Loch the group appears to thicken rapidly westward and to consist largely of massive, cross-bedded sandstone. Still further west the cross-bedded sandstone in turn gives way to flaggy sandstone with thin silty and shaly partings.

Area east of Brindister Voe

The succession in this area contains the following four lithological groups, listed from base to top:

Sandstones with pebbly bands and thin shaly partings (thickness 900 to 1500 ft (275–450 m); 2a on Plate X). The arenaceous bands in this subdivision resemble those below the Loch of Hostigates breccia (p. 74), which is taken as the highest unit of the basal beds. The rocks are mainly grey or pinkish in colour and range from pebbly grit to a very fine-grained quartzitic sandstone. They are almost continuously exposed on the north shore of the North Voe of Clousta where thin beds of purplish shale or slate, which make up between 3 and 5 per cent of the total sequence, alternate with the arenaceous beds. Both the upper and lower junctions of the slates are sharp, and slate chips are present in the sandstones. Pebbles of granite and quartzite are scattered throughout the sandstones but they are not sufficiently concentrated to form conglomerates. One impersistent pebbly horizon can, however, be traced for some distance on either side of South Loch of Hostigates.

Beds with well-rounded sand grains have been recorded throughout the sequence and with them are associated wind-faceted pebbles, chiefly of white vein-quartz. There are, however, few dreikanter pebbles and when these are found their edges are usually rather rounded. This seems to indicate that the wind faceting took place before they were transported by water to their present position.

Much of the sandstone and perhaps more especially that with well-rounded grains, has a pinkish tinge. This is mainly due to the colour of the grains themselves which include pink and rose-coloured quartz, reddish feldspar and pink felsite. In other beds there is a greenish tinge due to the presence of interstitial chloritic matter.

There is at least one locality on the northern slope of the North Voe of Clousta where tuffaceous sandstone is present, thus anticipating the volcanic rocks which are well developed higher in the sequence.

Medium-grained cross-bedded sandstones (thickness 500 to 800 ft (150–240 m), 2b on Plate X). The pebbly sandstone with shaly or slaty partings passes upwards into more massive strongly cross-bedded sandstone with even thinner shaly partings. This sandstone is grey to pink, medium-grained, usually rich in muscovite, and forms cosets up to 5 ft (1·5 m) thick, separated by thin beds (up to 4 in (10 cm)) of grey siltstone and silty mudstone.

On the shores of Muckle Head [297 582] which forms the headland between the North and South Voes of Clousta (Plate IX), the sandstone is trough-cross-bedded with individual troughs up to 3 ft 6 in (1 m) deep. The predominant trend of the troughs is north-north-east and the dip of foresets along the trough axes to the south-south-west. Many troughs have been affected by penecontemporaneous deformation which has resulted in the steepening and local overturning of their north-eastern or, in some cases, north-western slopes. The tops of the overturned troughs are usually truncated by planar erosion surfaces. Within this sequence some thinly bedded fine-grained sandstones with small-scale cross-stratification show convolute lamination. The sandstones with disturbed bedding underlie the Muckle Head Tuff (see below) and it is likely that earth tremors preceding its eruption were responsible for the deformation of the unconsolidated sediment (cf. Jones 1962, p. 238). Near the top of the sequence the sandstone becomes coarser and small subangular pebbles, up to $\frac{3}{4}$ in (2 cm) in diameter, composed almost invariably of felsite, become common. There are also thin lenses of conglomerate with pebbles, mainly of quartzite and felsite up to $1\frac{1}{2}$ in (4 cm) in diameter.

Muckle Head is formed from a lenticular mass of basaltic lapilli-tuff (possibly originally a volcanic cone) which has a maximum recorded thickness of 120 ft (36 m). The length of its present outcrop, on land, is 400 yd (360 m). In the lower part of the mass lenses of tuff are intercalated with beds of sandstone and conglomerate containing small subrounded pebbles of white and pinkish feldspar, quartz, felsite, microgranite and some deep red jasper. Certain conglomerate bands contain small pebbles of greenish siltstone, and in many bands the pebbles are set in a deep green clayey matrix (p. 82). The lowest tuffs contain irregular lapilli of altered glassy, vesicular basalt, which in many cases enclose small quartz grains. The presence of these grains suggests that the lava may have fallen into the unconsolidated ash as fluid droplets. The quartz grains could also have been picked up by the magma on its way up the vent. The basalt lapilli do not normally exceed $\frac{1}{2}$ in (1 cm) in diameter though larger fragments up to 2 in (5 cm) long have been recorded. The tuff also contains grains of quartz, feldspar, garnet, epidote and sphene (p. 82), some of which are well rounded. Fragments of quartz, feldspar and felsite become progressively less common near the top of the tuff cone. The matrix of the tuff is hard, argillaceous, dark green in colour and full of very small quartz grains.

Clousta Conglomerate (3 on Plate X). A thick band of conglomerate and pebbly grit can be traced as a fault-stepped outcrop from the shore south of Muckle Head eastwards for about 2 miles (3·2 km) almost to the Stead of Aithness (Plate IX). It attains a maximum thickness of 140 ft (43 m) and is composed of subrounded to subangular pebbles of vein-quartz, pinkish quartzite, felsite, granite and quartz-porphyry set in a sandy matrix. There are also some clasts of shale and siltstone. A high proportion of the coarser sand grains are composed of pink feldspar derived from granite and felsite. The pebbles average $1\frac{1}{4}$ in (3·2 cm) in diameter but may exceptionally reach 5 in (12 cm). They show no indication of sand blasting but the interstitial sand grains are in many cases well rounded.

The conglomerate contains a high proportion of unstable clasts which are readily destroyed by chemical and mechanical weathering (see Pettijohn 1957, pp. 254–5). It has abundant pebbles of granite, felsite and quartz-porphyry, which are comparable with the Vementry Granite and its associated minor intrusions (pp. 207–9). Apart from felsites, pebbles of acid igneous rocks are absent in the underlying sediments. As none of the north-north-east trending quartz-porphyry and porphyry dykes associated with the Vementry Granite cut the Old Red Sandstone sediments (Plate XXIV) it is thought likely that the intrusion of the Vementry Granite antedates the deposition of the Sandness Formation. The deposition of the Clousta Conglomerate may thus mark the stage at which the Vementry Granite or a similar late-Caledonian granite mass became uncovered by erosion in the source area.

On the Ness of Nounsbrough [294 574] there is a lenticular mass of conglomerate which is probably on a slightly higher horizon than the Clousta Conglomerate but has a similar pebble content. It attains a maximum thickness of about 100 ft (30 m) and contains, in the southern part of its outcrop, a proportion of tuffaceous material. Traced southward along the strike the conglomerate and interbedded pebbly sandstone give place to a series of acid lavas and tuffaceous sediments. The lavas are described on pp. 85 and 96.

Cross-bedded sandstone between Clousta Conglomerate and Clousta Basalt (4a in Plate X). The sediment between the Clousta Conglomerate and Clousta Basalt is mainly composed of thick medium- to coarse-grained, locally pebbly, sandstone with large-scale, mainly planar cross-stratification and with a predominant dip of foresets to south-south-east. Disturbed cross-bedding similar in type to that seen on Muckle Head (p. 78) is common.

At Aithness, in the extreme east of the area (p. 85) and on the east shore of Clousta Voe thin flows of basalt are interbedded with the sandstone. These are the lowest exposed flows of basic lava within the Sandness Group. Immediately below the Clousta Basalt there are a number of thin beds of tuffaceous sandstone containing a high proportion of basaltic detritus.

Area west of Brindister Voe

In the ground extending from Brindister Voe to the west coast of the Walls Peninsula it has not been found possible to recognize the subdivisions established further east. Conglomerates are absent and the scattered pebbles within the sediments are composed mainly of siltstone and mudstone of Old Red Sandstone type. The salient features of the sediments in the western area are (1) a gradual westward thickening from approximately 2700 ft (820 m) near Brindister Voe to 4000 ft (1220 m) at the longitude of Mousavord Loch and thence a more pronounced westward thickening to possibly 9000 ft (2750 m) at the longitude of Sandness Hill, and (2) the presence of a thick series of massive cross-bedded sandstones with virtually no intervening argillaceous beds. This series forms the rugged, hilly terrain which extends from Brindister Voe westward as far as Mousavord Loch and is up to 2500 ft (760 m) thick. Further west the sandstones become progressively more flaggy and micaceous, with much thinner cross-bedded sets and some shaly and silty partings.

Cross-bedded Sandstones (eastern facies). East of Mousavord Loch the series consists predominantly of grey to buff-coloured medium- to fine-grained sandstone. It contains beds of flaggy fine-grained micaceous sandstone which have alternate pale and dark grey bands with high concentrations of heavy minerals in the latter (p. 82). Siltstone bands are virtually absent in the lower and middle parts of the sequence.

Cross-bedded sets are up to 4 ft (1·2 m) thick and a high proportion of the foresets had an original sedimentary dip to between south and south-west. Disturbed and slumped cross-bedding is seen at a number of horizons north-east of Sulma Water, the most common structures being tight-crested anticlines overturned to the south-south-west. Intraformational conglomerates with shale and siltstone clasts are present throughout the sequence, but pebbles of meta-morphic or igneous rock are very rare.

Flaggy sandstone with siltstone partings (*western facies*). The westward transition from massive sandstone with predominantly large-scale cross-stratification to flaggy sandstone with subordinate siltstone bands takes place at approximately the longitude of Burga Water, but thick sets of massive, often cross-bedded and sometimes pebbly, sandstones have been recorded west of this line, particularly in the lower part of the sequence. Flaggy sandstones with subordinate trough- and planar-cross-bedded sets form the entire sequence exposed on the west shore of the Walls Peninsula between the north shore of the Bay of Deepdale and the Voe of Dale (Plate IX).

The most common rock type in this sequence is a hard fine-grained grey, greenish grey, pinkish, brown or locally purple, somewhat micaceous quartzose sandstone. This forms cosets ranging from 2 to 10 ft (60 cm–3 m) in thickness. The cosets are generally planar-bedded but there is a fair proportion of sets with both planar- and trough-cross-stratification. Disturbed cross-bedding has also been recorded. Many foresets in the cross-stratified sets dip to the south and south-east. The alignment of troughs suggests current movement from north-west to south-east. In the upper part of the sequence there are bands of medium-grained sandstone with small isolated pebbles of felsite and vein quartz. Some bands are slightly tuffaceous.

The beds of siltstone and fine-grained sandstone separating the sandstone cosets are up to 18 in (45 cm) thick, but in many cases much thinner. They are greenish grey to deep purple in colour. Many are tectonically deformed with microfolds, incipient cleavage and, in places, a vague lineation. Sedimentary structures are largely obliterated, but in the lower part of the exposed section (in the Bay of Deepdale) straight asymmetrical ripple-marks and sand-filled polygonal sun-cracks are common.

PETROGRAPHY

Though pebbly sandstones, conglomerates and breccias form an appreciable proportion of the sediments below the volcanic horizon, 80 to 90 per cent of the total rock is grey to buff, medium- to fine-grained sandstone with micaceous partings and with some thin dark laminae containing concentrations of heavy minerals.

The arenaceous rocks have the following characteristics:

Grain size. The specimens examined are poorly to very poorly graded and have an extreme range in grain size from 0·9 to 0·03 mm.

Shape of grains. The grains of the sandstone are normally subrounded to subangular (Plate XIII, fig. 1), but in some sandstones in the extreme west they are angular. The smaller grains are generally more angular than the larger ones and the feldspars are more angular than the quartzes, but in one specimen

(S 52736) the feldspar grains are more rounded than the quartz. Highly rounded grains, which may have been rounded by wind in the source area, have only been recorded in the basal sandstones at Sonso Ness [297 587] and along the south shore of North Voe of Clousta (S 49334). Many grains are strongly elongated, with the long axes mostly parallel to bedding, but in some instances there is a marked imbrication, with the long axes inclined at 20°–30° to the bedding.

Many quartz and feldspar grains have serrate margins due to the ingrowth of white mica and, less commonly, chlorite. The thin sections provide evidence for varying degrees of induration, such as frilling of grain boundaries, the partial resorption of margins of adjoining quartz grains (S 49334) and the development of authigenic sericite.

Composition of grains. All specimens examined have a relatively high proportion of feldspar grains and the ratio of quartz to feldspar clasts within the group ranges from 80:20 to 50:50 with a slight increase of the mean ratio from 65:35 in the eastern sector to 75:25 in the central and western sectors of the outcrop. This suggests a westward increase in the maturity of the sediment or a difference in the source areas. In many cases there is a marked difference in the proportions of quartz to feldspar in adjacent laminae, with a higher proportion of feldspar in the coarser laminae.

Quartz. The quartz grains of the medium- to fine-grained sandstone are in all cases devoid of strain shadows and cracks. Small liquid inclusions of the kind found in the quartz of the Sandsting Granite are absent, and solid inclusions are rare. In one specimen (S 30914) the quartz grains contain inclusions of rutile. Though the authigenic ingrowth of sericite and chlorite fibres into quartz grains is common in many specimens, the complete replacement of quartz is rare. In one specimen (S 30874) some quartz grains are partially replaced by carbonate.

Feldspar. Several distinct varieties of feldspar are present in varying proportions in nearly all thin sections. They are, in order of abundance:

1. Untwinned potash feldspar, generally clear or slightly cloudy, less commonly completely sericitized. In some specimens the feldspar shows rudimentary perthitic texture.
2. Plagioclase, fresh or slightly cloudy. The grains are somewhat smaller in size than the average grain size of the rock. Most plagioclase grains are oligoclase-andesine, though in one specimen albite-oligoclase has been recorded.
3. Microcline, generally fresh, less often slightly cloudy. Common in most slides but absent in some.
4. Microperthite, usually fresh.

The composition of the feldspar grains indicates that they could have been derived either from a granite-granodiorite complex or from metamorphic terrain. There appears to be some regional and stratigraphical variation in the relative abundance of the various types of feldspar grains, of particular note being the westward disappearance of plagioclase clasts. With the exception of two specimens (S 30905, 30914) the percentage of plagioclase is less than that of potash feldspar throughout the area and in the extreme west plagioclase is absent in all but the highest horizons.

Other Minerals. Scattered flakes of faintly pleochroic pale green to colourless muscovite commonly about 0·3 mm long are present in most sections. In the micaceous partings of the sandstone both muscovite and biotite are abundant.

Heavy mineral grains form up to 15 per cent of the total volume of grains in the dark laminae, which are in some cases up to 4 mm thick and are present throughout the sequence. They also occur as scattered grains, forming up to about 2 per cent of the total volume in many thin sections of the unlaminated sandstone, but are completely absent in others. The heavy mineral grains include, in approximate order of abundance, epidote, opaque minerals (ilmenite partly altered to leucoxene, hematite, pyrites and magnetite), sphene, apatite, allanite and tourmaline. Grains of garnet, which are present higher in the sequence (p. 99) have not been recorded. Epidote is extremely common in the dark laminae and common in the micaceous partings, but, in contrast to the sediments higher in the sequence (p. 99), it is virtually absent in the unlaminated grey fine- to medium-grained sandstones. Sphene and apatite are abundant in the dark heavy mineral laminae and in the micaceous partings, and they are also present in the unlaminated sediment.

Lithic clasts. Grains, ranging from 0·5 to 0·2 mm in diameter, of porphyritic felsite, basalt and andesite are abundant in the higher parts of the sequence, particularly in sediments associated with or adjacent to tuff and agglomerate lenses. In some bands (pp. 77–78) lava clasts form 10 to 30 per cent of the total volume. A number of thin sections of sandstone with igneous clasts also contain scattered grains with a small-scale mosaic texture which may be quartzite or silicified felsite.

Matrix and cement. The matrix of the sediments below the volcanic rocks forms between 15 and 25 per cent of the total volume of the rock. Two kinds of material can be distinguished in the matrix:

1. Authigenic chlorite and clay-mica which enclose the grains in a thin film, on average about 0·15 mm thick, and which in some cases penetrate into the quartz and feldspar.
2. Interstitial patches of carbonate, and more rarely, cryptocrystalline quartz. In some sliced rocks the interstitial carbonate is associated with finely disseminated mica (S 30874, 30918). The carbonate is usually coarsely crystalline, indicating that it formed as a chemical cement, but in several sections finely granular carbonate, which may be of detrital origin, is also present. In one thin section (S 30905) the rock contains over 30 per cent of matrix, some of which has clearly been formed by the alteration of feldspar grains.

Nomenclature. According to the classification of Pettijohn (1957, p. 291) feldspathic sandstones containing over 15 per cent of detrital matrix are termed feldspathic greywacke. In the case of these sediments, however, a proportion at least of the matrix is calcareous cement and of the rest a substantial part appears to be due to partial authigenic replacement of the mineral grains. It is therefore considered that the majority of specimens in this group are true arkoses. Some rocks contain an exceptionally high proportion of lava clasts, most of which may have been deposited directly as volcanic fall-out. These rocks are most aptly termed tuffaceous arkoses rather than subgreywackes, which would be the case if the clasts and detrital matrix had been incorporated through the erosion of nearby volcanic rocks.

CONDITIONS OF DEPOSITION

The main features of the strata between the basal beds and the volcanic rocks are as follows:

1. The westward transition from pebbly, strongly cross-bedded sandstone and conglomerate full of detritus of local origin, firstly to fine- to medium-grained cross-bedded sandstone with no extraformational conglomerates and only very rare siltstone partings, and then to flaggy sandstones with thin siltstone partings.
2. The relatively uniform direction of dips of foresets and trends of cross-bedding troughs indicating current movement from the north to north-east sector, though in the extreme west there is some evidence of current movement from the north-west.
3. The marked south-westward increase in the total thickness of sediment.

In the area east of Brindister Voe the beds contain a high proportion of coarse, apparently locally derived, readily weathered components which are generally fresh, and only a small proportion of sediment of shale or silt grade. Deposition probably took place close to the source area, probably at least in part, within alluvial fans. The character of the sandstone in the area between Brindister Voe and Mousavord Loch implies deposition by braided or straight rivers which did not have access to locally derived coarse detritus. The western, relatively flaggy, facies with its many thin grey and purplish siltstones, some with sun-cracks and ripple-marks, suggests deposition by rivers on a more extensive flood plain on which both channel and overbank deposits were preserved.

Clousta Volcanic Rocks and Associated Sediments

Though the first local manifestations of volcanism within the Sandness Formation appear fairly low in the sequence, the bulk of the Clousta Volcanic Rocks are confined to the upper half of the formation (Plate X). The lavas and tuffs form single flows or cones which, in many instances, are completely enclosed in sediment. The volcanic rocks belong to the calc-alkaline suite and are comparable with the Lower Old Red Sandstone extrusive rocks of Central Scotland.

The following extrusive rocks are present:

Basic and sub-basic lavas. These flows range in composition from pyroxene-andesite (p. 92) to basalt (p. 93). They usually form single highly vesicular flows, either closely associated with basaltic tuff, or entirely enclosed in sediment. East of Brindister Voe, individual flows range in thickness up to 200 ft (60 m) and have a maximum length of outcrop of 1½ miles (2·4 km). Between Brindister Voe and Sulma Water up to three flows with a total thickness of about 600 ft (180 m) are locally present. West of Sulma Water basic lavas are thin and impersistent, and only two outcrops of thin flows have been recorded west of the longitude of Mousavord Loch (p. 91).

Ignimbrites and acid lavas. A number of flows of ignimbrite are present in the Ness of Nounsbrough (Plate IX). They range in thickness from over 100 ft (30 m) to less than 10 ft (3 m). The thinner flows are intercalated with thin flows of altered basic andesite or basalt and with coarse acid tuffs. It is possible that some of the thin concordant acid bodies within the Sandness Formation are true rhyolite lava flows. One example is the thin sheet of fine-grained 'rhyolite' which crops out between Burga Water and Upperdale [196 531], in the western part of the area.

Basic tuffs and agglomerates. Coarse tuff and agglomerate which contains a higher proportion of basic than acid detritus forms a number of flat 'cones' which are interbedded with basic lava in the area between Galta Water and Brindister Voe. Fairly thick beds of lithic tuff with clasts of both basic and acid lava, set in a matrix with a vitroclastic texture, crop out just west of Brindister Voe and one such bed is exposed on the south shore of the Voe of Dale, in the extreme west of the area.

G

Acid lapilli-tuffs and agglomerates. These form two lenticular masses, probably originally cones, at least 1300 ft (390 m) thick, at Clousta and Aithness, in the eastern part of the area. A much flatter lenticular mass of acid tuff is present in the western part of the area between Burga Water and Upper Dale. Ignimbrite clasts are abundant in the coarse tuffs exposed between Sulma Water and Brindister Voe.

In addition to contemporaneous lavas and pyroclastic rocks, the Sandness Formation contains a number of roughly concordant intrusions of porphyritic felsite. The lenticular mass of felsite forming Smith's Hamar [252 548] just west of Sulma Water reaches a thickness of approximately 1400 ft (425 m) and the sill cropping out on the north shore of Voe of Dale attains possibly 500 ft (150 m). A number of felsite masses, like that cropping out on the west shore of Uni Firth, are strongly discordant. Many of the smaller felsite bodies are highly irregular in shape suggesting that they were intruded into loosely consolidated ashes and sediments.

AREA EAST OF BRINDISTER VOE

Volcanic Rocks and Intrusions

Clousta Basalt. The Clousta Basalt is the largest and most extensive flow of basic lava in the area east of Brindister Voe. Its outcrop forms the backbone of a prominent fault-stepped ridge which extends eastwards from the north shore of the Voe of Clousta to the west shore of the Loch of Clousta. The outcrop can be traced further east-north-eastward through a string of small islands in this loch.

The basalt is approximately 200 ft (60 m) thick between the western shore of the Loch of Clousta and the central part of its outcrop, but thins westwards to about 100 ft (30 m) on the shore of the Voe of Clousta and has not been found on the Ness of Nounsbrough. Near the western end of its outcrop the basalt is underlain by a thin bed of basic tuff but farther east it rests directly on sandstone which only locally contains small fragments of volcanic detritus.

Throughout the length of its outcrop the Clousta Basalt appears to consist of only one flow. It is usually vesicular throughout, even where its thickness approaches 200 ft (60 m). Its base contains well-developed pipe amygdales which are well seen on the north slope of the ridge, 720 yds (660 m) NNW of Clousta School. The top of the flow is in many places highly scoriaceous and usually has hollows and cracks filled with sand from above. It is best exposed at Little Head [297 576] on the north shore of the Voe of Clousta, where the top of the basalt is seen to be highly irregular with elongate bulbous protrusions aligned roughly east–west. The basalt is here overlain by 2 to 3 ft (60–90 cm) of purple mudstone which is full of large elongate vesicles. This is, in turn, overlain by up to 4 ft (1·2 m) of well-bedded reddish purple siltstone. The presence of vesicles in the mudstone suggests that in this area the basalt was either intruded into thin unconsolidated mud or that it was covered by mud immediately after its intrusion, before all volatiles had escaped. As neither mudstone nor siltstone shows signs of marked induration, the latter explanation seems more likely.

Aithness and Papa Little basalts. Three bands of vesicular basalt are present in the Aithness peninsula. The lowest of these is interbedded with massive cross-bedded sandstone and crops out near the north shore of Aithness. It can be traced for about 850 yd (770 m) along the strike from the head of the Stead of Aithness to The Rona, and may originally have been continuous with the basalt lava which traverses the island of Papa Little from its south-west shore north-north-eastward to the Walls Boundary Fault. In Aithness the flow is approximately 50 ft (15 m) thick at the south-western end of its outcrop, but thickens to about 100 ft (30 m) on the north-east shore of Aithness. In Papa Little it appears to be up to 150 ft (30 m) thick at the coast and is possibly thicker inland. Both the Aithness and Papa Little basalts are vesicular throughout. The latter is intensely sheared and amphibolized in the vicinity of the Walls Boundary Fault.

The higher basalts of Aithness are intercalated with agglomerate and tuff (pp. 85–86) but, though both are vesicular, only one appears to be a true lava flow. The other, thinner, bed which is exposed at intervals for 300 yd (280 m) seems to cut across the bedding planes of the tuffs and associated sandstones. The field characteristics suggest that it formed an intrusive sheet, probably in loosely consolidated country rock.

Ness of Nounsbrough Ignimbrites and Basic Lavas. The Ness of Nounsbrough contains a number of concordant sheets of fine-grained acid rock, which form several strong topographic features. The lower features exposed in the northern part of the peninsula appear to be sills of porphyritic felsite (p. 86). The coast section on the east shore of Brindister Voe, both north and south of the Head of Lahamar [287 564], contains four large and several smaller flows of pink to purplish ignimbrite, which bear a superficial resemblance to sills of highly porphyritic felsite, but are quite distinctive in thin section (p. 96). The lowest ignimbrite flow, which crops out on the coast 250 yd (230 m) N of the Head of Lahamar, is approximately 30 ft (9 m) thick and is separated from the Head of Lahamar ignimbrites by approximately 180 ft (55 m) of tuffaceous sediment with lenses of tuff. The Head of Lahamar is formed of three flows of ignimbrite, one of which is over 130 ft (40 m) thick, separated by thin beds of finely banded tuff and purple slate with lenses of limestone. Two further thin flows of ignimbrite, interbedded with acid lapilli-tuff and with four flows of basalt or andesite, crop out on the east shore of Uni Firth, some 450 ft (140 m) higher in the sequence. The basic lava flows average 30 ft (9 m) in thickness and have markedly vesicular tops and bases.

Both the ignimbrite and lava flows described above thin out in a north-easterly direction. The ignimbrite forming the Head of Lahamar thins very gradually and can be traced inland for a distance of 750 yd (700 m). All interdigitate north-eastward with the agglomerates and lapilli-tuffs of the Clousta area.

Lapilli-tuffs of Clousta and Aithness. The lenticular outcrop of lapilli-tuff, agglomerate and lavas, which extends from the east shore of Uni Firth to the south shore of Clousta Voe, is here termed the Clousta Tuff. It attains a maximum exposed thickness of possibly 1300 ft (400 m) and is exposed in numerous rocky knolls south of the Voe of Clousta and on the east shore of Uni Firth. The 'tuff' is well bedded and composed of roughly alternating layers of coarse lapilli-tuff and medium- to coarse-grained sandstone, in places with thin partings of buff or purple siltstone. The clasts in the lapilli-tuff normally range from $\frac{1}{4}$ to $1\frac{1}{2}$ in (6 mm–4 cm) in diameter but there are isolated angular blocks up to 18 in (45 cm) in size. Many of the fragments consist of pink sparsely porphyritic felsite and ignimbrite. Clasts of dark purple basalt or andesite are less common. In some exposures however the latter predominate over felsite.

The Aithness Tuff also reaches a maximum exposed thickness of 1300 ft (400 m) and its outcrop extends from the Loch of Clousta north-eastwards to

The Rona. It consists of acid lapilli-tuff with lenticular masses of agglomerate, which grades downwards into sandstone with thin beds of acid tuff. The composition of clasts is similar to that in the Clousta Tuff.

Relatively short impersistent flows of basic lava are present both in the Clousta and Aithness tuffs (pp. 85 and 94.)

Felsites. Sheets of felsite with feldspar phenocrysts are fairly common within the Clousta Tuff just south of the Voe of Clousta. Some of these are much brecciated and seem to grade into beds of tuff. Felsites with obscure field relationships appear to be interbedded with the coarse Aithness Tuff on the east shore of the Loch of Clousta. Elsewhere they show distinctly intrusive and transgressive features often with highly irregular margins. These characters suggest that intrusion took place under very shallow cover in loose ash.

Sediments intercalated with and overlying the Volcanic Rocks

The rocks overlying the horizon of the Clousta Basalt are extremely variable when traced along the strike. At the Ness of Nounsbrough and the Stead of Aithness, they are composed largely of pyroclastic and extrusive rocks (p. 85) but between these two areas the basalt is overlain by up to 2000 ft (650 m) of essentially non-tuffaceous sediment.

The lower 400 to 600 ft (120–180 m) of this sediment (4b in Plate X) is mainly massive medium- to fine-grained sandstone composed of thick sets with large-scale cross-stratification. Foresets of the planar cross-bedded sandstone dip, in the majority of cases, to the south to south-south-west sector. Small-scale disturbed cross-bedding is common near the base. The sandstone has a limited lateral extent, being confined to the same length of outcrop as the basalt.

The upper 1500 ft (460 m) of sediment (6 in Plate X) consists of flaggy sandstones interbedded with shales and siltstones containing thin irregular sandstone ribs. The section of these strata and some of the underlying beds, as seen on the north shore of Clousta Voe, between 350 and 680 yd (320 and 620 m) W of Clousta School [308 573] is shown in Fig. 8a. This sequence shows the following features:

1. The units composed predominantly of sandstone range in thickness from 3 to 25 ft (0·91–7·6 m). They are usually flaggy with planar-bedding and contain partings of purple or purplish grey siltstone or shale which are up to 3 in (7·6 cm) thick and 1 to 4 ft (0·3–1·2 m) apart. The upper surfaces of the sandstone ribs often have straight asymmetric ripple marks and both sandstone and siltstone are locally ripple-laminated. Desiccation polygons with sand-filled cracks are very common in the shale layers. Intraformational conglomerates are, however, absent.
2. Units composed of purplish shale, silty shale and siltstone with thin, often irregular, ribs of fine-grained sandstone range in thickness from 2 to 30 ft (60 cm–9 m). The total thickness of shale and siltstone usually exceeds that of the sandstone ribs which are from 1 in (2·5 cm) to, exceptionally, 2 ft 6 in (76 cm) thick. Ripple-cross-lamination is present in both sandstone and siltstone. Straight asymmetric ripple marks are seen on bedding surfaces, though the primary structures in the finer sediments are partially obscured by tectonic crinkling associated with a fracture-cleavage. Desiccation cracks are less common than in the siltstone partings within the flaggy sandstone. Penecontemporaneous disturbed lamination is seen at a number of horizons. True convolute lamination with the characteristic sharp-crested 'anticlines' and basin-shaped 'synclines' (p. 107) is absent, and the characteristic structures are overturned folds with local rupture and small-scale

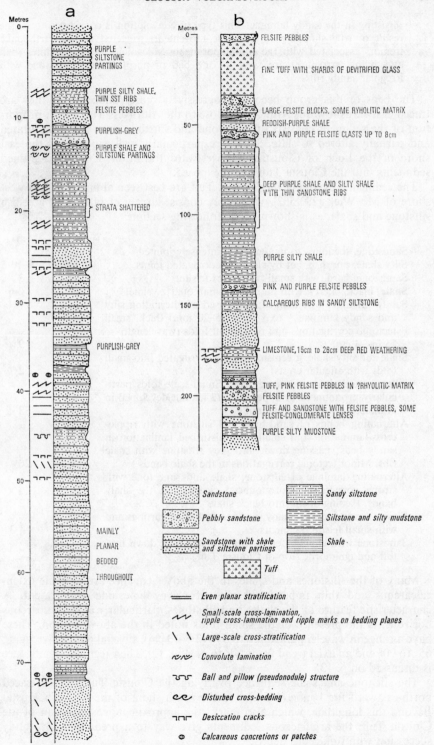

FIG. 8. *Sequences of strata interbedded with the Clousta Volcanic Rocks*

(a) Strata exposed on the north shore of Clousta Voe, between 320 m and 620 m west of Clousta School.

(b) Strata exposed at Voe of Dale.

thrusting in the sandy laminae. This type of convolution is considered to be the result of true slumping (down-slope movement) triggered off by earthquake tremors associated with the contemporaneous volcanicity. Pseudonodules (ball and pillow structure) are seen in one bed of purplish silty mudstone with irregular sandy ribs.

The beds of this group become progressively more tuffaceous as they are traced eastwards towards Aith Ness. Close to the Aith Ness Tuff a high proportion of the sediment consists of mudstone and siltstone, now highly indurated and partially altered to slate. Slates are particularly well exposed on the east shore of the Loch of Clousta. The westward transition of the fine-grained sediments into the Clousta Tuff is not exposed.

The sediments overlying the Clousta Tuff are best seen along the north-west shore of Mo Wick [293 562] where they consist largely of black and greenish siltstone and shale, as is shown in the following section:

	ft	(m)
Sandstone, medium- to fine-grained, intensely jointed ..	—	—
Silty shale, purplish, cut by close-set corrugated joints ..	30 to 35	9–10·5
Sandstone, hard, with small pebbles of felsite	10	3
Shale, black, cherty, with numerous small quartz veinlets ..	2	0·6
Sandy siltstone, well banded, composed of alternating silty and sandy laminae ½ to ¾ in (12·5–20 mm) thick, small tectonic corrugations and recumbent folds (wavelength = 1½ in—4 cm)	30	9
Shale and silty shale, greenish, strongly corrugated into small folds with angular crests	9	2·7
Siltstone and shale with sandy ribs. Small-scale folds particularly pronounced in upper 10 ft (3 m) but developed in argillaceous layers throughout	125	38
Alternating bands of sandstone and siltstone with ripple-cross-lamination and small-scale disturbed lamination in sandy beds; passing down into grey siltstone with shale ribs. Minor tectonic corrugations in the shale bands ..	85	26
Alternating laminae of siltstone, shale and sandstone with strong corrugations, pronounced cleavage in the shaly bands. Passing down into black shale	40	12
Flaggy sandstone with many siltstone and shale partings and with small felsite fragments	14	4·3
Sandstone with angular felsite fragments passing down into tuff and ignimbrite (top of Clousta Tuff)	—	—

Many of the siltstones and shales in the above sequence are to some extent calcareous and thin impersistent ribs of clayey limestone are present. A characteristic feature of the shales and siltstones, particularly at the calcareous horizons, is the presence of the small folds noted in the above section. These have a range in wavelength from 1 to 5 in (2·5–13 cm), an axial plunge averaging 65° to 90° and an axial trend from NE to N (Fig. 12). Their tectonic significance is discussed on p. 135.

The siltstone and shale sequence overlying the Clousta Tuff can be traced north-eastward for 1 mile (1·6 km) to the south shore of the Voe of Clousta. Beyond this longitude, which also marks the approximate eastern limit of the Clousta Tuff, the argillaceous sediments give way to a predominantly sandy succession with tuffaceous debris.

It is perhaps significant that both the Clousta and Aithness tuff 'cones' are in part, at least, associated with fine-grained sediment which both overlies and abuts against the tuff. The thick argillaceous and calcareous sediment at Mo Wick shows little sign of large-scale cross-stratification or other structures of a type associated with rapidly flowing water, and structures such as sun cracks which indicate deposition in very shallow water have not been seen. This suggests that it may have been deposited in the standing water of a lake or lakes. The thick irregular accumulations of pyroclastic material may have radically changed the topography of the area of deposition which was earlier controlled by a fluvial regime. By damming a river valley they could have been responsible for the formation of such a lake. The sandstone which overlies the fine-grained sediment at Mo Wick and Aithness suggests a return to a more normal fluvial regime.

AREA WEST OF BRINDISTER VOE

Sector between Brindister Voe and Smith's Hamar

Because of incomplete exposure and the presence of a number of strike-and-dip-faults within the area between Brindister Voe and Sulma Water, it is not possible to make a reliable interpretation of the stratigraphical relationships within the volcanic group in this area. A possible interpretation, shown in Plate X, suggests that there are five isolated lenticular flows of andesite and basalt within a sequence of coarse, predominantly basic, tuffs and tuffaceous sandstones.

In this area the group has an approximate thickness of 2000 ft (600 m) and is intruded by several lenticular masses of porphyritic felsite, the largest of which, the Smith's Hamar Felsite, extends from the south-east corner of Sulma Water westwards to Smith's Hamar (Plate IX).

Basic Lavas. The lavas which include both pyroxene-andesites and albitized basalts (p. 93) range in colour from pale grey to deep purplish grey, and range from fine-grained aphanitic to macroporphyritic with locally fluxion-banded plagioclase laths. They are commonly brecciated and some pass in places imperceptibly upwards into coarse agglomerate which, at one locality, 200 yd (180 m) E of the south-east corner of Sulma Water, is up to 30 ft (9 m) thick and passes in turn upwards into tuffaceous sandstone. The brecciated lava is in many places intensely veined by carbonate and both the spaces between individual lava fragments and vesicles within the lava are filled with sandstone or fine tuff.

Lapilli-tuffs with matrix exhibiting vitroclastic texture. In the area between Brindister Voe and the Loch of Hollorin there are several thick, somewhat lenticular, masses of compact lithic tuff with abundant clasts in the lower range of the lapilli-grade size, interbedded with basic lava and tuffaceous sandstone (Plate IX). The tuff is dark grey in colour with angular clasts of fine-grained dark grey lava predominating over somewhat smaller clasts of pink felsite and isolated crystals of pale pink potash feldspar, set in a fine-grained vitroclastic matrix, with a texture which is similar to that of the ignimbrites east of Brindister Voe. These tuffs pass westwards into lapilli-tuffs which have a sandy matrix, but the nature of the transition has not been investigated.

Lapilli-tuffs and agglomerates. In the area between Brindister Voe and Sulma Water, well-bedded lapilli-tuff and agglomerate forms a number of roughly lenticular masses some of which are closely associated with basic lava. In the eastern part of the area the matrix of the tuffs is partly vitroclastic but in the west it is entirely sandy, with sandstone partings separating individual tuff bands. The sandstones and shales between these tuffs contain small angular fragments of both basic lava and felsite.

In most pyroclastic deposits in this area there is a slight preponderance of clasts of basalt or andesite over those of ignimbrite and felsite. Fragments of baked sandstone and siltstone are present in certain bands. In the coarser deposits isolated angular blocks up to 1 ft (30 cm) long are set in a matrix of lapilli-tuff in which fragments range in size from $\frac{1}{4}$ to $\frac{1}{2}$ in (6–12·5 mm) or, exceptionally, 1 in (25 mm). Felsite and ignimbrite clasts are commonly angular to subangular and show an extensive range in size. Clasts of decomposed glassy basalt are more consistent in size (average 2·5 cm) and are generally rounded to subrounded, suggesting that they may have consolidated in flight following ejection.

To the west of Sulma Water one small isolated cone of coarse agglomerate is exposed on the north slope of Smith's Hamar (Plate IX). This contains many subrounded blocks of banded felsite up to 3 ft (0·9 m) long, set in a matrix of rhyolitic tuff.

Felsites. In addition to the Smith's Hamar Felsite which is a lenticular somewhat transgressive intrusion up to 1500 ft (450 m) thick, with a length of outcrop of $1\frac{1}{2}$ miles (2·4 km), there are a number of thin sill-like concordant felsite bodies in the vicinity of Sulma Water and Hamari Water [250 556] and some strongly transgressive felsite masses of highly irregular shape in the ground north of Loch of Hollorin [276 558] and on the west shore of Uni Firth [286 560]. As in the area east of Brindister Voe, these felsites are comparable in both texture and mineral composition with the felsite clasts in the pyroclastic deposits of the group. It is therefore reasonable to assume that these intrusions belong to the same general phase of volcanic activity as the somewhat earlier felsites from which the clasts were derived. The irregular shape of some of the bodies suggests that they were intruded into only partially consolidated sediments and tuffs.

Sector west of Smith's Hamar

Basic lavas and pyroclastics. In the area between Smith's Hamar and the longitude of Mousavord Loch there is a gradual westward decrease in the amount of contemporaneous basic rock, both lava and pyroclastic. Basic clasts occur as scattered pebbles and within discrete tuff bands which are not more than 18 in (45 cm) thick. In the lower part of the sequence these clasts are embedded in sandstone, in the upper part they occur mainly in mudstone and siltstone. West of Burga Water basic clasts are extremely rare.

The basic lavas exposed in the area west of Smith's Hamar are all relatively thin, never exceeding 50 ft (15 m), and are widely separated from each other by tuffaceous sediments (Plate X). Because of indifferent exposure these flows cannot be continuously traced along the strike, and it is not certain if the isolated outcrops are parts of one thin flow or represent a number of flows of restricted lateral extent. They are composed of relatively fresh fine-grained microporphyritic pyroxene-andesites and subordinate albitized basalts which, unlike the basalt flows further east, are almost everywhere devoid of vesicles along the junctions with the sediments. There is thus a possibility that some may be concordant minor intrusions rather than lava flows.

West of the longitude of Stanevatstoe Loch [216 545] there are only two records of basic lava within the volcanic horizon: in a small quarry 55 yd (50 m) ESE of the school house at Netherdale [181 523] where fine-grained purplish grey basalt overlies purple silty mudstone, and in an isolated exposure 150 yd (135 m) S of the Voe of Dale. The thickness and extent of these lavas are unknown.

Acid lavas, ignimbrites and concordant felsite intrusions. Between Burga Water and Littlapund [201 532] a concordant body of fine-grained acid rock up to 80 ft (24 m) thick underlies a coarse felsitic tuff and tuffaceous sandstone. This is a fine-grained pink to pale purplish rock which in thin section has the characteristics of felsite (p. 98). Because of the poor exposure of its margins it is not possible to determine if this is a concordant intrusion or a lava flow. Smaller concordant bodies of felsite are seen at higher horizons on the west shore of Burga Water and in the vicinity of Littlapund (Plate IX). Again their mode of origin is uncertain.

Along the south and south-east shores of Burga Water there are a number of thin beds to which the field description 'tuffaceous mudstone' has been given. Only one specimen from these beds has been sliced (S 30898) and its argillaceous matrix has a vitroclastic texture with relics of welded shards, suggesting that the rock is an ignimbrite.

The sill of sparsely porphyritic felsite which forms a continuous outcrop between the north shore of the Voe of Dale and Upper Dale [196 531] reaches a thickness of possibly 600 ft (180 m). Its upper junction is exposed on the north shore of the Voe of Dale, where it is seen to be intrusive but concordant with the overlying sediment. A number of small irregular felsite dykes, not usually exceeding 2 ft (60 cm) in thickness, cut the sediments underlying this sill.

Acid pyroclastic rocks and sediments. A thick bed of coarse lapilli-tuff which is here termed the Dale Tuff extends west-south-westwards from Burga Water to Upperdale. Near its eastern end it contains bands of agglomerate with felsite clasts normally up to 3 in (8 cm) in length, as well as some much larger blocks, set in a matrix of finer igneous detritus. There appears to be a gradual westward decrease in the size of component clasts coupled with a westward increase in the proportion of sandy matrix in the tuff. West of Upperdale the tuff gradually passes into a sandstone with abundant pebbles of felsite and some bands of tuff. This tuff and tuffaceous sandstone is intercalated with fine-grained argillaceous, silty and locally calcareous, sediments, which appear to thicken westwards towards the Voe of Dale, where the sequence of strata, thought to comprise the entire spread of volcanic rocks and associated sediments, is as shown graphically in Fig. 8b.

In this coast section the group has the following features:

1. Contemporaneous extrusive rocks are absent on the shore, though two exposures of basic lavas have been recorded just inland.
2. Agglomerates and lapilli-tuffs form only thin bands, which are closely associated with thick beds of coarse pebbly sandstone and conglomerate. The latter contain abundant subrounded boulders, cobbles and pebbles of pink and purple felsite, locally up to 2 ft (60 cm) long. There are several good exposures of channel-fill structures and of large-scale cross-bedding in the sandstone, suggesting that the beds were deposited by fast flowing rivers. The felsite clasts were probably derived by erosion from a nearby felsite agglomerate cone which may still have been active at the time and may have been the direct source of the thin beds of tuff and agglomerate.

3. The group contains a high proportion of reddish purple shale and siltstone and a few thin calcareous beds. The thickest argillaceous horizon consists of deep purple silty shale with thin ribs and irregular lenticles of sandstone and is 200 to 250 ft (60–75 m) thick. Several thinner beds of purple silty mudstone with thin ribs of tuffaceous sandstone occur in the lower part of the sequence.
4. Close to the top of the group there is a 100-ft (30-m) thick bed of very fine-grained pale purplish tuff, which contains shards of devitrified glass as well as quartz grains and fragments of felsite. Because of lack of exposures this tuff band cannot be traced far eastward from the coast.

There is a marked westward decrease in the total thickness of the beds containing the volcanic rocks. This is due, in part, to the westward decrease in the number and thickness of extrusive and pyroclastic bodies and, in part, to the westward transition of predominantly arenaceous beds into deep purple argillaceous sediments. Within the partially arenaceous sequence ripple-cross-lamination and desiccation cracks are present. The latter suggest deposition under shallow-water conditions with periodic drying out. The reddish colour of most of the fine-grained sediment might imply deposition in an arid environment free from reducing conditions, but the reddening may also be the result of iron staining from the products of weathering of basic lavas.

In the western part of the area the volcanic group passes upwards into grey micaceous flaggy sandstone with thin pebbly bands containing felsite clasts and thin partings of siltstone.

FLORA

Traces of plant remains have been found in two localities within the sediments interbedded with the Clousta Volcanic Rocks (Fig. 11, p. 86). Those on the east shore of the Loch of Clousta are unidentifiable graphitic streaks and patches. The specimens collected from the east shore of the Voe of Clousta, 30 yd (27 m) due west of Clousta School [309 573] have been examined by Prof. W. G. Chaloner who has identified *Psilophyton sp.*, and a minute spineless axis of cf. *Hostimella*. Chaloner stated that 'The plants do not give precise evidence of age, and *Hostimella* in particular is not decisive. The presence of *Psilophyton* however suggests a Lower or Middle Devonian age.'

PETROGRAPHY

Pyroxene-andesites

Pyroxene andesites are most abundant in the area west of Brindister Voe. They are generally sparsely porphyritic with feldspar phenocrysts of both square and lath-shaped outline ranging in length up to 4 mm. The andesites forming the highest flows just west of Burga Water are conspicuously porphyritic with abundant feldspar phenocrysts up to 7 mm long. Some of the lowest flows in this

---

EXPLANATION OF PLATE XI

A. West coast of Walls Peninsula, looking north from Mu Ness [166 524] towards Hill of Melby. Steeply inclined, predominantly planar-bedded sandstones of the Sandness Formation. (D 964).
B. Wats Ness, looking north at north shore of Trea Wick [172 507]. Steeply inclined, but relatively undeformed sediments of Walls Formation. (D 963).

A

Geology of Western Shetland (*Mem. Geol. Surv.*)   PLATE XI

B

area are non-porphyritic (S 30888–90). Phenocrysts of augite have been recorded in only one specimen (S 30913). The groundmass of most of the andesites consists of feldspar laths which make up 75 to 85 per cent of its total volume, ophitic or intersertal plates of augite and interstitial patches of finely fibrous chlorite and carbonate. Iron ore and pyrites occur as skeletal crystals, locally arranged in clusters. Other common accessory minerals are apatite, rutile and sphene.

The arrangement of the feldspar laths is in most cases decussate, but in the finer-grained specimens (S 30884, 30911) the laths are flow-aligned. Devitrified glass has been seen in the matrix of a number of specimens, and one such rock (S 30901) contains a radiating fibrous aggregate of crystallites of plagioclase, associated with finely granular iron ore.

The composition of both groundmass feldspars and phenocrysts ranges from calcic oligoclase to calcic andesine (S 30893, 30897, 30913). In many thin sections (e.g. S 30888–9) the feldspars are partially albitized but commonly contain irregular residual blebs of the original more basic plagioclase. Zoning has been recorded in only one specimen (S 30991) and in this it affects the outer rim only. In the area between Burga Water and Sulma Water the plagioclases of the highest flow appear to be more calcic (mid- to calcic andesine) than those of the lower flows (sodic- to mid-andesine).

Fresh augite is preserved in only about half of the specimens examined and no orthopyroxene has been recorded. In the highest flow of the Burga Water area (S 30897) augite forms nearly 20 per cent of the total volume of the rock and occurs mainly as ophitic crystals up to 1 mm in diameter. In most specimens, however, augite forms less than 15 per cent of the total volume. In some sliced rocks it forms sub-ophitic crystals, but in a number of slices it is intersertal or granular with grains not exceeding 0·3 mm in size. The pyroxene is commonly partly altered to chlorite.

The andesites are saussuritized to varying degrees, and some specimens (e.g. S 30893) contain scattered grains and irregular anastomosing veinlets of epidote. Euhedral blades of actinolite, which is pleochroic from yellow to bluish green, are present in some specimens, either within small chlorite–calcite veins (S 30913, 30987) or as a lining to and within calcite- and chalcedony-filled amygdales (S 30901, 30855). Amphibole crystals are only rarely found within the matrix of the andesites (e.g. S 30893).

Basalts

Basalt lavas form most of the flows in the areas east and immediately west of Brindister Voe. Only two basalt flows have been recognized in the Burga Water area. The more westerly basalts are highly porphyritic with phenocrysts of labradorite up to about 2·4 mm long (S 30885, 30902). The phenocrysts have corroded margins and some have a narrow reaction rim. Most basalts in the east are only sparsely microporphyritic, with plagioclase phenocrysts not normally exceeding 1·5 mm in length. The feldspar laths of the groundmass (also labradorite) generally show some flow alignment with intervening decussate zones. The basalts west of the longitude of Clousta contain sub-ophitic to ophitic plates of augite, which make up nearly 35 per cent of the total volume of the rock (S 30987, 30760, 30902). Specimens (S 30773, 30772) from flows north-east of Clousta, however, are uralitized to varying extents and the interstices between

TABLE 2

*Chemical analyses of sandstone, felsite sill and
basalt lava in Walls Sandstone*

	1	2	3
SiO_2	73·31	71·25	50·45
Al_2O_3	10·68	13·54	16·63
Fe_2O_3	0·30	0·45	3·79
FeO	2·37	1·91	5·50
MgO	1·18	0·99	6·95
CaO	2·97	0·20	7·10
Na_2O	3·29	1·84	2·35
K_2O	1·46	7·84	1·84
$H_2O > 105°$	1·21	0·94	2·80
$H_2O < 105°$	1·10	0·27	2·26
TiO_2	0·45	0·15	1·34
P_2O_5	0·14	0·01	0·31
MnO	0·05	0·05	0·12
CO_2	2·12	0·28	0·22
Allow for minor constituents	0·14	0·19	0·28
Total	99·77	99·91	99·94
*Ba	350 ppm	700 ppm	1000 ppm
*Co	< 10	< 10	20
*Cr	44	< 10	26
*Cu	10	< 10	20
*Ga	< 10	< 10	20
Li	12	8	40
*Ni	20	< 10	60
*Rb	n.d.	330	n.d.
*Sr	150	42	450
*V	54	10	220
*Zr	300	440	200
B	17	2	7
F	250	120	500

*Spectrographic determination   n.d. = not determined.

the feldspar laths are filled in part by chlorite and in part by aggregates composed of minute flakes of green mica and euhedral needles and blades of actinolite (Plate XIII, fig. 4). In the Clousta Loch and Aithness areas the secondary actinolite forms a very high proportion of the patchy intersertal aggregate which also contains abundant granular epidote and less than 5 per cent of chlorite. Throughout the area the original chlorite within the veins and vesicles has also been partly or completely altered to a green, highly pleochroic, fibrous aggregate of actinolite and biotite.

The development of secondary amphibole and biotite at the expense of pyroxene and chlorite becomes progressively more pronounced in a northeasterly direction and is most marked in the vicinity of the Walls Boundary

TABLE 2—(*contd.*)

NORMS

	2	3
Q	27·31	3·59
C	1·69	0·00
or	46·33	10·87
ab	15·57	19·89
an	0·93	29·40
di	0·00	3·04
hy	5·45	20·83
ol	0·00	0·00
mt	0·65	5·50
il	0·28	2·54
ap	0·02	0·72
Others	1·68	3·56
Total	99·91	99·94
Q	30·61	10·46
or	51·94	31·65
ab	17·45	57·89
Total	100·00	100·00
or	73·74	18·08
ab	24·78	33·06
an	1·48	48·87
Total	100·00	100·00
ab	94·38	40·35
an	5·62	59·65
Total	100·00	100·00

1. Sandstone, fine-grained, with calcareous matrix, 900 yd (820 m) E2°S of Efrigarth, Walls [277 503]. S 50128, Lab. No. 1992. Anal. J. M. Nunan and W. H. Evans, spectrographic work by C. Park. (*Summ. Prog.* 1967, p. 96).
2. Felsite sill cutting Walls Sandstone, hillside south of Longa Water, 800 yd (730 m) W37°S of West Burrafirth 'School' [263 567]. S 50129, Lab. No. 1991. Anal. J. M. Nunan and W. H. Evans, spectrographic work by C. Park. (*Summ. Prog.* 1967, p. 96).
3. Basalt lava flow, Water Head, west coast of Loch of Clousta, 850 yd (790 m) N 28°E of Grind [314 583]. S 50138, Lab. No. 1983. Anal. W. H. Evans, spectrographic work by C. Park. (*Summ. Prog.* 1966, pp. 76–7).

Fault. A similar development of amphibole and biotite has been recorded in basic minor intrusions in the northern part of the area described in this memoir (p. 244), a swell as in the area just east of the fault (F. May, personal communication) and in the basalt dykes of western Fair Isle (Mykura 1972, pp. 38–9). This alteration cannot be attributed to the progressively increased effects of shearing near the fault as the effects of shearing are confined to a much narrower strip adjoining the fault and also because the mineral assemblage in the amphibolized lava is not that of a basalt subjected to retrograde dynamic metamorphism. It is more likely to be due to the thermal or, more likely, late magmatic activity of a nearby or underlying major intrusive body.

In several thin sections of basalt chalcedony is observed both in the matrix and in epidote-chlorite veins. Amygdales generally contain calcite and chalcedony and they are in most cases bounded by an outer zone of acicular actinolite.

An unusual flow of autobrecciated macroporphyritic non-ophitic basalt, which occurs low in the volcanic sequence, crops out on the west shore of Burga Water (S 30917). This contains abundant phenocrysts, up to 5 mm in diameter, of sericitized plagioclase and euhedral crystals up to 2 mm in diameter of fresh augite set in an almost aphanitic groundmass. This lava also contains an appreciable proportion of quartz grains, which may have been incorporated during the passage of the lava over loose sand.

Ignimbrites

Head of Lahamar [289 564]. The greater part of the Head of Lahamar flows (Plate IX, p. 85) is composed of ignimbrite (S 53767) in which clasts that are clearly visible in hand specimen make up to 20 to 30 per cent of the total volume. The clasts consist mainly of subhedral to euhedral crystals of patchily kaolinitized untwinned potash-feldspar. There are also fragments of felsite and some flattened pieces of pumice which show axiolitic recrystallization of feldspar and quartz into crystallites up to 0·25 mm long and 0·04 mm thick. The matrix is formed largely of flattened shards on average 0·4 mm long and 0·03 mm wide aligned parallel to the bedding. Locally, however, shards display the original arcuate and Y-shaped structure with little flattening. There are also a number of narrow elongate zones in which the shards are twisted and distorted. These zones may either be narrow penecontemporaneous shear planes or passages used for the escape of residual gases. The shards show intense distortion and flattening both below and above the feldspar crystal-clasts. The matrix is largely recrystallized into an irregular granular mosaic of quartz and feldspar, with individual quartz patches exceptionally up to 0·045 mm in diameter. Axiolitic or spherulitic textures of the type common in the rhyolites of Papa Stour (pp. 167–8) are rarely developed. The purplish grey colour of the rock is due to a uniform dusting of hematite.

Certain parts of the Head of Lahamar flows (S 30750) are, however, composed of lapilli-tuff made up of 80 per cent of rounded clasts in the coarse tuff grade (5 to 6 mm diameter), set in a matrix made up of a fine-grained aggregate of carbonate and clay minerals with irregular patches of quartz and feldspar.

The highest thin flows of ignimbrite exposed at the Head of Lahamar, which are interbedded with basalt or andesite flows (p. 85), contain up to 40 per cent by volume of clasts. The latter consist of porphyritic and non-porphyritic felsites, basalts and andesites, euhedral feldspars, and some small scattered crystals of zircon. The matrix has some relict traces of flattened shards (S 30767) but is completely altered into an aggregate of clay minerals and sericite. One flow (S 51531) contains flattened inclusions with wispy flame-like ends. These are similar to the 'fiamme' structures of flattened particles of black devitrified glass, which have been recorded in many ignimbrites and tuff lavas elsewhere (Bersenev and others *in* Cook 1966, pp. 114–6).

The fabric of the flows in this area has been affected by subsequent shearing and small scale folding (p. 135) which has produced two distinct planar structures parallel to the limbs of the minor folds.

Burga Water. The thin flows of ignimbrite cropping out on the south and southeast shores of Burga Water (S 30898) consist of a compact dark grey rock with abundant subhedral to euhedral crystal clasts of pink potash feldspar, up to 2·6 mm in diameter, and rare fragments of altered andesite. Its groundmass is composed of flattened and, in places, intensely distorted shards, replaced by a fine mosaic of quartz and clay minerals with an irregular scatter of larger patches of quartz.

Voe of Dale. The flow of fine-grained 'tuff' near the top of the Sandness Formation just south of the Voe of Dale (p. 92) is a compact pink rock without the abundant crystal clasts of feldspar characteristic of the ignimbrites farther east. Clasts form a relatively small proportion of the rock and include rounded fragments of spherulitic felsite up to 2·8 mm in diameter, near-euhedral crystals of potash feldspar, scattered detrital grains of quartz and felsite together with rare iron ore, tourmaline and muscovite. The matrix is composed of a very fine-grained structureless aggregate of quartz, feldspar and clay minerals and has a dusting of hematite which gives it a red colour. Though much of its original texture is obliterated, there are some traces of structures comparable with those of the ignimbrites from other areas.

Acid tuffs and agglomerates

Clousta and Aithness. The megascopic textural variations within the Clousta and Aithness tuffs have already been described. The clasts consist predominantly of fine-grained acid rock, but in certain areas fragments of basic lava predominate over the acid ones.

Acid clasts within the Clousta Tuff include the three varieties of felsite described on p. 98, ignimbrite with flattened and welded shards, and feldspar crystals, which are either euhedral or broken but never markedly rounded. The matrix of the tuff consists of particles of quartz and feldspar ranging from 0·25 to 0·03 mm in diameter. The ratio of quartz to feldspar grains varies from 70:30 to 60:40. Accessory minerals include sphene, epidote, iron ore and muscovite.

The Aithness Tuff contains a higher proportion of clasts of basalt and a small proportion of pebbles of sediments. Grains of garnet have been recorded in the sandy matrix.

Area between Galta Water [248 544] and Loch of Hollorin [276 559]. A characteristic feature of the coarse tuffs in this area is the presence of a higher proportion of ignimbrite clasts than in the tuffs of the Clousta area. The ignimbrites (S 49343, Plate XIII, fig. 3) contain small fragments of both felsite and basic lava, as well as euhedral crystals up to 1·5 mm in diameter of both potash feldspar and quartz set in a matrix of shards which are only slightly flattened and locally retain their Y-shaped bifurcations. Small fragments of non-flattened pumice are also preserved.

Concordant felsite intrusions

The felsites which form concordant intrusions within the upper part of the Sandness Formation show a considerable variation in their petrographic character. Most felsites are porphyritic and the most common phenocrysts are clear potash feldspar, potash feldspar with radiating twin lamellae (see S 29918), microcline and microcline-microperthite. Many feldspar phenocrysts are partly kaolinitized. The phenocrysts vary greatly in size and can range up to 3 mm in diameter. In many thin sections they are arranged in clusters. Some felsites are very sparsely porphyritic (S 30906, 30916) or non-porphyritic (S 30913).

The matrix of the felsites varies considerably. In the Sulma Water (S 30906) and Smith's Hamar felsites (S 30922–3, 30944, 30974) it consists of a micrographic quartz-feldspar intergrowth full of patches containing slightly higher concentrations of quartz. There are also large irregular patches of clear quartz and smaller patches of chlorite with muscovite in the centres, as well as small euhedral grains of zircon. Matrices of this kind grade into the matrix found in the felsites from the Ness of Nounsbrough (S 30751) and Upper Dale (S 29974–5). In these it consists of small stumpy laths of potash feldspar, up to 0·025 mm long, set in poikilitic plates of quartz which are in optical continuity within areas extending over several millimetres. Where the feldspar laths are very minute, adjoining laths are in places joined to produce an irregular micrographic texture. Patches of chlorite and calcite are commonly found in this matrix.

In some decomposed felsite sills in the Ness of Nounsbrough (S 30976) the matrix of the rocks is altered to a very fine-grained aggregate of clay minerals with small irregular patches of quartz. This matrix is traversed by curved cracks which may be the remains of a perlitic texture in the original glass.

Two acid or intermediate intrusions near Upperdale (S 30585–6) consist of aligned laths of sodic oligoclase, which form 75 per cent of the total volume of the rock, set in a hematite-dusted chloritic matrix containing amoeboid patches of quartz.

Sediments associated with the Volcanic Rocks

Sandstones. Many of the sandstones associated with the volcanic rocks, particularly in the western and central sectors, contain small pebbles of felsite, and basalt or andesite. The pebbles are in most cases concentrated in bands or lenses, separated by bands of medium- to fine-grained arkose in which lithic clasts are small and less abundant. In the eastern sector, the sandstone overlying the Ness of Clousta Basalt is virtually devoid of volcanic detritus.

The sandstones of Clousta (Plate XIII, fig. 2) are medium- to fine-grained, well- to poorly-graded with angular to subrounded grains which range in diameter from about 0·5 to 0·08 mm. The ratio of quartz to feldspar grains ranges from 80:20 to 50:50, and averages 65: 35. The quartz grains are, with the exception of one specimen (S 52739), unstrained. The relative abundance of the various types of feldspar grains varies greatly in different specimens. In sliced rocks from the western and central zones the only feldspar that has been recognized is a slightly cloudy potash feldspar. This is accompanied by microcline and rare plagioclase in only one thin section from the central area. In the sandstone from the eastern sector, however, potash feldspars of various types and plagioclase are present, in one specimen (S 52739) in roughly equal amounts. In others, cleavage fragments of plagioclase predominate, but in two thin sections (S 30759, 49329) untwinned potash feldspar provides the most common feldspar grains. This distribution suggests that, whereas the detritus in the western and central sectors could have originated from a potassic granite, that in the eastern sector had its source in an area which may have included granite and possibly also members of a dioritic complex. The feldspars could also have been derived from metamorphic terrain.

Most of the sandstones interbedded with volcanic rocks contain a small percentage of heavy minerals which are particularly concentrated in the dark bands. The minerals are, in approximate order of abundance: epidote, sphene, iron ores, apatite, tourmaline and zircon. Large grains of garnet, peripherally altered to carbonate, have been recorded in only one thin section (S 30759). Clasts of devitrified glassy basalt are found in some of the lower sandstones (S 30759). They have an irregular outline and contain angular xenoliths of quartz which may have been incorporated in the lava during its extrusion through loose sand, or when the still plastic lava lapilli were embedded in the sand.

The proportion of matrix to clasts ranges from 10:90 (S 30886, 49329) to over 50:50 (S 30757), but is most commonly between 20:80 and 25:75. The matrix of most specimens consists of a thin greenish film of chlorite and subordinate clay-mica which mantles the grains. In some specimens there are large flakes of pale mica. Interstitial patches of carbonate are closely associated with the chlorite.

Local partial replacement of feldspar grains by sericite produces serrate margins in many grains. The effects of induration in these sediments are less marked than in the sediments below the volcanic rocks.

Siltstones, mudstones and subordinate limestone bands. The thick series of siltstone and shale with subordinate limestone exposed on the shores of Brindister Voe (p. 88) is locally deformed by the second major phase of folding which has affected the Walls Sandstone (p. 134). This has produced intense small-scale similar folding and a local axial plane cleavage (p. 135). The red shales and siltstones exposed on the shores of the Voe of Dale, in the extreme west of the area, are only slightly affected by folding. Small-scale crinkling producing a rudimentary lineation is developed in some of the thinner argillaceous bands along the shores of Galta Water [249 543] and strong slaty cleavage is present in some of the fine-grained sediments above the Aith Ness Tuff in the vicinity of the Loch of Clousta [316 583].

The ribs of argillaceous limestone interbedded with corrugated sediment on the east shore of Brindister Voe (S 52360) show little sign of intense deformation and there is no development of new minerals. In the limestone, irregular laminae of clay minerals with a small admixture of finely granular calcite are interbedded with laminae composed of calcite grains whose size is within the silt to fine sand range (0·015 to 0·005 mm). Angular quartz grains ranging from 0·16 to 0·07 mm, usually partially replaced by carbonate along the margins, are present in some thin layers in the limestone.

The fine-grained sediments rhythmically interbedded with sandstone around Clousta Voe show little sign of tectonic deformation or induration, but in the area around the Loch of Clousta these fine sediments are intensely indurated, with the development of new colourless mica. In this area the calcareous ribs consist of crystalline carbonate with abundant newly formed irregular grains of epidote and clinozoisite. These alterations may have resulted from the thermal or late magmatic effects of an underlying plutonic body which may also have been responsible for the formation of actinolite and biotite in the basalts (p. 93).

H

ENVIRONMENT OF DEPOSITION

The Clousta Volcanic Rocks belong to the calc-alkaline suite of igneous rocks, and the extrusive members of this suite consist of basalt, pyroxene-andesite and ignimbrite together with a high proportion of mainly acid tuff and agglomerate. Their petrographic characters are comparable with those of the Lower Old Red Sandstone extrusive rocks of Argyll and the Midland Valley of Scotland, though the proportion of pyroclastic deposits is here somewhat greater. All these volcanic provinces were formed during the post-tectonic phase of the Caledonian magmatic activity of the Scottish Mainland. In the Western Shetland area the explosive activity of gas-charged acid magma gave rise by explosion to the tuff-cones and probably also formed extensive sheets of ignimbrite. Though only small flows of ignimbrite are at present exposed within the series, the presence of abundant ignimbrite clasts in the tuffs and agglomerates suggests that ignimbrite flows were widely developed, and may have been present in the lowest horizons of the volcanic series.

Ignimbrites are commonly associated with calderas, and it is possible that the tuff-cones and ignimbrite flows exposed in the area were formed within or near the margins of calderas. Because of the steep dip of the strata and the presence of many faults of non-volcanic origin it is not possible to define the bounding faults of caldera basins, if such existed.

The volcanic rocks and fine-grained sediments are underlain by and, in many parts of the area, interbedded with, cross-bedded sandstones and rhythmic sequences which are very similar to the underlying fluvial deposits. These beds appear to have been laid down by braided and, in places, meandering rivers. It is probable that the lavas and pyroclastic rocks were extruded on river-plains bounded perhaps to the north or north-east by alluvial fans. The new landforms created by the extrusive rocks would greatly modify the existing drainage system, restricting the width of the flood plains of some of the existing rivers and damming up others to produce small lakes. Elsewhere again, the newly formed volcanic hills would be the source of local streams which produced alluvial cones full of coarse debris at their foot. This topography would be further modified by the possible development of calderas in which lake deposits could also have been formed.

REFERENCES

ALLEN, J. R. L. 1960. Cornstone. *Geol. Mag.*, **97**, 43–8.
—— 1965. A review of the origin and characteristics of recent alluvial sediments. *Sedimentology*, **5**, 89–191.
COOK, E. F. 1966. Editor. *Tuff lavas and Ignimbrites*. Amsterdam: Elsevier.
FINLAY, T. M. 1930. The Old Red Sandstone of Shetland. Part II: North-western Area. *Trans. R. Soc. Edinb.*, **56**, 671–94.
FLINN, D., MILLER, J. A., EVANS, A. L. and PRINGLE, I. R. 1968. On the age of the sediments and contemporaneous volcanic rocks of western Shetland. *Scott. Jnl Geol.*, **4**, 10–19.
JONES, G. P. 1962. Deformed cross-stratification in Cretaceous Bima Sandstone, Nigeria. *Jnl sedim. Petrol.*, **32**, 231–9.
MYKURA, W. 1972. Igneous intrusions and mineralization in Fair Isle, Shetland Islands. *Bull. geol. Surv. Gt Br.*, No. 41, 33–53.

PEACH, B. N. and HORNE, J. 1879. The Old Red Sandstone of Shetland. *Proc. R. Phys. Soc. Edinb.*, **5**, 80–7.

—— —— 1884. The Old Red Volcanic Rocks of Shetland. *Trans. R. Soc. Edinb.*, **32**, 359–88.

PETTIJOHN, F. J. 1957. *Sedimentary Rocks.* New York: Harper and Brothers.

SUMM. PROG. 1934. *Mem. geol. Surv. Gt Br. Summ. Prog. for 1933*

SUMM. PROG. 1935. *Mem. geol. Surv. Gt Br. Summ. Prog. for 1934*

SUMM. PROG. 1966. *Mem. geol. Surv. Gt Br. Summ. Prog. for 1965.*

SUMM. PROG. 1967. *Mem. geol. Surv. Gt Br. Summ. Prog. for 1966.*

TUDOR, J. R. 1883. *The Orkneys and Shetland. Their Past and Present State.* London.

Chapter 8

OLD RED SANDSTONE:
WALLS FORMATION

Introduction

THE OUTCROP of the sedimentary rocks belonging to the Walls Formation (p. 9) is bounded in the north by the Sulma Water Fault, in the east by the Walls Boundary Fault and in the south-east by the Sandsting Granite, which has intruded the sediments and produced a thermal aureole of variable width around its margin (pp. 229–233). As there are no stratigraphic marker horizons within the Walls Formation and as in some areas parts of the sequence may have been cut out or repeated by faults it is difficult to construct even an approximate succession or to make a realistic estimate of the thickness. Unless there are a number of major faults not recognized by the author which have repeated large parts of the sequence, the formation must be extremely thick, possibly attaining 30 000 ft (9150 m). In Plate XII a number of arbitrary horizons, taken to be 4000 to 5000 ft (1200–1500 m) apart, are marked 1 to 7, to give some idea of the possible stratigraphic relationships of the various outcrops within the area. If the structural interpretation shown on the diagram is correct, the oldest rocks within the formation crop out in the area around Clings Water [310 560] and Loch of Vaara [325 566] in the north-eastern part of the area, and the youngest preserved beds crop out along the axial region of the Walls Syncline near the head of Gruting Voe.

The enclaves of sediment within the Sandsting Plutonic Complex (p. 114, Plate XXV) are part of the Walls Formation, and the small outcrop of sediments bounded to the west by the Sandsting Granite, between Rea Wick and Roe Ness (Fig. 10, p. 112) is probably also part of this formation, though its stratigraphical position cannot be determined.

Lithology

The salient features of the Walls Formation are its extreme thickness and the uniformity of its lithology both throughout the sequence and along the strike within any given horizon. The succession is made up of rhythmic units, each composed of a sandy phase consisting of generally fine-grained grey to dark grey sandstone and a somewhat thinner phase of predominantly grey fine-grained thinly bedded sediment which consists of varying proportions of shale, siltstone and sandy siltstone, and in some cases limy shale and fine-grained limestone. The cyclic units range in thickness from $2\frac{1}{2}$ to 65 ft (0·75–20 m) and the thickness of the sandy phase within them varies from 2 to 60 ft (0·6–18 m).

Coarse-grained sandstones, pebbly sandstones and conglomerates are absent, as are sedimentary structures characteristic of deposition in very shallow water, such as sun cracks and fossil soils. Though planar and trough cross-bedded sets are present in the sandstone phases of the cycles, they are not usually developed throughout the coset. Individual cross-bedded sets do not often exceed 3 ft (1 m) in thickness. Strongly lenticular sandstones are absent. Some sandstones are laminated and dark laminae containing heavy mineral concentrates are common throughout the sequence.

The ratio of sandstone to shale and siltstone within the cycles varies in different parts of the sequence, and there appear to be roughly alternating groups of sand-rich and shale-rich cyclic units. The total thickness of a group of shale-rich cycles does not normally exceed 60 ft (18 m); the sand-rich groups are usually considerably thicker. The beds thought to belong to the lower part of the Walls Formation appear to contain a relatively high ratio of sandy rhythms in which the shale phase is either thin or absent. The sequences made up predominantly of sandstone form prominent hills and ridges with many rocky outcrops. Examples are the ridge extending from the Ward of Browland [268 515] north-north-eastward towards Brindister Voe, the high ground forming North Houlan [300 557] just west of Clings Water, and Brace Field [256 524] just north-west of the Loch of Browland. The well-exposed areas of high ground between Walls and the Voe of Browland and just east of the Voe of Browland consist largely of sandstone, but it seems likely that in these areas the shale phases of the rhythmic sequences have been cut out by shear along the bedding planes.

Good continuous sequences are rarely exposed within the lower part of the Walls Formation (i.e. in groups 1 to 3 of Plate XII). All exposures are inland or along the sheltered shores of voes, where the fine-grained sediments are usually obscured, giving the impression that sandstone forms virtually the entire succession. In the upper part of the formation there are many good, easily accessible shore sections. These sections, however, lie to a large extent in areas which have been strongly folded and where the fine sediments in particular are deformed by small-scale folds, crinkles and cleavage. Small sedimentary structures have thus been partially obliterated, deformed or obscured, and in some cycles the junctions between beds of contrasting grain size are shear planes. Good examples of such shear planes, which have produced little or no shattering of adjacent strata, can be seen on the north shore of The Peak [202 477] ¾ mile (1200 m) ESE of Braga Ness and along the coast between Ram's Head [182 498] and The Hamar [186 497] 1 mile (1610 m) ESE of Wats Ness.

COAST BETWEEN WATS NESS AND WESTER SOUND

The strata exposed along the shore between Wats Ness and Wester Sound consist of alternate groups of sand-rich and shale-rich cycles (Fig. 9). The groups of shale-rich cycles are 50 to 60 ft (15–18 m) thick, and the thicknesses of individual cycles (i.e. the vertical distance between successive bases of sand-stones) range from 2½ to 17 ft (0·75–5 m). The groups of sand-rich cycles are 80 to 150 ft (24–45 m) thick and individual cycles range from 12 to 65 ft (3·6–20 m).

Shale-rich Cycles

The rhythmic units containing relatively high proportions of fine-grained sediment have the following characteristics:

Sandstone Phase

The sandstone phase of the cycle usually has a smooth, sharply defined base. Irregularities due to the channelling of the underlying fine-grained sediment are seen in only a few instances and the maximum recorded extent of vertical down-cutting is 2 ft 6 in (0·75 m) (Fig. 9b at 23 to 25 ft (7–7·6 m) from top), though more often it does not exceed 3 in (7·6 cm). Fragments of siltstone incorporated in the basal part of the sandstone have been recorded in a few cycles at Gerdipaddle [185 497], ¾ mile (1200 m) E of Wats Ness. Some sandstone posts have a highly irregular base with lobate downward projections which resemble large pillow-like load casts. Most sandstones, however, have no recognizable bottom structures. In many cases this may be due to the obliteration of such structures by tectonic movement along the bedding. The sandstone cosets range in thickness from 2½ to 12 ft (0·75–3·6 m). Most are fine grained through-out, but many are finer grained at the top than the base. Many are massive and non-laminated and some are at least partly planar-bedded, with dark laminae full of heavy mineral concentrates (p. 117). Quite a number contain plane- and trough-cross-bedded sets usually 1 to 2 ft (0·3–0·6 m) thick, and there are sets with ripple cross-lamination. Neither type of cross-bedding is very common and, where present, it is only found in a part of the coset. Slumped and con-torted bedding are relatively common.

In a number of the sandstone beds exposed in the area around the Voe of Footabrough there is a three-fold zonation of sedimentary structures within the sandstone coset. This consists of:

1. a lower planar-bedded zone up to 3 ft (0·9 m) thick;
2. a disturbed zone 3 ft to 3 ft 6 in (0·9–1·1 m) thick; and
3. a thin upper flaggy zone.

The sandstone in the disturbed zone is laminated and the structures within it are similar to, but on a larger scale than, the convolute lamination described on p. 107. Individual folds or 'flame structures' are up to 18 in (45 cm) high and the distances between crests reach 2 ft 6 in (75 cm). In some instances the direction of overturning of adjacent anticlinal crests is inconsistent. Near the top of the disturbed zone the convolutions are on a much smaller scale and the disturbed beds grade upward into the upper evenly bedded zone. The latter contains some thin sets (up to 4 in (10 cm) thick) with small-scale planar cross-bedding, characterized by a consistent low dip of foresets to the south-west.

Though this three-fold sequence of structures is not often well developed there is everywhere a tendency for disturbed bedding and convolute lamination to be confined to the fine-grained sandstone in the upper half of the coset. Trough-cross-bedding with troughs up to 8 in (20 m) deep and 2 ft 6 in (75 cm) wide and an axial trend indicating movement from north–east to south–west has been noted in the lower, somewhat coarser-grained portions of some sandstones.

Though many sandstones become finer-grained upwards, graded bedding in unlaminated or evenly laminated sandstones has not been recorded. The tops of the sandstone cosets may either grade upwards into siltstone or may be sharply defined. There are even rare examples of eroded tops of sandstones, and in one section [184 497] a sandstone set with disturbed bedding has been partially removed by erosion and is now overlain by silty shale containing small irregular clasts of sediment.

Shale and Siltstone Phase

The fine-grained sediments between successive sandstone posts which make up the shale and siltstone phase range in thickness from a few inches to 15 ft

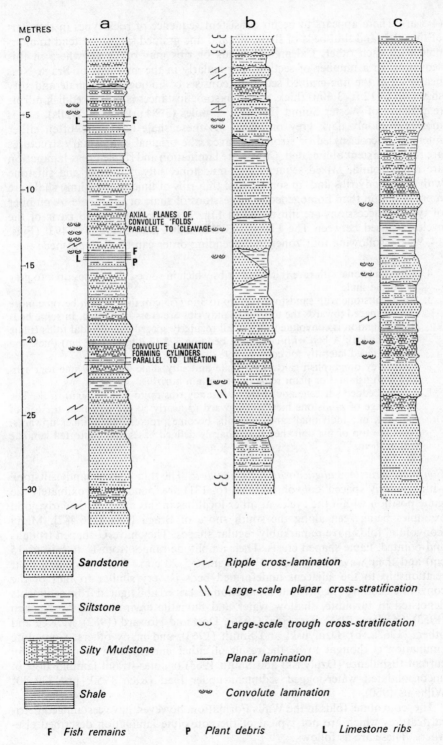

FIG. 9.   *Some characteristic sequences in the Walls Formation*

(4·5 m). There appears to be no consistent sequence of rock types in this part of the cycle and thin beds of shale, silt and fine-grained sandstone tend to alternate with each other. Calcareous bands or ribs may occur anywhere in this sequence. In a number of sections particularly in the area east of Braga Ness nearly all of the fine-grained sediment consists of homogeneous shale and silty shale, up to 12 ft (3·6 m) thick, which in some instances is red or reddish purple (e.g. Shore of Wester Sound [218 473], 2 miles (3·2 km) SW of Walls), but is more commonly dark grey or black. As these shale bands are often either intensely microcrinkled or strongly cleaved, the original sedimentary structures are to some extent obliterated. Convolute lamination and ripple-cross-lamination are rarely found. Mixed sequences of mudstone, silty mudstone and siltstone with thin sandy ribs and, in some cases, thin ribs of limestone or limy shale are more common than homogeneous successions of shale or silty shale. A number of typical successions are illustrated in Fig. 9. In the fine-grained parts of the cycles exposed between The Flaes [193 492] and Fidlar Stack [191 493] (Figs. 14, 9), the following four zones in descending order can be distinguished:

4. Silty mudstone with rare thin sandy ribs which in some cases passes up into grey or purple shale.
3. Sandy siltstone with sandstone ribs up to 3 in (7·5 cm) thick which become more closely spaced towards the top. The sandy ribs are cross-laminated. In some beds the lamination is convoluted into small regularly spaced cylindroidal folds (Plate XIV A and B). A bed with convolute bedding is 3 to 6 in (7·5–15 cm) thick and can be traced laterally for considerable distances.
2. Dark grey or purplish calcareous shale and silty shale with thin limestone ribs and with fish and/or plant remains on certain horizons.
1. Evenly bedded siltstone and shale with sandy ribs up to 3 in (7·5 cm) thick. The percentage of sandstone increases upward to nearly 50 per cent near the top of the unit, but individual sandstone ribs become progressively thinner upwards. Near the top of the unit they have sharply defined bases and contorted laminae within them.

Convolute structures aligned with tectonic fabric. The thin beds of sandy siltstone with regularly spaced 'convolute' folds (p. 131) are found in approximately the same position within the cycles at other localities in this area, particularly good examples being seen along the south shore of Braga Ness [196 482]. Many 'convolute' folds have remarkably regular shapes. They have U-shaped troughs and pointed, flame-shaped crests. Their amplitude ranges from 1½ to 6 in (4–15 cm) and their wavelength from 2½ to 8 in (6·5–20 cm). Their shape and their relationship to the adjacent undeformed beds is very similar to that of the convolute bedding or convolute lamination described and figured from sediments deposited in turbidite, shallow water and fluviatile environments by Kuenen (1953), Sanders (1956), Ten Haaf (1956), Dott and Howard (1962), McKee and others (1962a, b), Dzulynski and Smith (1963), and many others. Convolute lamination is thought to be the result of either current drag (Sanders 1956), current turbulence (Dzulynski and Smith 1963) or inter-stratal laminar flow in unconsolidated water-logged sediment under load (Rich 1950, pp. 729–30; Williams 1960).

The 'convolute' folds in the Walls Formation, however, have certain geometrical features which are not typical of the convolute lamination described elsewhere. These are as follows:

1. The axes of most of the 'convolute' folds recorded in the strongly deformed strata of this area are straight and parallel not only to the axes of the adjacent 'convolute' folds but also to the regional lineation of the rocks and to the axes of the minor folds and wrinkles which are of undoubted tectonic origin (p. 131).

2. The axial planes of the 'convolute' folds are parallel to the cleavage planes of the sediments in which they are formed (Plate XIVc). This is particularly striking on the steeply dipping limbs of the major folds, where bedding and cleavage are at an acute angle to each other. There is also a very consistent direction of overturning or 'recumbency' of the folds which is the same as the vergence of the minor drag folds (p. 128) found at various horizons in the same relative positions within the major folds.

3. Convolute lamination, as described by Ten Haaf (1956), is formed in bands of silt grade or of alternate laminae of fine sand and silt, which have been deformed into sharp-crested anticlines and rounded synclines, and whose deformation dies out gradually both upwards and downwards. Though many convolute structures in this area are of this type (Plate XIVb) there are many instances in which adjacent folds have become detached from each other at the anticlinal crest, so that the U-shaped troughs now form almost complete oval cylinders. In some cases the convolute folds pass laterally along the bedding into normal small-scale asymmetric folds. There are several instances in calcareous beds where the convolute layer does not pass upward without break into the overlying undistorted laminae, but is sharply truncated, the junction between the two layers being tectonic, not erosional.

The convolute lamination within the strongly deformed beds of the Walls Formation thus shares its geometric parameters with the tectonic structures of these beds, and it is difficult to envisage such a complete parallelism if the convolute structures are of purely sedimentary origin. The style of the convolute folds, however, leaves no doubt that they were formed when the sediments were still in a plastic state. If their formation is ascribed to current drag shortly after deposition there must have been some link between the direction of the currents and the geometry of the later tectonic structures, the currents moving either normal or parallel to the later regional lineation, depending on whether the 'convolute' folds were formed by current drag acting at right angles to their axes (see Sanders 1956) or by current turbulence along them (see Dzulynski and Smith 1963). In either case such a coincidence would suggest a strong topographic control of the currents either by the actively forming major folds themselves or by the margins of the basin of deposition which must have been parallel to the fold axes. If such were the case one would expect evidence of contemporaneous movement within the sediments, such as angular unconformities, mudflow breccias and coarse clastic sediments, none of which are present. The evidence from dips of foresets and trends of trough-axes in the cross-bedded sandstones suggests a predominant current movement from the north-east to the south-west, whereas the trend of the axes of the convolute folds in the vicinity of the Voe of Footabrough ranges from west to W10°S. This divergence of current directions could be explained by the supposition that the structures in the fine sediments were formed by currents flowing parallel to the axes of the trough, whereas the coarser sediments were deposited by lateral currents flowing down the sides of a westward plunging basin (see Kelling in Bouma and Brouwer 1964, pp. 76–85).

Rich (1950, p. 729) has described intra-stratal distortions in siltstones in the

Aberystwyth area, and has ascribed their origin to gravity sliding of the overlying sediments. Williams (1960, pp. 208–14) has suggested that most forms of
convolute lamination may result from lateral interstratal laminar flow due to the
liquefaction of certain laminae in a waterlogged unconsolidated sediment, the
flow pattern being determined by the differential movements of adjoining layers
which have remained solid. Within the Walls Formation the parallelism of the
geometry of the 'convolute' folds with that of the later tectonic folds, the sharp
tectonic upper contacts of some convolute horizons, and the observed transi-
tions to asymmetric similar folds favour the mechanism of interstratal laminar
flow, which was controlled either by the initial depositional slope or by pre-
contemporaneous flexuring along the incipient fold axes. The latter mechanism
presupposes either that the flexuring commenced very shortly after the affected
strata were deposited or that the convolute folds were formed a considerable
time after the deposition of the sediment under, possibly, a considerable
thickness of overburden. In the latter case it must be assumed that certain thin
beds in the sequence remained water-logged and in a state of loose packing, so
that they would be liable to liquefaction by shearing stress, even at a considerable depth. As is shown in Fig. 9a the layers with convolute lamination are both
overlain and underlain by beds of fine-grained impervious sediment, which
during the early stages of compaction could have acted as water seals, leaving
the thin silty layers waterlogged and preventing their lithification. The relationship between the geometry of the convolute folds and the regional cleavage
and lineation is discussed on pp. 128–134.

Disturbed bedding due to foundering of sandy laminae. Within the area under
description there are many irregular small-scale structures which may have
resulted from the foundering of thin sandy beds into the underlying mud. Good
examples are seen near the Point of Hus [197 482] where sandy laminae within
a 1-ft (30-cm) thick bed of silty mudstone form closely packed rolled-up ovoids
which average 6 in (15 cm) in length and breadth and 2 to 3 in (5–7·5 cm) in
height. Structures of this type are termed 'ball and pillow structures' (Potter and
Pettijohn 1963, p. 148) or 'pseudo-nodules' (Macar and Antun 1950). They are
thought to have been formed during periods when the mud was temporarily
liquefied, the actual break-up and foundering of the sandy laminae being due to
differential loading resulting from slight variations in the thickness of the sandy
beds. The foundering may have been initiated by earthquake shocks (*see*
Kuenen 1958). The disturbed sediment is usually overlain by non-laminated
medium-grained sandstone with a load-cast base, individual casts averaging 4 in
(10 cm) in depth and 9 in (23 cm) in width, and underlain by up to 3 ft 6 in (1 m)
of sandy shale which shows slight traces of penecontemporaneous disturbance.
The latter rests in one case on a 2 ft 6 in (0·8 m) thick bed of silty mudstone with
sandy laminae, which has several convolute horizons of the type described above
(p. 106).

Limestones and calcareous argillites

The rhythmic units shown in Fig. 9 illustrate well-developed shaly phases
within the cycles of the Walls Formation. In most cycles the shaly phase is
considerably thinner than those shown in the diagram and may have one or more
of the units missing. In many, both the calcareous and convolute units are
absent. There are, however, a number of cycles in which the calcareous sediments

are extremely thick and in some cycles almost the entire shaly phase consists of alternating layers of limestones and calcareous argillite. Details of a number of these are as follows:

1. *North shore of Wick of Watsness* [*176 503*], *730 yd (670 m) S20°W of Suther House.* In this section 30 ft (9 m) of calcareous sandy siltstone with pale and dark grey banding are intercalated with irregular ribs of clayey limestone. A number of the siltstone beds contain small-scale cross-lamination and some exhibit 'ball and pillow' structure (p. 108). The entire thickness is cut by an anastomosing network of calcite-filled fractures.

2. *Headland and south-east shore of Gorsendi Geo* [*177 503*], *860 yd (790 m) S5°E of Suther House.* This sequence is composed of 11 ft (3·4 m) of limestone and calcareous mudstone with several bands in which argillaceous limestone or calcareous mudstone is folded into asymmetric similar folds the axial planes of which are inclined at 64° to W30°N and the axes of which plunge at 39° to S30°W, which is parallel to the regional lineation.

The sequence is as follows:

	ft	(m)
Ribs of argillaceous limestone up to 1 ft (30 cm) thick with similar folds interbanded with evenly bedded limestone ..	6½	2
Shale, evenly laminated, unfolded	½ to 1	0·15–0·30
Limestone, massive, with shaly partings	4	1·2

The thick lower limestone is not folded, but the shale bands within it are either broken up into small fragments, or, near the base of the sequence, folded into very small asymmetrical folds with an alignment of fold axes and axial planes identical to that of the folds in the upper limestone.

This sequence is both overlain and underlain by sandstone.

3. *South-west shore of Ram's Head* [*182 498*], *1050 yd (960 m) S16°E of Suther House.* This is a relatively arenaceous sequence 10 to 11 ft (3–3·5 m) thick composed of intercalated bands of calcareous honeycomb-weathered sandstone and siltstone with one rib of massive limestone up to 10 in (25 cm) thick and several somewhat thinner ribs of clayey limestone. The limestone contains plant fragments preserved as a reddish film on the bedding planes.

4. *Sterling Geo* [*197 488*], *900 yd (820 m) SW of head of Voe of Footabrough.* The sediment exposed in Sterling Geo consists of calcareous shale with some ribs of calcareous siltstone, but no pure limestone. The entire thickness of sediment is deformed into plastic folds which plunge at 45° to S10°E thus making an angle of 60° with the trend of the regional lineation. The sequence is intensely veined by pink calcite.

5. *South shore of Braga Ness* [*203 477*], *100 to 200 yd (90–180 m) WNW of The Peak.* The calcareous sequence is here 25 to 30 ft (7·5–9 m) thick and composed of calcareous siltstone and shale. It contains a number of beds of sandstone up to 4 ft thick (1·2 m) with silty partings and small-scale ripple cross-lamination. There are several argillaceous limestones within the sequence, but no thick crystalline limestones. As in the limestone exposed on the east coast of Gorsendi Geo (p. 109) a number of the calcareous ribs show intense small-scale folding. Though individual folds are less regular than in the Gorsendi Geo section, their axes and axial planes are roughly parallel to the regional lineation and cleavage. There are several thin layers with 'convolute' folds the geometry of which is in accord with the geometry of the minor tectonic structures.

Sand-rich Cycles

The sedimentary features of the cyclic units which contain a high proportion of sandstone and relatively little siltstone and shale are as follows:

Sandstone Phase

Individual sandstone posts average 10 to 15 ft (3–4·6 m) in thickness but do, in rare cases, attain 60 ft (18 m). They are usually fine grained throughout. Most are planar-bedded and have flaggy partings and some dark bands with heavy mineral concentrates throughout the greater part of their thickness. Some sandstones, however, have well-developed trough-cross-bedding or, less commonly, planar-cross-bedding, usually but not always in the upper part of the posts. Trough-cross-bedded sets are in some cases 5 ft (1·5 m) thick, with individual troughs up to 25 ft (7·6 m) wide. Most are, however, less than 2 ft (60 cm) thick. Throughout the sequence there is a tendency for trough axes to have a south-west or west-south-west alignment and an inclination of cross-strata which indicates current movement from north-east or east-north-east. Sets with contorted (overturned) cross-bedding, in some cases up to 4 ft (1·2 m) thick, have been recorded in several of the sandstone posts.

Shale-siltstone Phase

The fine-grained sediments between the thick sandstone posts range in thickness from several inches to 4 ft (1·2 m). They consist of grey or purplish shale and silty shale, which may pass down into siltstone. Some have calcareous ribs, usually with plant remains, several inches thick, in their upper part. Small-scale cross-lamination is common in the siltstones but convolute lamination of the type described on pp. 106–7 is not usually developed. In many sections the fine-grained sediments between sandstone posts are intensely deformed by the shearing and small-scale folding and original sedimentary structures are largely obliterated.

VAILA SOUND AND VAILA

The strata exposed along the shores of Vaila Sound and Gruting Voe and on the Island of Vaila are partly within the belt of intense deformation and are affected in the north-eastern outcrops by two phases of folding (p. 133). They are also partly within the thermal aureole of the Sandsting Complex (Plate XXV). The strata within the thermal aureole do not contain penetrative tectonic structures such as lineation, cleavage and tight minor folds (p. 134), but they are in some areas cut by a large number of irregular shear planes, which tend to obliterate the junctions between sediments of contrasting grain size.

The major lithologic features in this area are the same as those of the strata exposed between Watsness and Wester Sound (pp. 103–110). The fine-grained phases within the shale-rich cycles, however, appear to be more homogeneous than in the Wats Ness area, consisting almost entirely of shale and silty shale, which is locally calcareous. These shale bands appear to increase in thickness both as the sequence is ascended and eastward along the strike.

Along the south shore of Vaila the cosets of predominantly planar-bedded, fine-grained sandstone up to 25 ft (7·6 m) thick alternate with thin beds composed almost entirely of dark grey or black evenly laminated shale. Somewhat higher in the sequence, along the west and north-west coasts of the island, the sandstones contain a higher proportion of sets with planar- and trough-cross-bedding as well as some sets with distorted or recumbently overfolded cross-bedding. Here the beds of fine-grained sediment are thicker, in one case reaching 12 ft (3·7 m). The latter are, again, composed almost entirely of shale and silty shale, with no trace of convolute lamination.

On the north-west shore of Wester Sound and Vaila Sound there are a fairly large number of cycles with thick units of fine-grained sediment, several of which are between 10 and 15 ft (3 and 4·6 m) thick. These consist predominantly of grey or purplish silty shale and shale with calcareous bands. Sandy laminae occur in only a small proportion of these beds, and convolute lamination has not been recorded. Most fine-grained sediments in this area are strongly cleaved, and some are affected by small-scale recumbent folding.

Along the east coast of Vaila and in the Whites Ness peninsula the shaly phases of rhythms are both thick and abundant. In most exposures they are made of black mudstone which has been hornfelsed so that the internal structures are largely obliterated. On Whites Ness, the peninsula between Vaila Sound and Gruting Voe, there are a large number of beds of black mudstone and silty mudstone between 6 and 15 ft (1·8 and 4·7 m) thick, but the details of their structure have been destroyed by thermal metamorphism and later shattering. On the east coast of Vaila, 850 yd (777 m) N of the southern end of Green Head (Vaila) there is an exceptionally thick development of 50 ft (15 m) of fine sediment with a thick limestone and calcareous mudstone near the middle. The sequence within this unit and the present mineral content of the component sediments is shown graphically in Fig. 25, p. 232. Both the thick mudstone and the limestone appear to thin out in a north-westerly direction and the probable horizon of this bed is represented on the north coast of Vaila by a 12-ft (3·7-m) thick bed of shale and siltstone.

GRUTING VOE, SELI VOE, SCUTTA VOE AND VOE OF BROWLAND

Along the coasts of Gruting Voe, Seli Voe, Scutta Voe and the Voe of Browland the fine-grained sediments are everywhere intensely puckered, cleaved and folded by two phases of deformation. On the shores of the Voe of Browland and in the northern part of Gruting Voe most junctions between fine sediments and sandstone are shear-planes.

The rhythmic units in this area are very similar to those of the upper part of the succession in Vaila Sound. The fine-grained sediment within these units is up to 11 ft (3·5 m) thick and consists mainly of thinly laminated shale and silty shale, but as all shaly units are tectonically distorted their true thickness cannot be established. It is therefore not possible to say whether the eastward thickening of the fine-grained beds continues into this area. Calcareous mudstone with limestone ribs is present in two localities:

1. On the north shore of Scutta Voe [278 498], 140 yd (128 m) WSW of Lee of Houlland School, where calcareous ribs up to 6 in (15 cm) thick are present in 5 ft (1·5 m) of intensely folded and cleaved black shale.
2. On the east shore of Seli Voe [293 484], 580 yd (590 m) NNE of Setter. Here almost the entire fine-grained phase of one cyclic unit appears to be made up of argillaceous limestone with thin partings of black shale. The limestone is affected by two phases of folding and has largely lost its original sedimentary characters.

Evidence for the presence of a very thick bed of black shale and silty shale with thin silty laminae is seen in a quarry cutting superficial deposits on the roadside [294 482] just east of Seli Voe (Chapter 18, p. 271). The drift in this quarry contains many fragments of lineated and crinkled black shale and dark grey siltstone. Although no rock is exposed sediments of this type may well underlie the entire quarry area and the shale band may be over 40 ft (12 m) thick.

In the sandstone posts in the Gruting Voe area sets with both trough- and planar-cross-stratification, between 1 and 2 ft (30–60 cm) thick, are present. Most sandstones, however, are planar-bedded and have thin flaggy partings. The area immediately north and north-east of Scutta Voe has an exceptionally high proportion of exposed rock. All outcrops consist of hard grey fine- to medium-grained sandstone which is so intensely and regularly jointed that it is in places difficult to distinguish joint planes from bedding planes. As in most other inland areas the shaly horizons are not exposed.

AREA BETWEEN LUNGA WATER AND WALLS BOUNDARY FAULT

The ground immediately south of the Sulma Water Fault, between Lunga Water [235 527] and the Walls Boundary Fault, has a fairly rugged topography with a large number of sandstone exposures but very few exposures in the fine-grained sediment. The sediments forming this area appear to be the oldest beds of the Walls Formation (p. 102, Plate XII) and they are folded and locally jointed and shattered by the north-east to north trending folds.

In the area south-west of Uni Firth [288 562] some of the higher ground is formed by relatively thick cosets of medium- and very locally even coarse-grained sandstone with relatively thick cross-bedded sets and some horizons with distorted cross-bedding. There are very few exposures of the fine-grained sediments and nothing is known of their thickness or composition.

In the area east of Uni Firth exposures are more patchy and the rocks are nearly everywhere broken or crushed. Nearly all outcrops are in indurated fine- to medium-grained sandstone which is in some cases intruded by irregular squirts of felsite. Cross-bedded sets are common throughout.

The presence of a fairly large number of roughly concordant felsite intrusions, petrographically comparable with the felsites associated with Clousta Volcanic Rocks, suggests a tentative correlation of these sediments with the highest exposed strata of the Sandness Formation. As there are, however, no close lithological similarities between these beds and the higher beds of the Sandness Formation and as there are here no basic lavas or pyroclastic rocks, such a correlation seems unlikely.

The sediments exposed along the shores of and in the vicinity of Bixter Voe consist of thin cosets of fine-grained dark grey, generally evenly-bedded sandstone alternating with thin beds of siltstone and silty shale. The latter are indurated and partly converted into slate. The sequence has been intruded by numerous concordant sheets of fine-grained graphic granite, and is shattered within a zone extending for 250 yd (230 m) from the Walls Boundary Fault.

REA WICK–ROE NESS PENINSULA

The one and a half mile (2·4 km) long outcrop of sedimentary rocks along the south-east coast of the Walls Peninsula between Rea Wick and Roe Ness

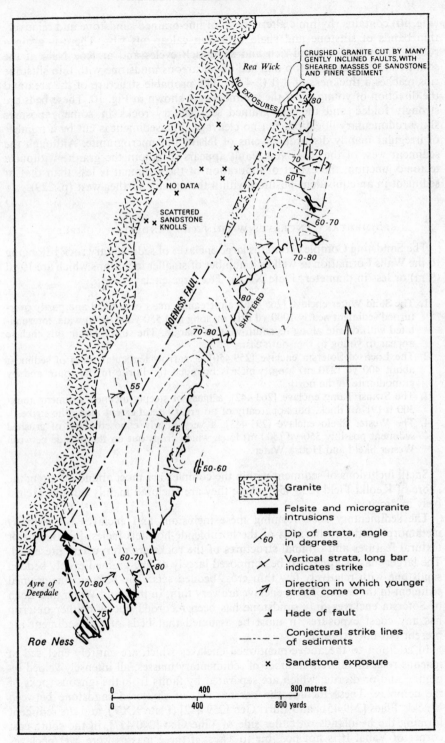

Fig. 10. *Geological sketch-map of the Roe Ness–Rea Wick area*

(Fig. 10) contains rhythmic alternations of fine-grained sandstone and relatively thin bands of siltstone and shale with thin calcareous ribs. There is again a rough alternation of sand-rich and shale-rich cycles and at Roe Ness at the southern end of the outcrop one bed of calcareous mudstone with thin siltstone ribs reaches a thickness of 8 ft (2·4 m). The probable structure of the area and the direction of younging of the sediments are shown in Fig. 10. These beds are strongly folded and the fine-grained sedimentary rocks in some exposures show rudimentary lineation, but no cleavage. The sediment is cut by a number of irregular mainly discordant veins of felsite and microgranite. Although the sediment west of the Roe Ness Fault appears to adjoin the granite without a tectonic junction, the extent of induration of the sediment is less than that of sediment in an equivalent position within the aureole farther west (p. 229).

SEDIMENTARY ENCLAVES WITHIN THE SANDSTING COMPLEX

The Sandsting Complex contains large enclaves of sedimentary rock belonging to the Walls Formation as well as a number of smaller inclusions which are 10 yd (9 m) or less in diameter (Plate XXV). The large enclaves are:

1. The Sand Water enclave [265 446], an irregular mass of vertical and partly over-turned sediment nearly 1000 yd (910 m) long and 550 yd (510 m) wide, interdigi-tated with diorite along its south-western margin. The strata within this enclave appear to young to the south-east.
2. The Loch of Sotersta enclave [259 450], a narrow lenticular mass of sediment about 400 yd (370 m) long, which is bounded by diorite in the south and by granodiorite in the north.
3. The Swinsi Taing enclave [263 442], a mass of steeply inclined sediment about 300 ft (91 m) thick, but apparently of no great lateral extent along the strike.
4. The Wester Skeld enclave [293 443], a very poorly exposed mass of crushed sediment possibly 550 yd (500 m) long, which crops out on the hillside between Wester Skeld and Housa Water.

Small inclusions of sediment within the complex are most abundant along the shore of Keolki Field [253 454] where they are enclosed mainly but not exclus-ively in diorite.

The sedimentary rock forming these inclusions and enclaves is thermally metamorphosed into hornfels of the hornblende-hornfels facies, but the major textural features and original structures of the rocks are everywhere preserved. The larger masses appear to be composed largely of fine-grained evenly bedded sandstone with relatively few thin cross-bedded sets. The beds of fine-grained sediment in the Swinsi Taing enclave are very thin. In the Sand Water and Loch of Sotersta enclaves only sandstone has been recorded but as neither outcrop has any coast exposures, it must be assumed that beds of fine sediment are present but not exposed.

In addition to the above-mentioned enclaves which are entirely enclosed in igneous rock there are a number of sedimentary masses, all intensely veined by granite and/or diorite, which are separated by faults from the igneous rocks of the complex. These include the narrow belt of shattered sandstone between Muckle Flaes [249 451] and Burri Geo [259 442] (Plate XXV), and the sediment forming the headlands on either side of Vine Geo [240 457], in the south-east corner of Vaila. It is not possible to assess if these masses were ever enclaves within the granite or if they are parts of the roof which collapsed during the final stage in the intrusive history of the complex (pp. 228–9).

Fauna and Flora

The first find of fossil plants within the Walls Formation was that by Peach and Horne in the area just north and east of Walls and in the hills between Gruting Voe and Bixter Voe (p. 5). Unfortunately the records of the exact localities of these finds have not been preserved.

Finlay (1930, p. 675) recorded plant remains from 'a bed of sandy shale below the Watsness Limestone', but he did not give the locality of this limestone, and it is now not possible to ascertain which of the limestones in the Watsness district he referred to. Finlay stated that Professor Lang examined the district for plant remains and appears to have found plant fragments at various unspecified localities. Lang described these plant remains as hostimellid in type.

In the course of the geological mapping of the Walls Sandstone by the Geological Survey during the recent revision mapping by the author a number of localities yielding both plant and fish remains were recorded. These localities are shown in Fig. 11 and briefly described below. The search for fossils in the Walls Formation has not been exhaustive and there can be little doubt that much new material remains to be found. The numbers in the list below refer to localities shown in Fig. 11. Localities 1 and 2 are in the Sandness Formation and described on p. 92.

3. *Trea Wick* [*173 507*], *600 to 615 yd (550–560 m) W15°S of Suther House*. Plant and fish remains from three horizons in dark grey, ochre-weathering, mudstone and siltstone. Fish remains include scales of *Gyroptychius microlepidotus* type and two fragments of articulated acanthodians which can be determined as belonging to the Cheiracanthidae and might be close to the genus *Cheiracanthus*.

4. *Loch of Watsness, south-west shore* [*175 506*], *450 yd (410 m) S45°W of Suther House*. Dark shale with plant fragments including *Hostimella sp*.

5. *Wick of Watsness* [*175 501*], *west shore, 450 yd (410 m) S45°W of Suther House*. Fine-grained grey sandstone with plant remains preserved as red film.

6. *Gorsendi Geo* [*179 510*], *810 yd (740 m) S3°E of Suther House*. Dark grey to purplish grey sandy siltstone with abundant plant debris. Plants occur above and within the limestone horizon. Only *Hostimella sp*. has been identified.

7. *Ram's Head* [*182 499*], *1200 yd (1100 m) S21°E of Suther House*. Dark grey siltstone interbedded with sandstone above the calcareous horizon. Rare fish scales of the *Gyroptychius microlepidotus* type.

8. *Coast between Fidlar Stack and The Flaes* [*192 492*], *1 mile 420 yd (2 km) S40°E of Suther House*. Fossils from two horizons (Fig. 9a). Pale purplish grey to purplish green calcareous silty mudstone with fish bones, including a climatiiform spine and scales of the *Gyroptychius microlepidotus* type. Also ?eurypterid remains and plant remains.

9. *Burn of Turdale* [*197 509*], *130 yd (120 m) E38°N of Turdale and 150 yd (140 m) E43°N of Turdale*. Grey fine-grained sandstone with beds full of plant remains including *Hostimella sp*. and some trace fossils or roots. (This is the only locality in the Walls Formation where trace fossils or root remains have been recorded.) Also dark grey to black shale with fish scales preserved as carbonaceous films.

10. *Voe of Littlure* [*208 477*], *620 yd (575 m) W23°N of outlet of Loch of Quinnigeo*. Grey mudstone with poorly preserved plant remains.

11. *Bridge of Walls* [*269 512*], *roadside 520 yd (475 m) S12°E of summit of Ward of Browland*. Flaggy micaceous sandstone with plant debris.

I

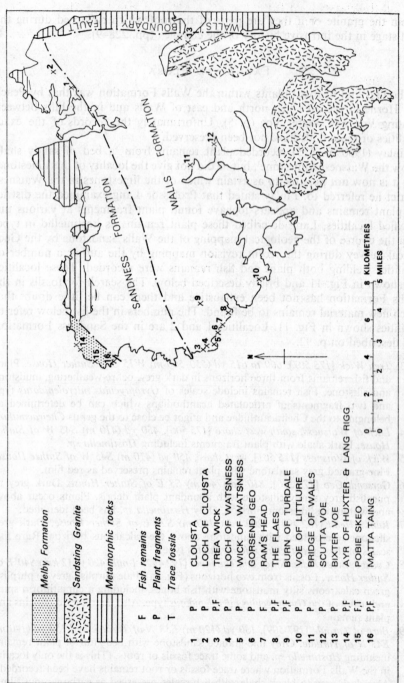

FIG. 11. Fossiliferous localities in the Old Red Sandstone of the Walls Peninsula

Melby Formation

Sandsting Granite

Metamorphic rocks

F   Fish remains

P   Plant fragments

T   Trace fossils

1	CLOUSTA	P
2	LOCH OF CLOUSTA	P
3	TREA WICK	P,F
4	LOCH OF WATSNESS	P
5	WICK OF WATSNESS	P
6	GORSENDI GEO	P
7	RAM'S HEAD	F
8	THE FLAES	P,F
9	BURN OF TURDALE	P,F
10	VOE OF LITTLURE	P
11	BRIDGE OF WALLS	P
12	SCUTTA VOE	P
13	BIXTER VOE	P
14	AYR OF HUXTER & LANG RIGG	P,F
15	POBIE SKEO	F,T
16	MATTA TAING	P,F

KILOMETRES

MILES

12. *Scutta Voe* [278 499], *north shore, 110 yd* (*100 m*) *W20°S of School at Lee of Houlland*. Grey silty mudstone with poorly preserved plant remains.

13. *Bixter Voe* [334 514], *north-east shore 650 and 850 yd* (*595 and 777 m*) *S of Bixter Post Office*. Grey siltstone with plant remains. Plants have also been recorded from the south shore of Bixter Voe, 1350 yd (1235 m) SE of Bixter Post Office.

The fish remains have been examined by Dr. R. S. Miles, who has reported as follows:

"The major portion of the collection comprises indeterminable Crossopterygian remains. They are principally scales of an irregular cycloid form, but also include some plates provisionally determined as from the exocranium (opercular, gular and branchiostegal bones). The scales are broadly of the '*Gyroptychius microlepidotus*' type in their surface morphology; a fact of little stratigraphical or systematic significance. Two fragments have the cosmine layer of the bone preserved, clearly indicating a member of the Osteolepoidea.

Two articulated acanthodians can be determined as belonging to the Cheiracanthidae, and might be close to the Scottish Middle Old Red Sandstone genus *Cheiracanthus*. A small fragment has been tentatively identified as the base of a spine of climatiiform type.

The fragmentary fish remains indicate a Middle Old Red Sandstone age for the Walls Sandstone, but do not permit a more precise statement on the matter."

W. G. Chaloner has identified the plants, but none of them has proved to be of diagnostic value in determining the age of these strata. Specimens of carbonaceous mudstone from two localities were submitted to Dr. B. Owens for spore determination, but no material suitable for identification, even at generic level, was obtained.

PETROGRAPHY

SANDSTONES

The characteristic features of the arenaceous sediments of the Walls Formation are the remarkable consistency in grain size within the 0·45 to 0·045 mm range, the absence of extraformational conglomerates, and the rarity of intraformational conglomerates. Individual grains in the sandstones are generally subangular to angular, poorly to very poorly graded, with corroded margins of quartz and feldspar grains due to partial replacement by the authigenic matrix minerals clay mica, chlorite and carbonate. The sandstones are very immature with the ratio of quartz to feldspar grains ranging from 55:45 to 80:20, and with up to 15 per cent of heavy minerals. Concentrations of heavy minerals also occur, as bands up to 3 mm thick, throughout the sandstone sequence (Plate XIII fig. 5). The most common detrital heavy mineral grains are epidote, sphene, iron ores and apatite. Rarer mineral grains include tourmaline, zircon, allanite and garnet. Lithic clasts are virtually absent.

The percentage of matrix and mineral cement in relation to the total volume of rock of the great majority of specimens ranges from 15 to 35, but reaches 50 in some of the laminae with high concentrations of heavy minerals. The matrix is partly of secondary origin, being formed by the partial or complete replacement of detrital grains, particularly the feldspars, epidotes and garnets. In the least altered sandstones the matrix consists of a thin, more or less continuous film of chlorite mantling the grains, and interstitial patches of granular calcite and chlorite.

According to the classification adopted by Pettijohn (1957, p. 291) most of the sandstones are feldspathic greywackes. As in many samples the matrix and mineral cement were partly formed by the replacement of grains, a proportion of the sandstones may have originated as fine-grained arkose.

Within the tightly folded zones (p. 128) and in the entire eastern part of the area, muscovite, chalcedonic silica, ?albite and epidote are extensively developed, and over a large part of the area the composition of the matrix indicates a low-grade dislocation metamorphism with pressure-temperature conditions ranging from the zeolite facies to the lower grades of the greenschist facies (Turner 1968, pp. 265–70). The thermal alteration of the sandstones within the aureole of the Sandsting Granite is described in Chapter 15, pp. 229–233.

In the following description the sandstones are divided into arbitrary strati-

---

EXPLANATION OF PLATE XIII

PHOTOMICROGRAPHS OF SEDIMENTARY AND VOLCANIC ROCKS OF THE WALLS SANDSTONE

FIG. 1.   Slice No. S 52737. Magnification × 20. Plane polarized light. Fine-grained flaggy sandstone, Sandness Formation, showing alternate quartz-feldspar and micaceous laminae. Scattered small grains of epidote throughout. West shore of Muckle Head. [297 581].

FIG. 2.   Slice No. S 52738. Magnification × 20. Crossed polarisers. Medium-grained arkose, Sandness Formation. Well-graded subangular to subrounded grains. The ratio of quartz to feldspar grains is 60:40. Matrix forms less than 10 per cent of total volume and is composed predominantly of carbonate. North shore of Voe of Clousta, 1225 yd (1100 m) WNW of Clousta School. [298 577].

FIG. 3.   Slice No. S 49343. Magnification × 40. Plane polarized light. Part of ignimbrite clast in lapilli-tuff in Clousta Volcanic Rocks, showing flattened and welded shards. Note the bending of shards around quartz clasts. Hillside, 710 yd (650 m) SW of western end of Loch Hollorin [267 552].

FIG. 4.   Slice No. S 30773. Magnification × 38. Plane polarized light. Basalt flow in Clousta Volcanic Rocks. Flow-aligned laths of sodic labradorite set in matrix composed largely of secondary amphibole with subordinate grains of epidote and a dusting of iron ore. Aithness peninsula, 220 yd (200 m) SE from north-west corner of peninsula [327 597].

FIG. 5.   Slice No. S 51496. Magnification × 16. Plane polarized light. Fine-grained feldspathic sandstone in Walls Formation with laminae of heavy mineral concentrates. Black grains are predominantly iron ore, other heavy mineral grains are apatite, sphene, epidote and tourmaline. North shore of Scutta Voe, 520 yd (475 m) WSW of Lee of Houlland [275 498].

FIG. 6.   Slice No. S 52748. Magnification × 100. Plane polarized light. Microfolded sandy siltstone, Walls Formation. Roadside, close to west shore of Loch Vaara [565 316].

FIG. 7.   Slice No. S 53696. Magnification × 100. Plane polarized light. Silty shale with F1 slaty cleavage (horizontal) refolded by F2 minor folds with incipient fracture cleavage developed along some fold limbs. Walls Formation. North shore of Scutta Voe, 520 yd (470 m) WSW of Lee of Houlland [275 498].

FIG. 8.   Slice No. S 53688. Magnification × 100. Crossed polarisers. Microfolded dark grey shale with axial-planar strain-slip cleavage inclined at 44° to bedding. West shore of Voe of Browland, 1620 yd (1480 m) S4°E of Browster [261 503].

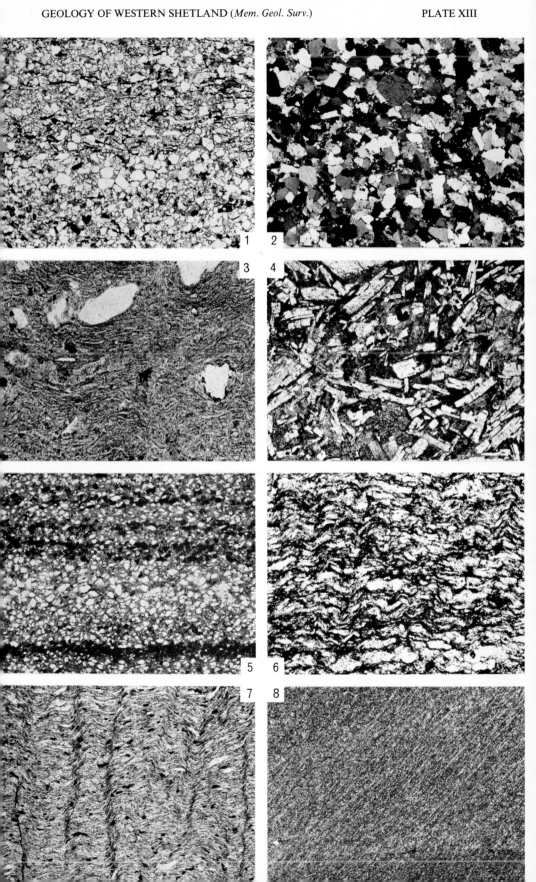

graphic units based on the structural interpretation shown in Plate XII. This brings out the slight petrographic differences within the sequence.

Lowest sediments (below band 3 on Plate XII). The sandstones are fine- to medium-grained, poorly graded with grain diameters ranging from 0·6 to 0·045 mm. Grains are angular to subangular, with the larger grains of some thin sections subrounded. Within the zone of tight north-north-east trending folds close to the Loch of Voxterby [260 533] quartz grains in a number of specimens have an undulose extinction and serrate interlocking margins. Small quartz grains are partly replaced by matrix minerals. In some cases a high proportion of quartz grains shows extreme elongation parallel to the bedding.

The ratio of quartz to feldspar averages 70:30 and ranges from 60:40 to 80:20. The feldspar clasts consist, in order of abundance, of untwinned potash feldspar, oligoclase-andesine, microcline and microperthite, the ratio of potash feldspar to plagioclase varying greatly in different specimens.

Thin bands of heavy mineral concentrations, ranging from less than 1 to 2·5 mm in thickness, are developed throughout the sequence, and nearly all specimens of pale sandstone examined contain over 10 per cent and in one specimen 20 per cent of dark minerals evenly scattered throughout. In the dark bands up to 50 per cent of the clasts are heavy minerals. These are, in order of abundance, epidote, sphene, apatite, clusters of opaque iron ore, tourmaline and rare clusters of allanite. One specimen (S 30975) from just south of the Sulma Water Fault ¾ mile (1200 m) W of Brindister Voe contains a high proportion of large garnet grains, altered along the joints to a chloritic aggregate. The garnets are present both in the dark bands and in the normal sandstone.

The amount of matrix and mineral cement in the specimens ranges from 10 to 15 per cent, rising to 20 per cent in the stratigraphically lowest specimens. The matrix is always somewhat more abundant in the dark bands and is variable in composition, depending on the extent of tectonic deformation. In the least altered sandstones it is composed of a green aggregate of chlorite and clay-mica coating the grains and interstitial patches of calcite. The specimens from the belts of intense folding (S 2746, 51494), have a fine-grained aggregate of green mica and a fine interstitial mosaic of silica and ?albite. In these specimens chlorite forms discrete irregular interstitial patches.

Sediments between bands (3) and (4) of Plate XII. The sandstones from this group are fine- to medium-grained and, with one exception, poorly graded. The grain diameters compare closely with those of the underlying sediments. They average 0·3 mm and range from 0·5 mm to 0·04 mm. Grains are subangular to subrounded but well rounded in one specimen (S 30900). The margins of the grains are either not corroded or only slightly corroded by the ingrowth of authigenic sericite. The ratio of quartz to feldspar grains in all but one specimen is close to 60:40 and the percentage of matrix ranges from 10 to 15 per cent in the normal sandstone, but is 30 to 40 per cent in the dark bands. The quartz grains are normally clear, but the samples from a locality ¼ mile (400 m) N of Lunga Water have quartz grains full of needles of rutile. The composition of feldspar grains is the same as that in the underlying sediments, with untwinned potash feldspar, which in this group is slightly cloudy, most abundant in many specimens, though in some (S 30872–3) sodic plagioclase predominates. Dark, rounded, mineral grains are present in all specimens and form up to 15 per cent of the total detrital grains (S 30900). Epidote, though forming up to 60 per cent of the total heavy minerals in one specimen (S 30900), is commonly less abundant than sphene and even apatite. Garnet forms abundant grains which are partially or completely broken down into chlorite in the dark bands of specimens from north of Lunga Water. The matrix consists of a thin, often discontinuous film of chlorite and clay-mica mantling most grains, together with interstitial patches of carbonate and chlorite and less commonly a mosaic of quartz and ?albite. Except in the heavy mineral bands, there is little evidence to suggest that a high proportion of the matrix was authigenically formed by the replacement of the detrital grains.

Strata between zones 4 and 5 of Plate XII. The sandstones in this group of strata are generally fine- to medium-grained with the average grain diameter slightly smaller than in the beds of the underlying groups, the maximum size ranging from 0·45 to 0·24 mm and minimum from 0·06 to 0·03 mm in different specimens. Some sandstones contain thin bands which are coarser and have grains exceptionally up to 0·8 mm in diameter. Grading is variable, most specimens are poorly graded but a number are moderately well to well graded. Grains are nearly always angular and within the strongly folded belt there are two cases with slightly serrate and interlocking margins. The grains of these sediments are, however, not as strongly deformed and serrated as the sandstones within the zone of the near-isoclinal north-east trending folds below band 3 (p. 119). There is a gradual increase in the serration and partial annealing of the grain margins as the aureole of the Sandsting Granite Complex is approached (p. 231).

The ratio of quartz to feldspar grains ranges from 60:40 to 80:20, with an average of 70:30. In only one specimen (S 52547) is feldspar virtually absent. Quartz grains are almost invariably free from inclusions. Among the feldspar grains untwinned slightly cloudy and, in some instances, sericitized, potash-feldspar predominates over or is roughly equal in amount to plagioclase (oligoclase-andesine), which is always clear. In some specimens (e.g. S 29928) potash feldspars form fresh angular cleavage fragments, suggesting a fairly local source. Muscovite occurs in scattered flakes or in clusters of stumpy plates and appears to be more abundant than in the lower sediments. Heavy mineral bands, up to 2 mm thick, appear to be less abundant than in the lower beds. Again, epidote, though still the most common heavy mineral in the dark bands, is rare or absent in many specimens of pale sandstone where sphene is by far the most abundant heavy mineral. Other minerals, in approximate order of abundance, are apatite, ilmenite-leucoxene, tourmaline, zircon and allanite. Clastic garnet has not been recorded in this group.

Matrix forms 5 to 30 per cent of the total rock volume and up to 40 per cent in the dark mineral bands. In nearly all specimens the matrix contains a high proportion (50 to 80 per cent) of carbonate which forms fairly large patches and has locally partially replaced quartz grains. The aggregate of chlorite and clay-mica forms a thin, often discontinuous rim around grains and has partly replaced some feldspars. Chalcedonic silica with chlorite forms irregular patches in several specimens. In some specimens in this group the matrix percentage is less than 10, and in others the matrix is almost entirely carbonate. Some of these rocks are thus not greywackes and it seems probable that the high proportion of matrix in the other specimens is largely of secondary origin.

Strata above band 5 of Plate XII. The beds above and just below band 5 appear to have a somewhat higher proportion of fine-grained sediments than the underlying groups, though the grain size of the sandstones is much the same. Average grain diameters range from 0·4 to 0·03 mm. Coarse-grained sandstones with a maximum grain diameter of 0·8 mm are very rare. Though most specimens are poorly sorted with subangular to angular grains there is a somewhat higher proportion with moderate to good sorting than in underlying sediments.

The composition of the clasts shows a quartz-feldspar ratio ranging from 60:40 to 80:20 with a possible slight westward increase in the proportion of feldspar. The feldspar clasts are in most cases predominantly untwinned potash feldspar with subordinate plagioclase (oligoclase-andesine) and rare microcline and microperthite. The latter is, however, fairly abundant in specimens from the Watsness district. Heavy minerals appear to be marginally less abundant than in underlying sediments, and epidote appears to be largely confined to the dark bands. Within the normal pale sandstones sphene is most abundant, in places forming 80 per cent of the total heavy mineral grains. Apatite is common throughout; tourmaline, iron ore and allanite are rare. Garnet has been recorded within this group only in a calcareous siltstone from the Watsness district (S 50821).

The percentage of matrix and mineral cement within this group is very variable, ranging from less than 5 to 40 per cent. In some specimens from the Watsness area it consists only of a thin discontinuous film of chlorite and small interstitial patches of carbonate. Further east within the highly deformed sandstones, muscovite is more abundant than chlorite and a quartz-albite aggregate as well as calcite forms interstitial patches.

Rea Wick–Roe Ness peninsula. The sandstones east of the Roeness Fault (pp. 00–00, Fig. 10) are comparable with the non-hornfelsed sandstones with a relatively high percentage of calcareous cement from the main outcrop. The only sliced specimen (S 51542) from this area is slightly better graded than most sandstones from the Walls Formation, the grain size being within the limited range of 0·25 to 0·08 mm. The grains are angular with serrate margins, and quartz grains are partially replaced by carbonate. The ratio of quartz to feldspar clasts is 70:30 and of the feldspar grains approximately 80 per cent are potash feldspar with irregular blotchy inclusions of albite and carbonate; the remainder is clear sodic plagioclase. Muscovite forms abundant thin flakes and heavy minerals consist of abundant small rounded grains of apatite, scattered granules of iron ore, and rare olive-green tourmalines. Mineral cement forms over 20 per cent of the volume of the rock and consists essentially of finely granular calcite with scattered flakes of sericite.

Stratigraphic and regional variations within sandstones. The above petrographical descriptions deal with a thickness of possibly 30 000 ft (9000 m) of sediments which show remarkably little petrographic variation either stratigraphically or geographically. Nearly all specimens of sandstone are poorly graded, have subangular clasts, and contain a high proportion of readily altered minerals, which include up to 40 per cent of relatively fresh feldspar and varying proportions of epidote. Stratigraphic variations, which may not be statistically significant, are as follows:

1. There is a slight decrease in average grain size as the sequence is ascended, and a slightly higher proportion of subrounded to rounded grains near the top.
2. There is a fairly marked decrease in the amount of detrital epidote within the pale sandstones in the upper part of the sequence as well as a less marked decrease in the frequency of dark laminae of epidote-rich heavy mineral concentrates.
3. Garnet-bearing sandstones are confined to the lower part of the sequence and are found mainly near the northern margin of the area.
4. There is a slight but inconsistent upward decrease in the percentage of matrix. As a high proportion of the matrix was, however, formed by the partial or complete replacement of clasts by sericite, calcite and a quartz-albite aggregate during the intense folding of the beds the relationships of the original matrix content to the stratigraphy are largely obliterated.
5. There appears to be no obvious connection between the relative abundance of the various types of feldspar clasts (p. 120) and the stratigraphy. In the higher part of the sequence, however, there appears to be a marked westward increase in the percentage of perthite.

SILTSTONE AND MUDSTONE

Siltstones and mudstones make up between 10 and 20 per cent of the total sediment in the upper half of the Walls Formation. Only a limited number of thin sections are, however, available.

The siltstones (S 29932, 29937) consist of alternate quartz-rich and muscovite-rich laminae ranging, in thickness, from 0·5 to 1·5 mm. Feldspar grains are rare or absent in many of the specimens. Muscovite forms thin flakes normally up to 0·25 mm long, and in some specimens these are associated with flakes of chloritic material. There are also thin laminae of very fine-grained brownish aggregate of chlorite and mica. Heavy minerals are relatively rare in sediments of silt grain-size, but some bands with rounded epidote grains and rare sphenes (S 29937) have been recorded. Minute grains and flecks of iron oxide are common throughout. The matrix of most specimens is a mixture of calcite, chlorite, and clay-mica, with the calcite content in the sandy laminae of certain specimens (S 52550) reaching 40 per cent.

The fine-grained sediments have clearly recorded the effects of the two major episodes of folding. In the area where only the first fold phase is recognized, they have small-scale folds with wavelengths up to 0·3 mm and a strong axial-planar slaty cleavage. In the areas affected by the two periods of folding the early slaty cleavage, characterized by the development of greenish pleochroic micas, is re-folded into regular asymmetric folds with a newly developed strain-slip cleavage (p. 136).

LIMESTONE AND CALCAREOUS ARGILLITE

The limestones and calcareous mudstones and siltstones are generally finely banded and consist of alternate laminae of calcite-rich mudstone and calcite-rich siltstone, which contain varying proportions of grains of other minerals such as partially replaced quartz, feldspar, sphene and epidote, as well as fine flakes of muscovite and granules and films of carbonaceous material. Coarse-grained limestones, composed entirely of crystalline carbonate, are absent.

The calcareous deposits are very prone to mechanical deformation and many show intense small-scale folding or, in extreme cases (S 50803), complete disruption of the fabric. In some thin sections (e.g. S 50821) the laminae of mudstone and shale within the folded calcilutite are broken up into discrete highly irregular fragments ranging up to 7 mm × 1 mm in size, which in many cases show evidence of some degree of initial folding. In the extremely deformed bands (S 50803) the calcite-rich mudstone is discordantly folded and broken into irregular fragments. The latter are cemented by irregular branching veins of carbonate composed of parallel and, in many cases, slightly bent calcite fibres. These are in turn cut by a suite of often branching parallel veinlets of crystalline calcite, which shows no sign of deformation.

Tectonically undeformed calcareous mudstone is composed of laminae of calcite grains, which are elongated parallel to the bedding and enclosed in a thin film of carbonaceous material. Thin laminae of carbonaceous debris, which tend to split laterally, are interdigitated with laminae of pure carbonate rock, which are up to 0·4 mm thick. Most laminae contain scattered grains of quartz, and subhedral highly poikilitic crystals of authigenic pyrite up to 0·35 mm in diameter. The inclusions in the pyrite crystals are invariably calcite. Calcite also forms a narrow rim around these crystals, and in one case, the pyrite is completely enclosed by an euhedral crystal of calcite.

The limestone bands are particularly sensitive to the thermal alteration by the Sandsting Granite and the mineral assemblages developed in the limestones exposed on the east coast of Vaila are described in Chapter 15, pp. 232–3 and shown in Fig. 25, p. 232.

Environment of Deposition

The sediments of the Walls Formation bear some resemblance to the higher members of the Sandness Formation exposed along the west coast of the Walls Peninsula (p. 80). The latter are believed to be river deposits laid down on alluvial plains, and they consist of fining-upward cycles in which channel sandstones are overlain by relatively thin overbank deposits. The fine-grained beds contain sun-cracks, asymmetrical ripple marks and local red beds, all of which are evidence for deposition on flood plains (p. 83). The beds of the Walls Formation also consist of rhythmic units composed of alternating sandy and shaly members, but they lack many of the features associated with river deposits.

The sandstones are almost uniformly fine-grained and contain no basal pebbly lenses or scattered pebbles throughout them. Though cross-bedded sets are present in many sandstones, they usually form only a part of the total coset, which also contains finely banded planar-bedded sets and massive unbedded sets. Channelling at the base of sandstones is rare, and some sandstones have lobate load-cast bases. All the sandstones contain a very high proportion of fine-grained matrix, and though some of this matrix was formed by the authigenic replacement of grains after lithification, the rocks do not have the character of typical 'clean' fluvial sandstones. Perhaps most significant of all is the remarkable uniformity of the sandstones throughout their entire outcrop and over a vertical distance of possibly 30 000 ft (9000 m). This uniformity is in marked contrast to the great vertical and lateral variation in the lithology of the predominantly fluvial Sandness and Melby formations.

The fine-grained phases of the rhythmic units in the Walls Formation show none of the characteristics of fluvial overbank deposits. Desiccation cracks are unknown and asymmetrical ripple marks are rare. The red beds are not of the type associated with topstratum beds. Many of the deposits are calcareous and some display certain of the characteristics found in the lacustrine flagstones of Orkney (Fannin 1970) and Caithness (Crampton and Carruthers 1914). The fine lamination of some bands and the presence of thin lime-rich and carbonaceous laminae associated with authigenic pyrite (p. 122) are obvious examples. The fish genera found in these beds are also of the type associated elsewhere with a lacustrine environment. In the flagstones of Orkney and Caithness, however, the laminated lacustrine beds are usually underlain and overlain by predominantly fine-grained sediment with sun-cracks and other structures which are diagnostic of a shallow water or mud-flat environment that was periodically exposed to sub-aerial conditions. As these features are completely absent in the Walls Formation, it is probable that it was deposited, not in shallow lakes of the Orkney–Caithness type, but in a rather deeper body of inland water. The great regularity in the direction of minor structures, such as the 'convolute' folds, could mean that the shape of the basin of deposition exercised a strong control over the directions of the currents within it (but see p. 132). If this hypothesis were to be accepted, it should be possible to use the alignment of sedimentary structures to get some idea of the approximate shape of the basin. The basin must have subsided fairly rapidly, as subsidence must at all times have kept pace with the deposition of a very considerable thickness of sediment.

The sandstones interbedded with the Middle Old Red Sandstone lacustrine flagstones of Orkney and Caithness are generally regarded as shallow-water deposits laid down either in the distributary channels of deltas which periodically invaded the shallow lakes, or in channels of rivers which traversed the dried up lake bed. The Walls Formation sandstones, however, bear a completely different relationship to the fine sediments, as they themselves appear to have been laid down in relatively deep water (but see below, p. 124). Though they are not marine sediments they can in some respects be compared with sandy flysch deposits which are generally believed to have been laid down in or on the slopes of marine basins. The Walls Formation has the following characteristics in common with flysch deposits:

1. The sandstones are 'muddy' and have a high proportion of clay grade material in the matrix.
2. The sequence displays the rhythmic repetition of sandstones and fine-grained sediments which is characteristic of most flysch successions.
3. The sandstones have sharply defined bottom surfaces on which sole markings are in some instances preserved. Sole markings are, however, less common than in typical flysch deposits.
4. Many sandstone cosets become finer-grained upwards. Good graded bedding of the type seen in many turbidites is, however, uncommon.
5. Fine-grained sandstones and sandy siltstones are generally well laminated and many beds show convolute lamination which is a characteristic, but not exclusive feature of flysch deposits. Slump bedding and other forms of contorted bedding are common in the sandstones.
6. There is no rapid variation in the overall composition of the sediments, either laterally or vertically.
7. The alignment of directional sedimentary structures, particularly in the fine-grained sediments, appears to be constant over considerable distances.
8. There is no evidence of sub-aerial or very shallow water conditions.

The sandstones of the Walls Formation, however, show a fair amount of large-scale cross-stratification which is only rarely found in flysch sandstones (Dzulynski and Walton 1965, p. 178) and which could not easily have been formed in deep water. There is also no evidence in the Walls Formation for the occurrence of large-scale slumping, nor are there any mudflow breccias or conglomerates of the type recorded in some flysch deposits. Most flysch sandstones are thought to be turbidites, which have been deposited by turbidity currents. The idealized complete turbidite unit should show the following four structural sub-units (Bouma 1962, pp. 49–51; Walker 1965): (1) lower unit with graded bedding and no cross-lamination; (2) unit with parallel lamination; (3) unit with ripple-cross-lamination, and (4) upper unit with current lamination. Some rhythmic units in the Walls Formation (Fig. 9 and pp. 104–6) have similar subdivisions, but it is doubtful if many of the sandstones in the formation can genuinely be classed as turbidites.

The origin and mode of deposition of the sediments of the Walls Formation is thus still open to speculation. Nor can anything be said at this stage about the size or shape of the original basin of deposition, which must have occupied a considerably larger area than the present outcrop of the Walls Formation.

REFERENCES

BOUMA, A. H. 1962. *Sedimentology of some Flysch Deposits: A Graphic Approach to Facies Interpretation*. Amsterdam: Elsevier.
—— and BROUWER, A. 1964. (Editors). *Turbidites*. Amsterdam: Elsevier.

CRAMPTON, C. B. and CARRUTHERS, R. G. 1914. The Geology of Caithness. *Mem. geol. Surv. Gt Br.*

DOTT, R. H. Jr. and HOWARD, J. K. 1962. Convolute Lamination in non-graded Sequences. *Jnl Geol.*, **70**, 114–21.

DZULINSKI, S. and SMITH, A. J. 1963. Convolute Lamination, its Origin, Preservation and Directional Significance. *J. sedim. Petrol.*, **33**, 616–27.

—— and WALTON, E. K. 1965. *Sedimentary features of Flysch and Greywackes.* Amsterdam: Elsevier.

FANNIN, N. G. T. 1970. The sedimentary environment of the Old Red Sandstone of Western Orkney. *Ph.D. Thesis, University of Reading* (unpublished).

FINLAY, T. M. 1930. The Old Red Sandstone of Shetland. Part II: North-western Area. *Trans. R. Soc. Edinb.*, **56**, 671–94.

KUENEN, PH. H. 1953. Significant features of graded bedding. *Bull. Am. Ass. Petrol. Geol.*, **37**, 1044–66.

—— 1958. Experiments in geology. *Trans. geol. Soc. Glasg.*, **23**, 1–28.

MACAR, P. and ANTUN, P. 1950. Pseudonodules et glissement sous-aquatique dans l'Emsien inferieur de l'Œsling (Grand Duché de Luxembourg). *Annls. Soc. géol. Belg. Mem.*, **73**, 121–50.

McKEE, E. D., REYNOLDS, M. A. and BAKER, C. H. 1962a. Laboratory studies on Deformation in unconsolidated sediments. *U.S. Geol. Surv. Prof. Paper*, **450 D**, 151–5.

—— —— —— 1962b. Experiments on intraformational recumbent folds in cross-bedded sands. *U.S. Geol. Surv. Prof. Paper*, **450 D**, 155–60.

PETTIJOHN, F. J. 1957. *Sedimentary Rocks.* New York: Harper and Brothers.

POTTER, P. E. and PETTIJOHN, F. J. 1963. *Palaeocurrents and Basin Analysis.* Berlin: Springer Verlag.

RICH, J. L. 1950. Flow markings, groovings and interstratal crumplings as criteria for recognition of slope deposits, with illustrations from Silurian rocks of Wales. *Bull. Am. Ass. Petrol. Geol.*, **34**, 717–41.

SANDERS, J. E. 1956. Oriented phenomena produced by sedimentation from turbidity currents and in subaqueous slope deposits. *J. sedim. Petrol.*, **26**, 178.

TEN HAAF, F. 1956. Significance of Convolute Lamination. *Geologie Mijnb.*, **18**, 188–94

TURNER, F. J. 1968. *Metamorphic Petrology.* New York.

WALKER, R. G. 1965. The origin and significance of the internal sedimentary structures of turbidites. *Proc. Yorks. geol. Soc.*, **35**, 1–32.

WILLIAMS, E. 1960. Intra-stratal flow and convolute folding. *Geol. Mag.*, **97**, 208–14.

Chapter 9

WALLS SANDSTONE—STRUCTURE

THE SEDIMENTARY and volcanic rocks of the Walls and Sandsting formations have been affected by two periods of intense folding (Plate XII). The axes of the earlier folds (F1) trend in an east to north-easterly direction, while those of the later folds (F2) have trends which range from N25°E to N20°W. The more intense folding is confined to two intersecting belts, the limits of which are shown in Fig. 12. Both belts contain zones in which the finer-grained sediments exhibit well-developed minor structures, such as cleavage, small-scale folds, and various types of lineation. The geometric relationships of these structures correspond to those of the major folds. In the coast sections along the shores of Gruting Voe and the Voe of Browland the cleavage and lineation associated with the two periods of folding are seen to intersect each other and produce small-scale interference structures.

The minor structures produced by the two periods of folding do not extend into the area in which the beds are indurated or hornfelsed by the Sandsting Granite, and the evidence suggests that granite intrusion preceded the folding. Radiometric determinations give an age of 360 m.y. for the date of empiacement of the granite. This suggests that the folding, which on fossil evidence (p. 117) was certainly later than at least part of the Middle Old Red Sandstone, took place not earlier than late-Middle or early-Upper Old Red Sandstone times.

STRUCTURES ATTRIBUTED TO THE FIRST PHASE OF FOLDING (F1)

The first phase of folding which affected the Walls Sandstone produced an east to east-north-east trending synclinorium with a northern limb in which the strata are consistently steep and only locally flexured, a complex hinge zone containing small, locally isoclinal, folds arranged *en echelon*, and a fairly steep southern limb. Cleavage, lineation and minor folds are confined to the fine-grained sediments in the axial zone and southern limb of the synclinorium.

STRUCTURAL UNITS OF THE SYNCLINORIUM

Northern Limb

The northern limb of the synclinorium is formed by the outcrop of the Sandness Formation, and is separated from the hinge zone by the Sulma Water Fault (Plate XII; Fig. 13). This fault appears to have acted as a hinge fault and has a considerably greater southward displacement at the western end of its outcrop than in the east.

The dip of the sedimentary and volcanic rocks of the Sandness Formation is, over the greater part of the area, inclined at between 45° and 80° to east-south-east, steepening to vertical and shifting in trend to north-north-east on the island of Papa Little, in the extreme east of the area. In the area between Sandness Hill and the head of West Burra Firth the northern limb of the synclinorium

126

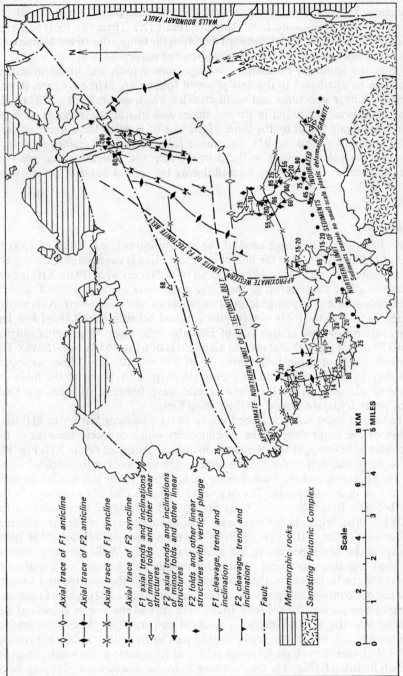

Fig. 12. Location of tectonite belts and prominent linear and planar structures in the Old Red Sandstone of the Walls Peninsula

contains a parasitic flexure which forms the Djuba Water Syncline and the complementary Mousavord Loch Anticline (Plate IX). These two folds converge and die out westwards; in an eastward direction the hinge zone of the Mousavord Loch Anticline passes into a tectonic movement plane.

Within the Sandness Formation cleavage, minor folds and linear structures which can be attributed to the first phase of folding are extremely rare. Poorly developed linear structures and small crinkles which are inclined at 25° to 30° to the south-west are found in the red shales and siltstones associated with the Clousta Volcanic Rocks on the shore of the Voe of Dale, at the extreme western end of the outcrop (pp. 91–92). The minor folds within fine-grained sediments adjoining the volcanic rocks in the Burga Water and Brindister Voe areas all appear, however, to have been formed during the second phase of deformation (pp. 88–89).

Hinge Zone

Major structures. The hinge zone of the synclinorium has a width of 1500 to 1650 yd (1400–1500 m) but the belt of intense folding is confined to the southern half of this zone and is only 440 to 880 yd (400–800 m) wide (Plate XII). In the western third of the outcrop of the hinge zone there are a number of second-order folds whose axes plunge in most instances to west-south-west. As is shown in Figs. 14 and 15 the folds of this order exposed between Ram's Head and The Flaes, just north-west of the Voe of Footabrough, have wavelengths ranging from 120 to 222 yd (110–200 m) and amplitudes of 65 to 165 yd (60–150 m). The interlimb angle of these folds varies from 55° to 110° (i.e. 'close' to 'open' according to the classification of Fleuty (1964, p. 470). Some of the folds are westward plunging monoclines and several have irregular disharmonic folds (third order folds) developed on their steep limbs.

The second order folds described above lie on a more or less horizontal limb section of the hinge zone of the synclinorium which connects the axes of the Scarvister Syncline and the Watsness–Browland Anticline (Plate XII, Fig. 13). The style of the first order fold becomes tight and eventually isoclinal in an easterly direction, and the second order folds appear to die out east of the head of the Voe of Footabrough. The Scarvister Syncline dies out eastwards in the area between Braga Ness and Vaila Sound and is replaced *en echelon* by the Walls Syncline, which is the main and most southerly axis of the synclinorium further east (Plate XII). East of Vaila Sound a steep to vertical shear plane appears to be developed along the axial plane of the Walls Syncline. Similar east–west trending movement planes, not usually recognizable on the ground, appear also to be present within the northern limb of the syncline and seem to have cut out considerable portions of the sequence (Plate XII). These faults may be parallel to the limb of the fold. In the area around the Voe of Browland and Gruting Voe the Walls Syncline is isoclinal and its axial plane is inclined at between 65° and 80° to the north, so that part of its northern limb is inverted.

The Watsness–Browland Anticline is also at its tightest at the longitude of the Voe of Browland (Fig. 13, C-C') where it forms a westward plunging near-monocline with a gently inclined northern limb and a nearly vertical southern limb. The anticline becomes progressively more open both west and east of this area.

In the west and north-west of the Walls Peninsula all folds plunge to the west

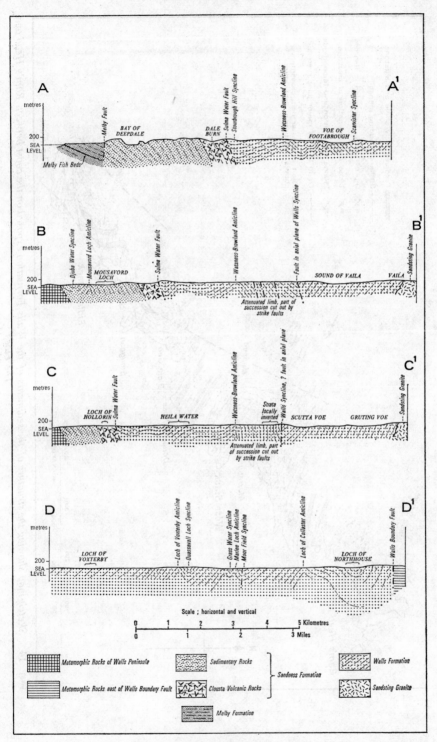

FIG. 13.   *Horizontal sections illustrating the major structural features of the Old Red*
Sandstone of the Walls Peninsula
Lines of sections are marked on Plate XII.

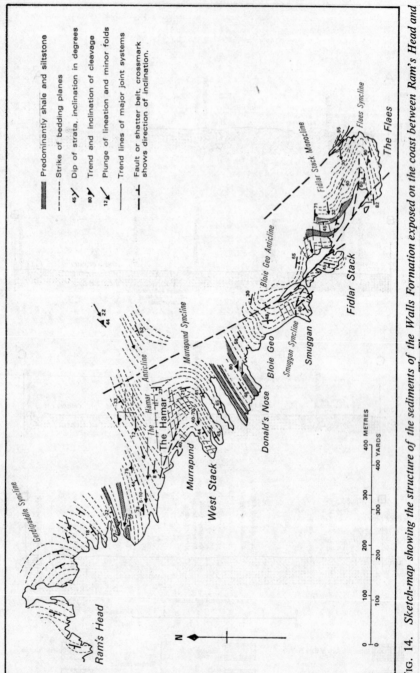

Fɪɢ. 14. *Sketch-map showing the structure of the sediments of the Walls Formation exposed on the coast between Ram's Head and The Flaes*

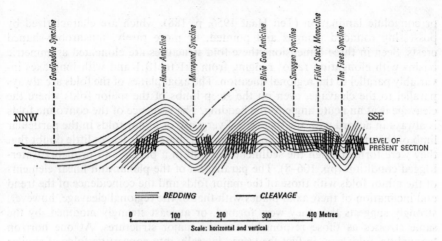

NNW   SSE

LEVEL OF
PRESENT SECTION

=== BEDDING   —— CLEAVAGE

0   100   200   300   400 Metres
Scale: horizontal and vertical

FIG. 15.   *Idealized horizontal section showing the structural pattern of the folded area along the coast between Ram's Head and The Flaes*

or west-south-west, but in the ground between the Voe of Footabrough and Gruting Voe the plunge of the fold axes and concordant small-scale linear structures is in places to the east. This reversal in plunge may be partly due to re-folding along the north–south trending second folds (F2).

The east-north-east trending F1 folds do not appear to extend into the extreme eastern part of the Walls Peninsula, where exposure is very poor. It is likely that these folds may originally have continued eastward but that they are now difficult to recognize owing to the effects of later north–south trending folding. *Cleavage, jointing and lineation.* An axial plane cleavage which locally obscures the bedding is present in many shales and siltstones within the belt of intense F1 folding and a corresponding refracted fracture cleavage affects the inter-vening sandstone bands. The trend and inclination of the cleavage is everywhere sub-parallel to the axial surfaces of the folds (Figs. 14 and 15). In most instances the cleavage is a true slaty cleavage, characterized by the parallel alignment of phylloblastic minerals either along the limbs of the minor corrugations (e.g. S 52549) or along the axial planes of the minor folds (e.g. S 52550). In other examples the cleavage planes are less regular and in extreme cases the cleavage surfaces have a phacoidal (lens-shaped) pattern (cf. Elliston 1963, pp. C11–C13).

Many fine-grained sediments, particularly the calcareous beds and the thinly interlaminated shales and sandstones, exhibit intense microcrinkling. This pro-duces a strong lineation (Plate XV, A) which is invariably parallel to the plunge of the axes of the major folds. Linear structures are of varying sizes, ranging from very fine striations on bedding surfaces of sandy sediments to the axes of small folds with wavelengths up to 1 cm.

'Convolute' minor folds. There are numerous horizons, particularly in the area between The Flaes and Fidlar Stack, just west of the Voe of Footabrough (Fig. 14) and close to Point of the Hus [196 482] on the south-west shore of Braga Ness, at which small-scale 'convolute' folds are developed in silty or inter-laminated shaly and sandy beds. These structures are described and their origin discussed on pp. 105–8 and they are illustrated in Plate XIV. They resemble the sedimentary structures known as convolute bedding (Kuenen 1953, pp. 1056–8)

K

or convolute lamination (Ten Haaf 1956, p. 188), which are characterized by possessing rounded troughs and pointed, or more rarely, mushroom-shaped crests. Seen in three dimensions these fold structures are elongated asymmetric basins with elongation ratios ranging from 5:1 to 10:1 and with long axes invariably parallel to the regional lineation. The axial planes of the folds are always parallel to the cleavage, even on the steep limbs of the major folds where the cleavage is at an acute angle to the bedding. The vergence of the convolute folds is always in accord with the theoretical vergence of minor folds in the particular limb of the major fold. The shape and style of these folds leaves little doubt that they were formed when the sediment was still in a plastic and probably water-logged condition (pp. 106–8). The parallelism of the planar and linear elements of the minor folds with those of the major folds and the coincidence of the trend and inclination of their axial planes with that of the regional cleavage, however, strongly suggests that they were formed, or at least strongly modified, by the same stresses as those responsible for the major structures. At one horizon 'convolute' folds can, in fact, be traced laterally into concertina folds of similar amplitude.

The above reasoning could suggest that the major folds, cleavage and lineation of this area were all initiated at a time when the finer-grained sediments were still in a plastic or waterlogged condition. Evidence for the formation of slaty cleavage in unconsolidated waterlogged sediments has been brought forward by Maxwell (1962, pp. 284–302), who, in the Martinsburg Slates of the Delaware Water Gap area of New Jersey and Pennsylvania, has recorded the presence of small sandstone dykes, which branch out from sandstone beds along the cleavage planes of the adjacent slates. Maxwell concluded that the sandstone beds were unconsolidated and waterlogged at the time the cleavage was formed and has shown that in these slates the cleavage is due to the presence of aligned plates of illite, and not sericite or muscovite which are the characteristic platy minerals of slates formed under metamorphic conditions. He suggested that the cleavage in the slates was formed in waterlogged sediments by deformation which produced a series of similar folds under very low-grade or non-metamorphic conditions, at depths probably not exceeding 12 000 ft (3658 m) and at no greatly elevated temperatures. He considered that the slaty cleavage resulted from the upward expulsion of water from the sediments which led to the rotation, upward transport and consequent parallel orientation of the component minerals. Further evidence for the formation of cleavage in unconsolidated sediments has been assembled by Elliston (1963, pp. C11–C13), who in Tennant Creek, Tasmania, has recorded fragments of slate with a strongly developed, often bent or curved cleavage set in a 'quickstone' (intrusive sandstone) breccia, which must have formed when the sandstone beds were still in a plastic state.

More recently Williams and others (1969, pp. 421–5) have recorded a poorly developed axial plane cleavage within small penecontemporaneous folds in the Devonian Bunga Beds exposed on the south coast of New South Wales, Australia. These folds differ from the convolute folds of Western Shetland in that the

---

EXPLANATION OF PLATE XIV

A. to C. Fidlar Geo, south-west shore of Walls Peninsula [190 494]. Thinly bedded siltstone and mudstone of Walls Formation, showing relationship of convolute lamination to cleavage. (A—D 956, B & C by w.m.).

orientation of their axes and axial planes is very variable and quite unrelated to the major tectonic elements. Though the style of the individual folds is indistinguishable from that of many tectonic folds, the sum of their relationships has convinced Williams and his co-authors that they are convolute laminations formed by load deformation of a semi-fluid sediment, and that the slaty cleavage was formed at the same time as the folds by the rotation of pre-existing inequant mineral grains.

In the present area no work has yet been done to determine the composition of the phylloblastic minerals which impart the slaty cleavage on the fine-grained sediment. It is possible that the cleavage may, as in the Martinsburg Slates and the Bunga Beds of New South Wales, have formed before the sediments were consolidated in a manner similar to that postulated by Maxwell. There are here, however, no sandstone dykes branching from sandstone beds out along the cleavage of the finer sediment, which would have suggested that the sandstones were still 'mobile' when the cleavage was being formed. It is more likely that in this area the close accord of the geometric parameters of the 'plastic folds' with those of the undoubted tectonic structures resulted from a continuous homoaxial deformation which gave rise to a series of stages in the development of the tectonic pattern of the sediments. These stages may be summarized as follows:

1. During and shortly after deposition the environment was unstable and repeated earthquake shocks produced beds with pseudo-nodules and probably also some unaligned convolute laminations.
2. The compressive stresses which eventually led to the formation of the F1 synclinorium commenced while the finer-grained sediments within the highest part of the Walls Formation, as now exposed, were still plastic and waterlogged. This led to the formation of the aligned 'convolute folds' in certain mixed fine-grained bands and probably also to the regular deformation of previously formed convolute laminations and pseudo-nodules. Sandstones were apparently no longer plastic at this stage, as no 'liquidized' sandstone dykes occur. Minor folds of convolute type have not been recorded in the lower part of the Walls Formation, suggesting that part of the formation may already have been lithified when the compressive movements commenced.
3. With continued compression all the water was expelled and the fine-grained beds were lithified. After this stage, when the large-scale folds were developing, the same continuing stresses produced the cleavage and lineation in the fine sediments, the small-scale folds in the calcareous shales, limestones and in some mixed sediments, and the complex joint systems in the sandstones.

In many localities within the intensely deformed areas the boundaries between sandstones and fine-grained sediments are shear planes. Movement planes are also developed along or close to the axial planes of folds both in the south-western coastal areas and to a much greater extent in the areas around Walls and the head of Gruting Voe (Fig. 13, sections B–B^1, C–C^1). These shear movements may be the latest manifestations of the first (F1) orogenic phase.

In the most intensely folded part of the hinge zone around Walls and especially at the head of Gruting Voe, lineation, cleavage and intense 'internal folds' within the fine-grained beds are everywhere developed. The style of F1 minor folding at the head of Gruting Voe is illustrated in Plate XV, B and C. The folds are complex sharp-crested zig-zag folds with several orders of fold size (cf. Ramsay 1967, pp. 354–5) which appear to have formed after the rock was lithified. Folds of the 'convolute' type have not been recognized in this area. The F1 slaty cleavage in the shales is here very pronounced and is characterized by the complete alignment of all micaceous and elongate minerals (Plate XIII, figs. 6–8). Though the trend of the lineation and the axes of minor folds in this area is more or less parallel to the regional strike of the beds, the amount and direction of their plunge varies greatly with short distances (Fig. 12). Although some of the variations could have developed during the F1 folding, most are thought to be due to the refolding of the F1 folds by the second (F2) phase of earth movement.

Southern Limb

The southern limb of the F1 synclinorium is composed of evenly bedded sediments dipping at 40°–90° to between north-east and north. The limb is almost vertical in the area just south of Walls. Fine-grained sediments within the limb are cleaved and lineated. The linear features are either minute straight corrugations or intersections of bedding and cleavage. Small-scale folds are relatively rare and 'convolute folds' have not been recorded east of Vaila Sound.

Neither cleavage nor lineation are seen within the aureole of the Sandsting Granite and the beds which contain these structures farther west become strongly jointed and shattered as they enter the aureole. These features are best seen in the peninsula south of Walls and on the east shore of Gruting Voe. This transition suggests that the fine-grained sediments within the aureole were already hornfelsed when they were subjected to the regional compressive stresses, and were no longer sufficiently plastic to form small-scale folds or crinkles or to take on a cleavage. The presence of small folds and linear structures in granite dykes and scapolite veins outwith the thermal aureole indicates that folding was still in progress after the latest intrusions and veins associated with the Sandsting Complex had formed.

STRUCTURES ATTRIBUTED TO SECOND PHASE OF
FOLDING (F2)

Major Folds. The folds attributed to the second phase of folding have axial trends ranging from north-north-east to north-north-west (Fig. 12). The main belt of folding is of limited width and extends from the area between the Voe of Clousta and Brindister Voe south-south-westwards to the Voe of Browland where it abuts against vertical strata, previously folded by the F1 movements. Folds with the same general trend but with more open styles occur south-west of this belt and extend to Braga Ness. Fairly open north-north-west trending folds are also present in the area between the main fold belt and the Walls Boundary Fault where the strata are generally steeply inclined and have a

---

EXPLANATION OF PLATE XV

A. Fidlar Geo, south-west shore of Walls Peninsula [190 494]. Strongly lineated siltstone and mudstone of Walls Formation (W.M.).
B. East shore of Gruting Voe [275 493]. Intense F1 folds in flaggy sandstone and siltstone of the Walls Formation (W.M.).
C. East shore of Gruting Voe [275 493]. Close-up of intense F1 folding in fine-grained sandstone and siltstone of Walls Formation (W.M.).

predominantly north-north-westerly trend. This area has, however, been affected by later movements, possibly associated with the Walls Boundary Fault (p. 236) and because of indifferent exposures it is here not possible to assess the style of the F2 folds and associated structures.

The north-north-east trending folds within the main F2 fold belt have inter-limb angles of 60° to 70° and wavelengths from $\frac{1}{2}$ to $1\frac{1}{2}$ miles (800m–2$\frac{1}{2}$km). The traces of individual fold axes are usually of limited length ($1\frac{1}{2}$ to 4 miles; 2·5–6·5 km) and adjacent folds are arranged *en echelon*. Because of discontinuous exposures and the rarity of exposures in fine-grained sediments it has not been possible to determine the tectonic pattern of this area with any accuracy. *Minor folds, cleavage and lineation.* Though the major folds of the main F2 fold belt are confined to a belt bounded by the Sulma Water Fault in the north and the crest of the Watsness–Browland Anticline in the south, the minor structures produced by this period of folding extend both northwards into the area occupied by the Sandness Formation and southward into the vertical strata exposed on the shores of Gruting Voe.

Small-scale folds and corrugations are developed in the siltstones, shales and interlaminated sandy siltstones associated with the Clousta Volcanic Rocks and they are well seen on the shores of Brindister Voe. They also occur in rare inland sections between Brindister Voe and the Voe of Clousta and westward as far as the shores of Burga Water (Plate XII). The size of individual folds ranges from 4 in (10 cm) wavelength and 2 in (5 cm) amplitude to 10 in (25 cm) wavelength and 8 in (20 cm) amplitude. The interlimb angles range from 70° to 120° and the crests of the folds are predominantly angular. The fold axes plunge in the direction of the regional dip and their plunge is thus invariably steep. The axial planes of the folds are generally perpendicular to the regional strike. Slaty cleavage has not been recognized in the corrugated beds within the Sandness Formation described above but is common in crinkled and lineated fine-grained sediments of this formation farther east, along and close to the east shore of the Loch of Clousta [319 584].

Within the belt of intense folding between the Sulma Water Fault and the Watsness–Browland anticline, exposures of fine sediments are relatively rare and sandstones are generally intensely jointed. Only minute wrinkles have been recorded in the shales of this belt, but all the fine-grained sediments have a cleavage, which is particularly pronounced in the area west and north-west of the Voe of Browland, where the zone of intense F2 folds intersects the northern margin of tight F1 folding (Plate XII). Lack of exposures in the fine-grained sediments has, however, not permitted a meaningful analysis of this cleavage.

Within the zone of vertical, inverted and steeply inclined strata exposed on the shores of the Voe of Browland and Scutta Voe, F2 minor folds, corrugations and cleavages have refolded or cut the generally more pronounced F1 minor structures. As in the Sandness Formation to the north (p. 88) the F2 linear structures plunge down the dip of the vertical or steeply inclined regional bedding planes. The plunge of the F2 fold axes is either vertical or inclined at angles greater than 70° to east or west. The F2 folds are much more open than the tight F1 folds and they are usually gentle puckers with wavelengths of 1 to 3 in (2·5–8 cm). Larger open folds with wavelengths up to 3$\frac{1}{2}$ ft (1 m) and rare smaller linear F2 structures form corrugations with wavelengths of less than 5 mm, and in extreme cases they appear as fine striations on bedding planes. In many instances these have largely obliterated the smallest linear F1 structures.

The F2 microcrinkling has produced an incipient cleavage in the fine-grained sediments, and the refolding of the F1 slaty cleavage is well seen in thin section. In one specimen (S 53688) the first cleavage, which is sub-parallel to the bedding, has been puckered into asymmetric microfolds and an incipient second cleavage is formed by the alignment of mica flakes along the limbs of the fold. The aligned plates form belts up to 0·02 mm thick and 0·045 mm apart which are parallel to the axial planes of the folds. This new cleavage is a strain slip cleavage and makes an angle of about 44° with the original slaty cleavage (Plate XIII, fig. 8). In the siltstones (S 53696) the F1 slaty cleavage, characterized by aligned mica flakes and strongly elongated and aligned quartz grains, has been corrugated into folds of two orders of magnitude. (Major corrugations: $\lambda = 6$ to 7 mm, minor corrugations: $\lambda = 0·15$ to 0·2 mm). The microfolds of the lower order have wide rounded hinge zones and very short steeply inclined straight limbs (Plate XIII, fig. 7). The alignment of micas along the short straight limbs has produced an incipient strain slip cleavage. In the sandstone (S 51495) it has not been possible to differentiate between the effects of the two phases of folding. In some of these the individual quartz (with highly undulose extinction) and feldspar grains now have diamond-shaped outlines and the surrounding fine-grained micaceous matrix is intensely corrugated and imbricated.

An interesting feature of the F2 fold system in the Walls Sandstone is the varying plunge of the fold axes. In the central sector of the fold belt, between the Sulma Water Fault and the axial plane of the Watsness–Browland Anticline, where the post-F1 dip of the bedding was more or less horizontal, the axes of the major F2 folds are roughly horizontal. In the vertical or steeply dipping belts to north and south of this belt, the plunge of the axes of the F2 minor folds is steeply inclined and more or less parallel to the true dip of the strata. In these areas there are no large F2 folds. The axial plunges and varying magnitudes of F2 folds appear to have been determined by the F1 structure of the area. Thus in the ground where the strata were gently inclined large folds with horizontal axes could form, but in the belts of vertical strata compression parallel to the strike could only produce minor crinkles with steeply plunging axes. It is also likely that during the F2 phase considerable lateral movement took place along shear planes trending parallel to the axial planes or limbs of the F1 folds. This would account for the different amounts of crustal shortening in the various sectors (p. 133).

COMPARISON WITH OTHER AREAS OF LATE-CALEDONIAN FOLDING

The Walls Sandstone is unique among the Lower and Middle Old Red Sandstone sediments of the Northern Isles and Scotland in that it is the only sediment which:

(a) may have been laid down in a fresh-water flysch basin, rather than a fluviatile or shallow water lacustrine environment;
(b) has a total thickness of over 30 000 ft (9000 m) and
(c) has been involved in two periods of intense late-Devonian folding, which also affected the underlying metamorphic basement.

The first of these tectonic phases produced major folds the trend of which is west-south-west. This is virtually at right angles to the Caledonoid trend of the metamorphic rocks in the greater part of eastern Shetland and to the strike of the steeply inclined Middle and Upper Old Red Sandstone sediments of south-east Shetland, which is more or less north–south.

East to east-north-east trending basins of Lower and Middle Devonian sediments occur along the west coast of Norway between Trondheimfjord and Sognefjord (Holtedahl 1960, pp. 285–93; Brynhi 1963; Nilsen 1968). These basins contain great thicknesses of strata, 20 to 25 km having been estimated by Kolderup (1927, p. 41) for the Hornelen basin, though Brynhi (1963, p. 24; 1964, p. 385) considers that the centre of deposition within this basin was displaced progressively eastwards and that the true thickness of sediments at any point in the basin may not exceed 5000 m. Estimated thickness for some of the other basins are as follows: Håsteinen 1000 m+ (Kolderup 1927, p. 28), Solund 5200 m (Nilsen 1968, p. 14). The sediments filling these basins are fluvial deposits, consisting of conglomerate, sandstone and basal breccias. It is believed that the present outcrops are separate depositional basins and that the Nordfjord–Sognefjord area must have undergone vertical tectonic movements during the Devonian period, producing block-faulted areas and east–west trending graben structures with marked intermittent uplift of the adjoining areas (Nilsen 1968, pp. 98–9). The belief that the Old Red Sandstone basins were involved in late-Devonian large-scale thrusting (Holtedahl 1960) has not been substantiated by later research, but evidence obtained from an outcrop of Devonian rocks near Röros in Eastern Norway indicates that in this area Devonian strata have been strongly folded together with the metamorphic basement (Holmsen 1963). The Devonian earth movements in Norway have been attributed to the last phase of the Caledonian Orogeny, which has been termed the Svalbardian Phase by Vogt (1928).

Earth movements of Middle and Upper Devonian date have also affected the Old Red Sandstone sediments of East Greenland (Haller *in* Raasch 1961, pp. 179–85; Bütler 1959, pp. 179–81). In the Moskusoksefjord area Bütler distinguished five phases of earth movement during the Middle and Upper Devonian. Some of these were essentially differential vertical block movements, but the later phases also involved the thrusting of basement blocks over the Old Red Sandstone sediments and the local folding of the sediments. Haller showed that over a greater part of East Greenland the Devonian earth movements were first tensional, producing an extensive graben-like basin bounded by north-north-west trending step-folds. The fracturing was followed by compressional phases, which gave rise to two fold belts trending respectively north-west and north-east, as well as to associated areas of superficial thrusting. As in Western Shetland, the fold belts are linked with the syntectonic intrusion of granitic magma.

In Norway the late-tectonic earth movements were not of the same intensity as those affecting the Walls Sandstone, but in Greenland there were local belts where the basement rocks were remobilized and rejuvenated by chemical and thermal action (Haller *in* Raasch 1961, p. 180). In the two areas both the tensional and compressional phases of the earth movement appear to have been confined to zones of limited extent. In Shetland the two phases of folding in the Walls Sandstone affected neither the lower Givetian Melby Formation nor the upper Givetian sediments of Eastern Shetland. Though the present near-juxtaposition of the three outcrops of Old Red Sandstone deposits in Shetland may have been brought about by lateral transcurrent movements along the Melby and Walls Boundary faults (p. 266), it must be assumed that the phases of folding recorded in the Walls Sandstone were either of very restricted areal extent or were completed before the deposition of the Melby Sandstone.

It is possible that the motive force for these localized tectonic episodes was provided by the forcible emplacement of the Western Shetland granite-diorite complexes. The extent of the folding within the Walls Sandstone, however, suggests that the folding is not causally connected with the granite emplacement. The folding may have been preceded by differential vertical block movement which was responsible for the formation of a restricted basin in which the great thickness of Walls Sandstone was deposited. A similar history is reported from Greenland, and there is a marked similarity between the Devonian tectonic history of East Greenland, West Norway and West Shetland, all of which record various aspects of the final Svalbardian phase of the Caledonian Orogeny.

References

BRYNHI, I. 1963. Relasjonen mellom senkaledonisk tektoniki og sedimentasjon ved Hornelens og Håsteinens devon. *Norg. geol. Unders.*, **223**, 10–25.

—— 1964. Migrating basins on the Old Red Continent. *Nature, Lond.*, **202**, 384–5.

BÜTLER, H. 1959. Das Old Red-Gebiet am Moskusoksefjord. *Meddr. Gronland*, **160**, Nr. 5, p. 188.

ELLISTON, J. 1963. Gravitational sediment movements in the Warramunga Geosyncline in WARREN, C. S. (Editor). *Syntaphral Tectonics and Diagenesis: A Symposium.* University of Tasmania, Hobart.

FLEUTY, M. J. 1964. The description of folds. *Proc. geol. Ass.*, **75**, 461–92.

HOLTEDAHL, O. 1960. (Editor). Geology of Norway. *Norg. geol. Unders. Småskr.*, **208**.

HOLMSEN, P. 1963. On the tectonic relations of the Devonian complex of the Røragen area, east-central Norway. *Norg. geol. Unders.*, **223**, 127–38.

KOLDERUP, C. F. 1927. Hornelens devonfelt. *Bergens, Mus. Årb.* 1926. *Naturvidensk. Raekke.*, Nr. 6.

KUENEN, PH. H. 1953. Significant features of graded bedding. *Bull. Am. Ass. petrol. Geol.*, **37**, 1044–66.

MAXWELL, J. C. 1962. Origin of slaty and fracture cleavage in the Delaware Water Gap Area, New Jersey and Pennsylvania. *Mem. geol. Soc. Am.* (Buddington Volume), 281–311.

NILSEN, T. H. 1968. The Relationship of Sedimentation to tectonics in the Solund Devonian District of south-western Norway. *Norg. geol. Unders.*, **259**, 1–108.

RAASCH, G. O. 1961. (Editor). *Geology of the Arctic.* Toronto.

RAMSAY, J. G. 1967. *Folding and Fracturing of Rocks.* New York.

TEN HAAF, F. 1956. Significance of Convolute Lamination. *Geologie Mijnb.*, **18**, 188–94.

VOGT, T. 1928. Den Norske Fjelkjedes revolusjons-historie. *Norsk. geol. Tidsskr.*, **10**, 97–115.

WILLIAMS, P. F., COLLINS, A. R. and WILTSHIRE, R. G. 1969. Cleavage and penecontemporaneous deformation structures in sedimentary rocks. *Jnl Geol.*, **77**, 415–25.

Chapter 10

OLD RED SANDSTONE:
MELBY FORMATION

INTRODUCTION

THE ROCKS of the Melby Formation occupy an area of approximately one square mile (2·6 sq. km) in the north-west corner of the Walls Peninsula (Plate XII, Fig. 16). They are separated from the sediments of the Sandness Formation and the Walls Peninsula metamorphic rocks by the north-east trending Melby Fault. For part of its course at least, this fault is a reversed fault. It is inclined at 60°–70° to the south-east at Hesti Geo, where it is bounded on its north-western side by a 300-yd (270-m) wide zone of intensely folded, sheared and somewhat indurated sediment. The possible transcurrent nature of this fault is discussed in Chapter 17.

Though there are a number of gaps in the shore section along the north coast both west and east of Ness of Melby, there is sufficient evidence (p. 147) to support the conclusion reached by both Finlay (1930, p. 676) and Knox (Summ. Prog. 1934, pp. 72–3) that the rocks cropping out north-west of the Melby Fault form a single structural and stratigraphic unit. If the large-scale faulting within the outcrop of the formation is confined to the narrow zone bounding the Melby Fault the sequence within the Melby Formation must be similar to that as shown in Fig. 17. This suggests that the total exposed thickness may be about 2500 ft (760 m). At the top of the formation there appear to be two groups of flows of silicified rhyolite or ignimbrite, the lower of which thins out southwards. It is possible, however, that the two outcrops of rhyolite are parts of the same group of flows repeated by a fault trending sub-parallel to the Melby Fault.

The lower part of the sequence contains two thick bands of pale grey sandy siltstone and shale with carbonate-rich ribs and nodules, which have yielded fish remains and abundant plant fragments and are known as the Melby fish beds. Though Knox (Summ. Prog. 1934, p. 73) considered that only one fish bed is present in the Melby area, the field evidence leaves no doubt that there are two fish beds separated by just over 300 ft (90 m) of predominantly arenaceous sediment. Watson (1934, pp. 74–6) compared the fish from Melby with the Middle Old Red Sandstone fish faunas of Orkney, Caithness and other parts of Scotland and concluded that one of the Melby genera, *Pterichthyodes*, occurs elsewhere only at the Achanarras horizon of Caithness and in the Stromness Beds of Orkney which contain the Sandwick Fish Bed (Wilson and others 1935, p. 14). Most of the other forms found at Melby are also present at these horizons. The Melby fish beds have thus been roughly equated with the Sandwick and Achanarras fish beds, which are thought to be of basal Givetian age (Westoll 1937, p. 38; 1951, table iii; Waterston *in* Craig 1965, p. 291).

Both Finlay (1930, p. 676) and Knox (Summ. Prog. 1934, pp. 72–3) believed that the two groups of rhyolite flows exposed at Melby are the equivalents of the rhyolites of Papa Stour and that the Melby sandstones are stratigraphically below the Papa Stour volcanic rocks. Flinn and others (1968, p. 15) have, however, suggested that such a correlation is not justified and that there may be no stratigraphic connection between Papa Stour and the Mainland. The present author believes that the correlation of the Melby and Papa Stour volcanic rocks is a reasonable one.

The small islands in the Sound of Papa, Holm of Melby and Forewick Holm, are formed of sediments and igneous rocks which are considered by the author to belong to the Melby Formation. Forewick Holm consists of two flows of rhyolite separated by tuffaceous sandstone and is described in the chapter dealing with Papa Stour (p. 169). The Holm of Melby is formed of flaggy, in part calcareous sandstones, pebbly tuffaceous sandstones and one lenticular flow of basalt (p. 153). There is no direct evidence bearing on the stratigraphical relationships between these beds and the rocks of Papa Stour and Melby. They may form a link between the basalts and tuffaceous sediments underlying the Papa Stour rhyolites and the sediments below the Ness of Melby rhyolites.

STRATIGRAPHY

The sediments below the Melby rhyolites can be divided into the following two lithological groups (Fig. 17):

(a) A lower group approximately 600 ft (180 m) thick. This contains the two fish beds in a predominantly arenaceous, fluvial, sequence which was deposited by currents from the west or west-north-west.

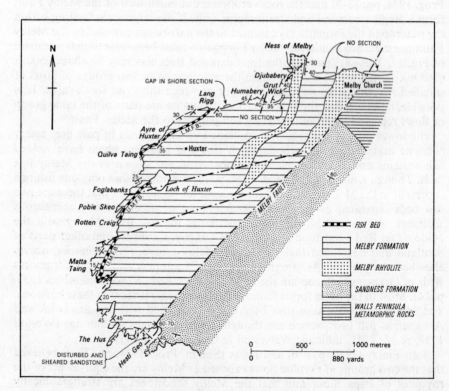

FIG. 16. *Geological sketch-map of the Melby Formation on Shetland Mainland*

(b) An upper group containing some thick beds of pinkish feldspathic sandstone with clasts of pink rhyolite and many plant fragments, as well as thick beds of purplish sandy siltstone. This group appears to have been deposited by currents from the east-north-east.

The two groups are separated by a gap in the shore exposures 610 yd (570 m) NE of Huxter. South-west of Huxter the boundary between the two groups is taken at the top of the Upper Melby Fish Bed.

STRATA BELOW LOWER MELBY FISH BED

The strata below the Lower Melby Fish Bed are exposed in an excellent, easily accessible section on the shore north of Huxter, between the Ayre of Huxter and Lang Rigg. The sequence is shown graphically in Fig. 18. Two units are present as follows:

1. *The Lower Unit* is 65 ft+ (20 m+) thick and composed of soft reddish brown-weathering medium-grained feldspathic sandstone with thin subordinate siltstones. The cross-bedded sets in the sandstone range in thickness from 2 to 10 ft (0·6–3 m). Cross-bedding is predominantly planar and most foresets are inclined to the east-south-east (Plate XVI, A). Trough-cross-bedded sets are rare but such as there are have a marked ESE–WNW alignment of trough axes. The foresets are strongly laminated, with individual laminae etched out by weathering and emphasized by the presence of thin pale greenish bands. Carious weathering is well seen at certain horizons. Ripple-marked fine-grained sandstones and siltstones with sand-filled sun cracks are rare. The only thick bed of fine-grained sediment occurs 45 ft (14 m) below the top of the unit and consists of a pale purple siltstone up to 2 ft 2 in (0·67 m) thick, which has an irregular eroded top and is overlain by 1 ft (0·3 m) of intraformational conglomerate. The greater part of this unit consists of medium- to coarse-grained channel-fill sandstone, beds of finer-grained sediment, probably representing overbank deposits, having been largely removed by channel scour.

2. *Upper Unit.* The reddish sandstone passes abruptly upward into pale buff, ochre-weathering medium- to coarse-grained calcareous sandstone. At the Ayre of Huxter the latter is 30 ft (9 m) thick and contains planar cross-bedded sets ranging in thickness from 2 ft 6 in to 7 ft (0·75–2·1 m). The cross-bedding near the base and top of the unit is contorted with the lamination in the upper third to half of the set forming semi-recumbent convolute folds with sharply crested anticlines and rounded synclines. These folds are invariably overturned to the east. The cross-bedded sandstone passes up into finer-grained, generally unlaminated, sandstone alternating with thin beds of silty shale and sandy siltstone with ripple cross-lamination, ripple-marked bedding planes and some thin beds with convolute lamination, which in some cases is truncated by the overlying planar-bedded sediment. The outcrop of the Upper Unit is repeated on the shore at Lang Rigg, where there is a marked increase in the number of siltstone partings and where sets with disturbed cross-bedding are rare.

LOWER MELBY FISH BED

Though there is considerable lateral variation within the Lower Melby Fish Bed, which crops out at the Ayre of Huxter and at Lang Rigg, the following three lithological units can be recognized throughout (see Fig. 18):

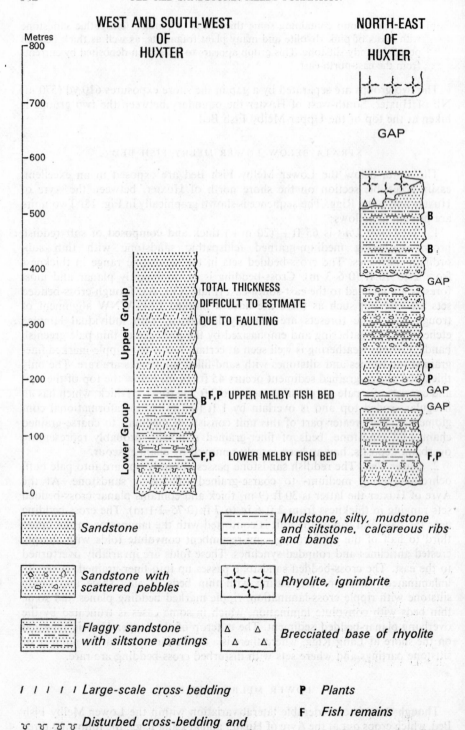

FIG. 17. *Melby Formation: probable successions*

1. A lower unit, 9 to 10 ft (2·7–3 m) thick composed of evenly laminated pale grey siltstone and silty mudstone with thin sandy ribs, particularly at the base. This is characterized by the abundance of sand-filled desiccation cracks. Plant remains are present at Lang Rigg.

2. A central unit 10½ ft (3·2 m) thick composed of pale to dark grey or black, locally bituminous, fissile shale alternating with unlaminated mudstone with irregular carbonate-rich ribs and nodules. Nodules are up to 1 in (25 mm) high and 3 in (75 mm) long. Ribs do not exceed 1·4 in (35 mm) in height and are up to 6 ft (2 m) long. The shale passes down into silty shale with fewer dark carbonate-rich ribs. The sediment throughout is intensely sheared and jointed, and a number of joint planes contain a black bituminous substance. Plant remains are abundant throughout this unit, fish remains are mainly from the calcareous nodules. Thin veinlets of fibrous carbonate traverse the lower part of the sequence.

3. The upper unit is up to 15 ft (4·5 m) thick and has a variable lithology composed mainly of interlaminated siltstone, sandy siltstone and, near the base, mudstone and silty mudstone. There are also lenticular masses of sandstone and sandy siltstone. Certain horizons contain finely disseminated plant debris and some plant stems aligned in an E20°S direction.

The lithological pattern within the Lower Melby Fish Bed shows some resemblance to the pattern within the lacustrine phases of the cycles in the Stromness Flags of Orkney (Fannin 1970) and the Caithness Flagstones (Crampton and Carruthers 1914, pp. 91–7). No strict parallels can however be drawn. The sun-cracked, laminated sediments of the lower unit appear to represent shallow-water mud-flat deposits which were periodically exposed to the atmosphere and dried out. The base of the central unit may mark the sudden onset of deeper water lacustrine conditions, when thinly laminated sediments with alternating carbonate-rich and bituminous laminae were laid down. Deposition may have taken place in a lake with temperature distribution and water circulation controlled by a tropical climatic regime (see Rayner 1963; Fannin 1970, pp. 228–31; Bradley 1929). The carbonate concretions which enclose many of the fish remains at Melby are, however, less common in the Orkney and Caithness fish beds. The upper unit with its thin lenticular masses of sandstone, which may represent small sand-filled channels, may indicate a period when the lake was again becoming shallower and the influx of clastic material brought into the lake by river and stream deltas was on the increase.

At the Ayre of Huxter the Lower Melby Fish Bed is affected by a number of minor east-north-east trending folds, which vary from open synclines at the north-eastern end of the exposure to recumbent folds locally passing into small thrusts. There are also a number of minor north-north-east trending vertical faults and some thrusts along the bedding planes.

STRATA BETWEEN LOWER AND UPPER MELBY FISH BEDS

West shore. The 270 to 300-ft (80–90-m) of predominantly arenaceous beds between the lower and upper fish beds are exposed in the cliffs between Quilva Taing and Pobie Skeo. Four groups of strata with distinct characteristics have been recognized:

1. *The lowest group* consists of about 45 ft (14 m) of flaggy sandstone with fairly closely spaced siltstone partings and some thick intercalations of siltstone

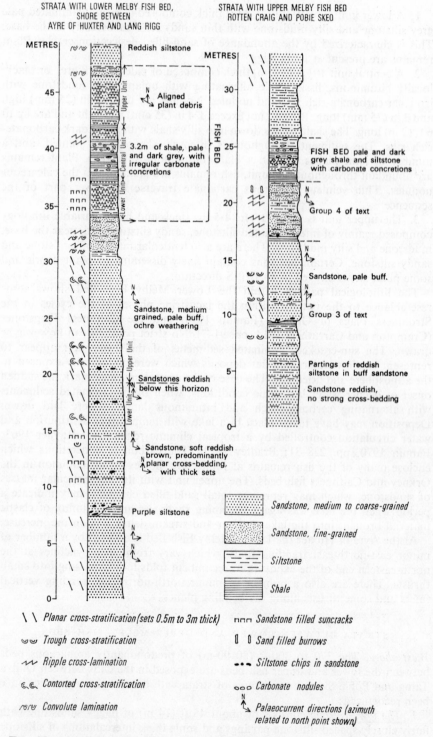

FIG. 18. *Sections of the Lower and Upper Melby Fish Beds and adjacent strata*

and sandy siltstone. At Quilva Taing the 6 ft (2 m) of strata just above the fish bed are convoluted into semi-recumbent folds which are up to 3 ft (0·9 m) high, have north-west trending axes and are overturned to the north-east. South of Quilva Taing the flaggy beds pass upwards into fine-grained sandstone with posts 2 to 3 ft (0·6–0·9 m) thick separated from each other by thin purplish siltstone partings. A number of these sandstone beds are partially reddened. Convolute lamination is common in the finer-grained sediments of this group.

2. *The second group* consists of about 180 ft (55 m) of reddish brown, medium-grained sandstone with fairly thick predominantly planar cross-bedded sets, and relatively few siltstone partings.

3. *The third group* is 50 to 55 ft (15–17 m) thick and consists mainly of buff medium- to fine-grained generally planar-bedded sandstone. This contains some sets with planar cross-bedding, in which the foresets dip consistently towards the east to east-north-east sector. In the lower part individual sets are separated by lenticular beds of purple siltstone and silty mudstone, some of which reach a thickness of 4 ft (1·2 m).

4. *The highest group* is 15 ft (4·5 m) thick and is composed of flaggy sandstone interbedded with finer sediment. Ripple cross-lamination is common throughout. Beds of siltstone and shale up to 2 ft 6 in (0·75 m) thick contain calcareous nodules similar to those in the overlying fish bed. Near the top of the unit small near-horizontal sand-filled tubes are present in certain sandy siltstone laminae. Individual tubes are round in cross section, approximately 2 mm wide, straight or slightly curved, and up to 20 mm long. These tubes may have been formed by burrowing invertebrates.

North Shore. On the north coast, between Lang Rigg and Humabery, the Lower Melby Fish Bed is overlain by an incomplete sequence approximately 250 ft (75 m) thick, composed mainly of medium-grained brownish-weathering sandstone interbedded at intervals with thick beds of siltstone and flaggy sandstone. Near the middle of the sequence these fine-grained sediments contain macerated plant debris. They are generally poorly laminated. Some beds have ripple-marked surfaces, convolute lamination and local signs of bioturbation. Beds of sandy siltstone with ripple cross-lamination and convolute lamination are also present at the base of the sequence. A feature of the sandstones is the gradual increase in the percentage of feldspar grains as the sequence is ascended. Trough- and planar-cross-bedded sets up to 2 ft (0·6 m) thick are scattered throughout this predominantly planar-bedded sequence. The cross-bedding suggests that current directions were still predominantly from the west.

The beds in the northern outcrops have the following features which are not seen further west:

(a) A slightly higher proportion of fine-grained sediment, but a complete absence of red siltstone and mudstone partings.
(b) A higher proportion of pink feldspar grains in the sandstones.

UPPER MELBY FISH BED

The Upper Melby Fish Bed is almost continuously exposed along the cliff-top between Foglabanks and Rotten Craig. The fish-bearing carbonate-rich shales and flags exposed on the south shore of Matta Taing, 1600 yd (1460 m) SW of

Huxter, appear to be the most southerly representatives of this fish bed. The sequence of the fish bed at Pobie Skeo is as follows (see Fig. 18):

	ft	in	m
Flaggy sandstone with siltstone partings..	—	—	—
Silty shale, pale grey, evenly bedded with carbonate-rich ribs and nodules. Fish remains in calcareous bands..	4	0	1·22
Shale, dark grey, thinly bedded, somewhat contorted; some calcareous bands are nearly black and slightly bituminous; carbonate nodules with fish remains throughout .. .. 1 ft 6 in to 2		0	0·45–0·6
Siltstone and sandy siltstone, slightly calcareous with thin ribs and elongated nodules of carbonate ..	5	0	1·52
Shale and silty shale, dark grey, calcareous, with ribs of limestone .. .. 1 ft 6 in to 2		0	0·45–0·6
Siltstone, pale grey, evenly bedded, slightly calcareous .. .. 6 in to 0		9	0·15–0·22
Flaggy sandstone with thin bands of sandy siltstone with ?invertebrate burrows (p. 145)	—	—	—

As in the lower fish bed this sequence varies somewhat along the strike, and lenticular masses of sandstone are locally present.

STRATA ABOVE UPPER MELBY FISH BED

The sequence of strata above the Upper Melby Fish Bed cannot be readily determined. Along the south crop between Matta Taing and Hesti Geo the strata are cut by a number of minor faults and close to Hesti Geo they are affected by intense irregular folding and faulting. On the north shore between Lang Rigg and Ness of Melby there is a 60 yd (55 m) wide gap in the shore section at the probable horizon of the Upper Melby Fish Bed. This gap may contain one or more faults. It is not possible to correlate the northern and southern exposures of this group.

South Crop

The strata exposed between Matta Taing and Hesti Geo consist almost entirely of brownish feldspathic medium- to fine-grained sandstone with some large-scale cross-stratification in its lower part. On the east face of The Hus a 5 to 6 ft (1·5–1·8 m) thick bed of grey, probably calcareous shale and siltstone is interbedded with the reddish sandstone. As the Melby Fault is approached the sandstone becomes progressively more indurated and, close to the fault, the bedding is largely obliterated. As this shore section is virtually inaccessible, no plant remains or rhyolite fragments have been recorded.

EXPLANATION OF PLATE XVI

A. South shore of Sound of Papa, 350 yd (320 m) N5°E of Huxter [175 574]. Planar cross-bedded purple sandstone underlying Lower Melby Fish Bed in Melby Formation. (D 891).

B. & C. South shore of Sound of Papa, 150 yd (140 m) W of Melby House (184 577). Purple sandy siltstone with irregular buff sandstone laminae with ?bioturbation structures. (W.M.).

A

B

C

North Crop

Between the gap in the exposures along the north coast, 600 to 650 yd (550–600 m) NE of Huxter, where the Upper Melby Fish Bed would be expected to crop out, and the point where the Melby Fault cuts the coast, 700 yd (640 m) E of Melby, the exposed section of sediments and rhyolites has two major gaps, breaking the exposed sequence into three blocks (Fig. 16).

Humabery Shore. The most westerly block is exposed over a distance of 430 yd (400 m) along the coast at Humabery. It contains a lower group of relatively soft pebbly sandstones and an upper mixed group.

1. The lower group is approximately 270 ft (80 m) thick and is made up of feldspathic sandstone with a matrix of ochreous or purple clay. The sandstone contains pebbly bands and lenses with clasts of pink rhyolite and some quartz, and chips of greenish siltstone and mudstone. The pebbly beds are most abundant in the lowest and highest 70 ft (20 m) of the sequence. Near the top, the sandstone contains some thin pink tuffaceous laminae. Plant fragments are present throughout, and are particularly abundant near the base of the group where they are preserved as coaly ribs up to 4 in (10 cm) long and $\frac{3}{8}$ in (10 mm) wide. Cross-bedded sets up to 3 ft (0·9 m) thick are common throughout the sequence and their foresets dip consistently to the west to south-west. Sets with disturbed cross-bedding occur near the middle of the sequence and there are a number of irregular dome-shaped structures which may have originated as sand volcanoes.

2. The upper mixed group of sediments is approximately 250 ft (75 m) thick, and contains beds of reddish purple siltstone and silty mudstone with irregular sandy laminae. These beds are 20 to 30 ft (6–9 m) thick, and they alternate with beds of pink to reddish purple, fine-grained, predominantly planar-bedded sandstone. Pebbly sandstones containing scattered pebbles of rhyolite are present at a number of horizons. There are vague suggestions of bioturbation in some of the purple sandy siltstones and flaggy sandstones. Convolute lamination is present in some of the purple siltstones.

Ness of Melby and Djubabery. The Ness of Melby Rhyolite is underlain by the sediments which form the reefs at Djubabery. Approximately 350 ft (100 m) of strata are exposed, and the sequence is as follows:

	ft	m
Irregular base of Melby Rhyolite		
7. Sandstone, grey, micaceous, predominantly ripple-laminated, with lenses, up to 4 in (10 cm) thick, of pebbly sandstone with rhyolite clasts	20 to 40	6–12
6. Purplish brown sandy siltstone with disturbed laminae, ?bioturbated	6	2
5. Sandstone, grey, predominantly planar-bedded, with scattered pebbles of pink rhyolite and lenses of rhyolite grit. Some silty laminae with symmetrical ripple marks	150	45
4. Purplish brown sandy siltstone with highly disturbed laminae of fine-grained sandstone and siltstone. The sandstone laminae have lobate bases and there are detached tubes of sandstone in the siltstone (Plate XVI, figs. B and C). These structures may be invertebrate burrows	80	25

L

	ft	m
3. Flaggy sandstone, consisting of alternate bands of sandstone and sandy siltstone 2 ft (60 cm) thick. Small-scale cross-lamination throughout	30	9
2. Purple sandy siltstone, evenly bedded, composed of alternate laminae of purplish grey, fine-grained sandstone and deep purple silty shale	25	7·5
1. Sandstone, hard, fine- to medium-grained with pebbles of rhyolite up to 4 in (10 cm) in diameter and pebbles of basic lava	40	12

MELBY RHYOLITES

Ness of Melby. The base of the rhyolite forming the Ness of Melby is highly irregular and the underlying sandstone is distorted by blocks and tongues of rhyolite which have pushed downwards into unconsolidated sediments for distances of up to 2 ft (60 cm). The basal 15 to 20 ft (4·5–6 m) of the rhyolite is brecciated, consisting of rhyolite blocks up to 2 ft (60 cm) long, set in a pink rhyolite matrix. Both the breccia and the overlying rhyolite contain irregular cracks and cavities filled with a greenish sandy sediment. The estimated maximum thickness of the Ness of Melby rhyolite is 230 ft (70 m), and there is some inconclusive evidence that it may consist of several flows. The rhyolite is purple when fresh and pale pink, greenish, brownish ochre or mottled on weathered surfaces. It contains relatively sparse pink euhedral phenocrysts of feldspar and, in many exposures, is strongly banded, the banding being in many instances near-vertical and perpendicular to the inclination of the flow. The top 10 to 12 ft (3–4 m) of the rhyolite consist of breccia composed of fragments and blocks of rhyolite ranging up to 3 ft (0·9 m) in size set in a sandy matrix. In its upper 5 ft (1·5 m) the rhyolite contains numerous small vesicles lined with crystalline quartz and a dark green mineral and filled with calcite.

Only a few feet of sediment are exposed above the Ness of Melby rhyolite. These consist of fine- to medium-grained planar-bedded sandstone with scattered rhyolite clasts up to ½ in (12 mm) in diameter.

Melby Church. The shore exposures north of Melby Church are composed entirely of strongly banded, variably porphyritic rhyolite. As there is no indication of the dip of this rhyolite no estimate of its thickness can be made. In a number of exposures the rhyolite has a pyroclastic aspect, with small rounded clasts of darkish rock set in a rhyolitic matrix.

FAUNA

Additional fish remains to those recorded by Watson (1934) were collected from the Melby Fish Beds by Mr. P. J. Brand in 1966. Some of the original specimens and the later collection have been identified by Dr. R. Miles who recorded the following forms:

Upper Melby Fish Bed:

Matta Taing, 1540 yd (1400 m) S37°W of Huxter [166 561]:
   Cheiracanthus sp., *Coccosteus cuspidatus* Miller *ex.* Agassiz M.S., *Gyroptychius?*, *Mesacanthus sp.*

Coast north of Matta Taing, 1140 yd (1035 m) S42°W of Huxter [167 563]:
   Coccosteus cuspidatus, *Mesacanthus sp.*

Pobie Skeo, 810 yd (610 m) W40°S of Huxter [168 567]:
   Cheiracanthus sp., ?Coccosteus cuspidatus, Homostius milleri Traquair.

Lower Melby Fish Bed:

Quilva Taing, 450 yd (410 m) W6°N of Huxter [171 572]:
   Coccosteus cuspidatus, Dipterus valenciennesi Sedgwick and Murchison,
   Gyroptychius agassizi (Traill).
Ayre of Huxter, 290 yd (265 m) W16°N of Huxter [173 573]:
   Coccosteus cuspidatus, Dipterus valenciennesi, Glyptolepis cf. *leptopterus*
   Agassiz, *Gyroptychius agassizi.*
Lang Rigg, 490 yd (440 m) N34°E of Huxter [177 576]:
   Coccosteus sp., Pterichthyodes sp.

Dr. Miles agrees with the conclusion reached by Watson (p. 139) that this fauna can be equated with that of the Sandwick Fish Bed of Orkney and the Achanarras Limestone of Caithness. The age of this part of the Melby Formation is thus most probably basal Givetian.

PETROGRAPHY

SANDSTONES

The sandstones of the Melby Formation are poorly graded arkoses which contain up to 40 per cent feldspar clasts, a small, but variable proportion of lithic clasts and, in most cases, virtually no detrital matrix.

Lower Group. The reddish brown sandstone below the Lower Melby Fish Bed (Fig. 18) is a relatively soft, poorly graded arkose (S 49338, 50674, Plate XVII, fig. 1) composed of laminae of coarse-grained relatively quartz-rich sandstone alternating with laminae of finer-grained, feldspathic sandstone. The former have a strongly bimodal grain-size frequency, with large rounded grains within the size range 1·3 to 0·5 mm, composed of 80 per cent quartz and 20 per cent feldspar grains, set in a groundmass of smaller angular grains averaging 0·15 mm in size. The medium-grained sandstone and the finer laminae which alternate with laminae of coarse-grained sandstone are composed of subrounded to subangular grains, ranging from 0·6 mm to 0·07 mm in diameter with a quartz-feldspar clast ratio ranging from 60:40 to 50:50. As in the case of all other sandstones of the Melby Formation the feldspar clasts contain a high proportion of slightly cloudy to fresh untwinned potash-feldspar, up to 20 per cent sodic plagioclase, a small proportion of very fresh microcline and subordinate microperthite. There appears to be no significant variation in the percentage distribution of the various types of feldspar in the sandstones throughout the group. Lithic clasts form 5 to 10 per cent of the total volume and are composed mainly of unstained silicified felsite and altered basic lava. Heavy mineral grains are never abundant. They include garnet, which is always present, apatite, tourmaline, zircon and allanite. The matrix is composed of a thin film of limonite-stained argillaceous material which incompletely mantles the grains.

The buff-coloured sandstone (S 50673) which lies between the reddish brown sandstone and the Lower Melby Fish Bed does not have the alternating coarse and fine laminae of the underlying sandstone. It is a relatively well-graded arkose with subrounded to well-rounded grains and a grain diameter ranging up to 0·8 mm. The quartz-feldspar ratio is 60:40, with quartz forming most of the larger grains. Muscovite is abundant and 5 per cent of the clasts are composed of fine-grained argillaceous sediment. The other lithic clasts are silicified felsite.

Of the sandstones between the two Melby Fish Beds only two thin sections have been examined (S 51501, 49339). Both specimens are medium- to fine-

---

EXPLANATION OF PLATE XVII

PHOTOMICROGRAPHS OF MELBY FORMATION AND PAPA STOUR VOLCANIC ROCKS

FIG. 1.   Slice No. S 49338. Magnification × 16. Crossed polarisers. Pink medium-grained sandstone below Melby Fish Bed, Melby Formation. Feldspathic sandstone with bi-modal grain size distribution. Quartz-feldspar ratio 70:30. Among large subrounded grains quartz predominates. Accessory grains are garnet, zircon, tourmaline and apatite. Lithic clasts are composed mainly of altered acid lava and form less than 10 per cent of the total grains. Most grains are covered by a thin reddish film of iron ore. South shore of Sound of Papa, 340 yd (310 m) N of Huxter [174 575].

FIG. 2.   Slice No. S 30602. Magnification × 31. Crossed polarisers. Thin flow of basalt within tuff sequence in Melby Formation. Ophitic basalt with rare phenocrysts of sodic labradorite. Vaguely flow-aligned laths of calcic andesine are partly enclosed in ophitic augite. Matrix is a deep olive-green amorphous aggregate. Holm of Melby, west coast [191 586].

FIG. 3.   Slice No. S 54285. Magnification × 32. Plane polarized light. Poorly welded or non-welded tuff near base of Ness of Melby rhyolite. Partially flattened devitrified glass shards and small potash feldspar plates and laths, set in matrix of microlite rods. North-west corner of Ness of Melby, 240 yd (220 m) NW of Melby House [185 580].

FIG. 4.   Slice No. S 30944. Magnification × 31. Plane polarized light Coarse ophitic dolerite with plates of cloudy plagioclase set in ophitic pyroxene. Papa Stour, 560 yd (500 m) SSE of Skerry of Lambaness, 1850 yd (1690 m) NW of Gardie [167 620].

FIG. 5.   Slice No. S. 30930. Magnification × 32. Crossed polarisers. Spherulitic rhyolite. Spherulites are composed of radiating fibres of brownish-stained potash feldspar. Small patches of quartz between adjoining spherulites. Papa Stour, south shore of West Voe, 550 yd (500 m) NW of Gardie [176 612].

FIG. 6.   Slice No. S 30931. Magnification × 32. Plane polarized light. Spherulitic rhyolite. Spherulites composed of tightly packed clusters of irregularly radiating laths of orange stained potash feldspar, set in large interstitial areas of clear quartz. Quartz forms a small central nucleus in some spherulites. Papa Stour, east shore of West Voe, 920 yd (840 m) NW of Gardie [175 616].

FIG. 7.   Slice No. S 30933. Magnification × 31. Crossed polarisers. Spherulitic rhyolite showing two contrasting types of spherulites. The large spherulites consist of radiating fibres of quartz and potash feldspar and are set in a groundmass of small near-spherical spherulites of consistent size (with black cross). Papa Stour, Doun Hellier, 1220 yd (1100 m) NNE of Gardie [183 620].

FIG. 8.   Slice No. S 30962. Magnification × 16. Plane polarized light. Porphyritic rhyolite, with stumpy euhedral plates of slightly kaolinized potash feldspar, set in an irregular banded matrix of microlites of orange-stained potash feldspar and irregular patches of quartz. Papa Stour, south-west coast, close to Shepherd's Geo, 800 yd (730 m) SW of Bragasetter [165 592].

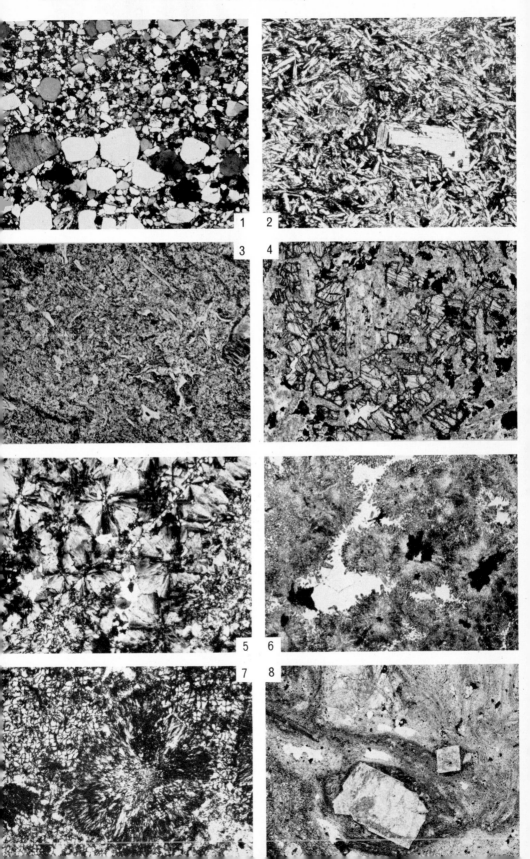

grained arkose (quartz-feldspar ratios 50:50 and 60:40) with grains ranging from 0·45 to 0·1 mm. Grains are predominantly subrounded in S 49339 and angular in S 51501, and in both cases there is slight induration with grain margins in places interlocking. In both cases matrix is virtually absent.

Upper Group. The sediment above the Melby Fish Bed exposed on the coast north-east of Huxter is characterized by the presence of pebbly bands with clasts of fine sediment and pink felsite. The thin section examined (S 50672) is a medium-grained poorly graded, but not bimodal, arkose, in which the larger grains are well rounded and the smaller grains angular. Grain size ranges from 0·7 to 0·14 mm. The quartz-feldspar ratio is 55:45 and most quartz grains contain liquid inclusions similar to those in the quartz of the Sandsting Granite. Lithic clasts form approximately 20 per cent of the total and are composed of unstained felsite and fine-grained sediment. Of accessory minerals well-rounded grains of garnet are relatively abundant; tourmaline and apatite are rare. As in the under-lying sandstone detrital matrix and cement are virtually absent. The sediment is fairly indurated and some grains have serrate margins.

The sandstone exposed along the south-west coast between Matta Taing and Hesti Geo (S 52744) shows evidence of induration as well as mechanical deformation. The quartz grains have undulose extinction and irregular conchoidal cracks and adjacent quartz grains have serrate margins. The quartz-feldspar ratio of this sandstone is 65:35 and there is a higher proportion of microperthite grains than in the other sandstones of the Melby Formation. Lithic clasts, mainly of felsite, form up to 10 per cent of the total volume, and large garnet grains are relatively abundant. As in the other sandstones of the Melby Formation the argillaceous matrix forms only a minute portion of the total volume.

RHYOLITES

The petrographic characters of the rhyolites from Melby have been referred to by Flinn and others (1968, p. 16) who state that they show vague signs of being an accumulation of fragments of welded tuff and that in one specimen there are abundant quartz phenocrysts, which are rare in the rhyolites of Papa Stour. They also mention one specimen, which, they consider, might be an altered and reddened very fine-grained andesite.

Only a small number of rhyolites have been examined by the author, and no assessment of the mode of formation of the entire thickness of the two 'flows' can thus be made. Only the specimen collected from the lower part of the Ness of Melby Rhyolite has the vitroclastic texture of welded tuff (S 54285). This rock contains abundant phenocrysts of euhedral patchily red-stained potash feldspar as well as subrounded xenoliths of non-silicified rhyolite and baked sediment, set in a matrix composed of only partially flattened glass shards, many with a characteristic Y-shaped outline, and some larger pumice fragments. The latter contain spherical or slightly flattened vesicles (Plate XVII, fig. 3). Under crossed polars the matrix is completely silicified into a fine mosaic aggregate of silica. Axiolitic and spherulitic textures are absent. In addition to the silicified pumice fragments there are also a number of highly elongated fragments, up to 10 mm long, of less silicified pumice, many of which have the ragged, serrated outlines characteristic of 'fiamme'. The subangular xenoliths of red-stained rhyolite are not completely silicified, but do contain irregular patches of silica. They are irregularly banded and contain an anastomosing network of opaque microlite rods.

The specimens from close to the top of the Ness of Melby Rhyolite (S 29972, 30582) have no vitroclastic textures. They are in part strongly porphyritic with patchily red-stained euhedral to subhedral phenocrysts of potash feldspar. Quartz phenocrysts have not been recorded. The matrix consists in one case (S 30582) of strongly flow-aligned microlites of hematite-dusted feldspar, together with irregular patches composed of a fine-grained mosaic of quartz. Though the rock appears to be spherulitic in hand specimen, there are no true spherulites made up of radiating fibres. The spherical red-stained masses have no consistent internal structure and they are separated from each other by irregular sheaths of mosaic-quartz.

A specimen from the Melby Church outcrop (S 29971) contains both feldspar and quartz phenocrysts set in a partially silicified matrix, which contains no textures that would suggest a welded tuff origin. Many of the quartz crystals are partially corroded.

CONDITIONS OF DEPOSITION

The sediments of the Melby Formation appear to be predominantly of fluvial origin. In the lower half of the sequence, up to the Upper Melby Fish Bed, the foresets of most planar cross-bedded sandstones dip predominantly to E10°S and there is a marked east-south-easterly trend of the axes of trough-cross-bedded units. This suggests a prevalent current direction from the west or west-north-west. In the upper part of the sequence the pebbly sandstones have cross-bedded units with an equally consistent dip of foresets to the west-south-west, suggesting that the direction of the currents depositing these sediments was from the east-north-east. The change in current direction coincides with the appearance of abundant pebbles of pink rhyolite and rarer pebbles of basic lava in the sediment, and it is suggested that the topography and drainage pattern of the area was suddenly altered by outpourings of volcanic rocks and by associated earth movements which created an area of high relief to the east-north-east.

A high proportion of the sediment below the Upper Melby Fish Bed consists of medium-grained, predominantly planar-cross-bedded sandstone with a uniform inclination direction of the foresets and a low percentage of fine-grained sediments. The sedimentary structures and petrographic characters (pp. 141–5) of the sandstones suggest that they were largely laid down in the channels of straight or braided rivers (Allen 1965, pp. 164–5). Most of the original overbank deposits appear to have been eroded by the migrating channels, leaving only thin lenticular masses of purple siltstone, but providing material for the isolated lenses of intraformational conglomerate and the scattered siltstone clasts within the sandstones. The strata just below both fish beds and a proportion of the beds between them are fine-grained flaggy ripple-marked sandstones interbedded with siltstones and shales. These beds were laid down by relatively slow-moving currents, probably in river channels and crevasses and as overbank deposits on an alluvial plain.

The two fish beds show some resemblances to the lacustrine phases of the Orkney (Fannin 1970) and Caithness (Crampton and Carruthers 1914) flagstone cycles (see p. 143) and they were probably deposited in shallow, but extensive lakes. Fannin (1970) has shown that the lacustrine beds of the Sandwick Fish

Bed Cycle are very much thicker than the equivalent beds within the other cycles in the Stromness Flagstones of Orkney and that in western Orkney the Fish Bed lake may at one stage have had a depth of 164 ft (50 m). The Sandwick Fish Bed therefore represents a major transgression and it is just possible that one of the Melby fish beds was laid down in the same extensive lake as the Sandwick and Achanarras Fish Beds. Such a concept assumes that the lower members of the Melby Formation are river deposits laid down near the north-western margin of the flat depositional basin which, towards its centre, contained the Orkney–Caithness lakes. The implications of this concept have a far-reaching effect on our understanding of the structural and palaeogeographical relationship of the Melby Sandstone to the other Old Red Sandstone deposits in the Shetland Islands. This relationship is discussed in Chapter 17.

The higher sediments of the Melby Formation exposed along the coasts of Humabery and Djubabery differ from those below the Upper Melby Fish Bed in that the sandstones contain pebbly beds with clasts of acid and basic lavas, abundant chips of siltstone and large fragments of coaly plant material. The sandstones are interbedded with considerable thicknesses of reddish brown sandy siltstones, some of which show evidence of extensive disturbance, either by burrowing animals or by the mechanical break-up and foundering of sandy laminae. If the beds cropping out on this shore are the equivalents of those exposed along the west coast between Matta Taing and The Hus, there is also a considerable lateral facies change within the group. These data suggest that the rivers which deposited the sandstones were fairly short and had relatively steep profiles. They carried angular clasts of the newly erupted lavas and deposited them in alluvial fans near the foot of the volcanic hills which appear to have lain to the north-west of the present area. Such hills may have been formed by the earliest members of the extrusive rocks which later formed the Papa Stour–Eshaness Volcanic Series. The formation of this high ground could be responsible for the diversion and possible ponding of the old established drainage system, and the thick beds of purplish brown siltstone and sandy siltstone may have been laid down either as overbank deposits or in ephemeral lakes within the ponded areas. The presence of large pieces of plant debris in the pebbly sandstones could suggest that the newly formed volcanic areas provided a fertile environment for the growth of large Old Red Sandstone plants.

HOLM OF MELBY

The Holm of Melby (Fig. 19) is a small island in the Sound of Papa, 650 yd (590 m) NE of the Ness of Melby, with a geological structure which does not tie in with the structural pattern of the sediments and lavas of the Melby area. It may be separated from the latter by an east–west trending fault.

The sequence, as seen on the west and north coasts of the island, consists of a basal series of pale flaggy sandstone interbedded with and underlain by beds of pale grey carbonate-rich siltstone and silty mudstone. This is succeeded by a series composed mainly of tuffaceous sandstone with scattered clasts of acid lava and with thin lenses of rhyolitic agglomerate and lapilli-tuff. The tuffaceous sandstone contains several thin flows and one fairly thick (up to about 35 ft [10 m]) lenticular flow of basic lava. The latter forms a prominent west-north-west trending ridge in the centre of the island but thins out south-eastwards. In

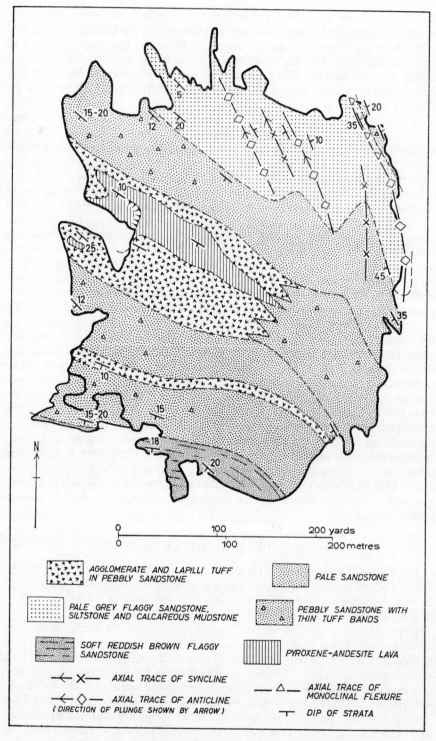

FIG. 19. *Geological sketch-map of the Holm of Melby*

the upper part of the sequence the tuffaceous sandstone becomes reddish brown, fine-grained and in part thinly bedded with reddish silty laminae.

The lapilli-tuff bands within the tuffaceous sandstone are water-deposited and usually have strongly erosive bases. They consist of clasts of pink to purplish felsite or rhyolite, up to $\frac{1}{2}$ in (12·6 mm) in size, together with rarer clasts of dark fine-grained or glassy lava and greenish-weathering fine-grained sediment. The clasts are, in some cases, strongly imbricated and are invariably set in a matrix of fine-grained greenish sandstone, which normally forms less than 20 per cent of the total volume.

The tuff bands and igneous clasts become less abundant and largely disappear south-eastward along the strike, indicating that the source of igneous material must have lain to the north or north-west. It is possible that the upper part of the Holm of Melby sequence is the stratigraphic equivalent of the basalt and tuffaceous sandstone underlying the rhyolites of Papa Stour (p. 158), and the pebbly sandstone beneath the rhyolite of Melby.

The sediment exposed along the north-east side of the Holm of Melby is folded into a series of gentle folds plunging gently to the north-north-west (Fig. 19).

Petrography. The lapilli-tuff (S 54286) is composed of subrounded clasts of pink to purplish acid rock, ranging up to $\frac{1}{2}$ in (12·7 mm) in length, with sub-ordinate clasts of altered fine-grained basic lavas, set in a matrix which does not exceed 20 per cent of the total volume. The matrix is composed in part of calcite, and, in part, of fine-grained sandstone with angular quartz grains. The acid clasts are composed of rhyolite and sub-acid lavas, with a very wide range in texture and grain size. The finest grained and most common types are flow-banded patchily silicified, red-stained, sparsely porphyritic rhyolite containing very fine aggregates of feldspar microlites and minute rods of iron ore. Many of the rhyolite clasts are partly or completely silicified and a number have spherulitic texture. Some rounded clasts are single spherulites of red-stained rhyolite. Less abundant clasts include microgranite composed of feldspar laths set in a clear quartz matrix, sub-basic lava composed entirely of flow-orientated feldspar laths set in a turbid matrix, and altered basic lava.

The basic lava (S 30602, Plate XVII, fig. 2) is a fine-grained strongly ophitic pyroxene-andesite, composed of closely spaced highly ophitic augites commonly about 0·8 mm in diameter but reaching 1·5 mm. The pyroxenes enclose laths of calcic andesine, which occupy 40 per cent of the volume of the rock. The latter average 0·17 mm in length and show a vague fluxion banding. Outside the pyroxenes the plagioclase needles are embedded in a deep olive-green structure-less base containing abundant grains of pyrite. In the altered vesicular lava (S 30601) fresh pyroxene is not preserved. The vesicles are lined with chlorite and filled with calcite, sometimes containing a central mass of chlorite.

REFERENCES

ALLEN, J. R. L. 1965. A review of the origin and characteristics of recent alluvial sediments. *Sedimentology*, **5**, 89–191.

BRADLEY, W. H. 1929. The varves and climate of the Green River epoch. *Prof. Pap. U.S. geol. Surv.*, **159** E, 87–110.

CRAIG, G. Y. 1965. Editor. *The Geology of Scotland.* Oliver and Boyd.

CRAMPTON, C. B. and CARRUTHERS, R. G. 1914. The Geology of Caithness. *Mem. geol. Surv. Gt Br.*

FANNIN, N. G. T. 1970. The sedimentary environment of the Old Red Sandstone of Western Orkney. *Ph.D. thesis, University of Reading* (unpublished).

FINLAY, T. M. 1930. The Old Red Sandstone of Shetland. Part II: North-western Area. *Trans. R. Soc. Edinb.*, **56**, 671–94.

FLINN, D., MILLER, J. A., EVANS, A. L. and PRINGLE, I. R. 1968. On the age of the sediments and contemporaneous volcanic rocks of Western Shetland. *Scott. Jnl Geol.*, **4**, 10–19.

RAYNER, D. H. 1963. The Achanarras Limestone of the Middle Old Red Sandstone, Caithness, Scotland. *Proc. Yorks. geol. Soc.*, **34**, 1–44.

SUMM. PROG. 1934. *Mem. Geol. Surv. Gt Br. Summ. Prog. for* 1933.

WATSON, D. M. S. 1934. Report on Fossil Fish from Sandness, Shetland. *Mem. geol. Surv. Gt Br. Summ. Prog. for* 1933, pt. 1, 74–6.

WESTOLL, T. S. 1937. The Old Red Sandstone Fishes of the North of Scotland, particularly of Orkney and Shetland. *Proc. geol. Ass.*, **48**, 13–45.

—— 1951. The vertebrate-bearing strata of Scotland. Rep. *XVIII Int. Geol. Congr.*, pt. 11, Great Britain 1948, 5–21.

WILSON, G. V., EDWARDS, W., KNOX, J., JONES, R. C. B. and STEPHENS, J. V. 1935. The Geology of the Orkneys. *Mem. geol. Surv. Gt Br.*

Chapter 11

MIDDLE OLD RED SANDSTONE
VOLCANIC ROCKS AND SEDIMENTS OF
PAPA STOUR

INTRODUCTION

THE ISLAND of Papa Stour (Plate XVIII) consists of a virtually unfolded series of rhyolites, basalts, tuffs and sandstones which resembles the lower part of the Esha Ness Volcanic Series of north Shetland. It has, in fact, long been assumed that the volcanic rocks of Papa Stour and Esha Ness form part of a single suite of Middle Old Red Sandstone volcanic rocks (Finlay 1930, p. 681; Summ. Prog. 1935, p. 69; Flinn and others 1968, pp. 11–15). The Papa Stour rhyolites have also been correlated with the rhyolites of Melby the base of which may be about 1000 ft (300 m) above the Upper Melby Fish Bed (p. 142).

The greater part of the island is formed of rhyolite, which reaches a thickness of over 300 ft (90 m) and forms the upper part of the Papa Stour Volcanic Series. The rhyolite was considered to be an intrusive sill by both Geikie (1879, p. 420) and Peach and Horne (1884, p. 371), but was shown by Finlay (1930, p. 681) to be a thick lava flow. Wilson (Summ. Prog. 1935, p. 69) stated that the mapping of the Geological Survey showed that there are, in fact, two rhyolite flows separated by a bed of tuff and agglomerate, which locally attains a thickness of 40 ft (12 m). The lower flow was thought to be 80 to 100 ft (24–30 m) thick, and the upper, the top of which is not seen, over 200 ft (60 m).

The lower rhyolite rests on an undulating, eroded surface (Plate XVIII). In the western part of the island it overlies a variable thickness of tuff and tuffaceous sandstone which in many places has been completely removed by erosion. The tuff is underlain by several flows of basalt which forms the lowest exposed formation on Papa Stour. In the central section of the island the rhyolite rests in most places directly on basalt or basaltic rubble, but in the south-east tuffaceous sandstone which reaches a thickness of over 100 ft (30 m) at Housa Voe separates the two lava types.

The generalized sequence and the approximate thickness of the various groups is as follows:

		Thickness	
		ft	m
R2	Upper Rhyolite	280+	85+
T2	Inter-rhyolitic Tuff and Agglomerate	8–80+	2·4–24+
		(average 30	9)
R1	Lower Rhyolite	0 to 150	0–40
T1	Lower Tuff and tuffaceous sandstone (western half of island)	0 to 40	0–12
SST	Sandstone with tuffaceous bands (eastern half of island)	0 to 100+	0–30+
B	Basalt (up to 4 flows seen)	80+	24+

The index letters are those shown in Plate XVIII and Fig. 20.

BASALTS

FIELD RELATIONSHIPS

The basic lavas of Papa Stour are intensely weathered basalts and dolerites which vary in texture from medium- to coarse-grained in the central parts of the flows to aphanitic in the scoriaceous tops. Along the northern shores of the island only one very thick flow appears to be exposed, but on the south shore up to four flows with a total exposed thickness of 80 ft (24 m) have been recorded. As the base of the basalts is nowhere seen the total thickness of the group is unknown. Individual flows have in most places thick scoriaceous upper zones and in some exposures they are vesicular throughout. Vesicles are filled with chalcedony, calcite, baryte and locally, zeolites. Agates with cores of baryte are common in some exposures. The cracks and hollows of several flows are filled with pink or purplish sandstone and the topmost parts of some flows consist of irregular fragments of scoriaceous basalt embedded in a sandy matrix. The upper surface of the basalt series is highly variable. At the north-west shore of Housa Voe, where the basalt is overlain by sandstone, the lavas are cut by a clean erosion surface which is inclined at an angle of 40° to the dip of the flows. At the north end of Aesha Bight on the other hand, vesicular basalt is overlain by 15 ft (4·5 m) of basaltic rubble composed of subangular basalt blocks up to 3 ft (0·9 m) in diameter embedded in a matrix of reddish brown sandstone (Plate XX, B).

Plate XVIII shows the distribution of the basalt outcrops on Papa Stour, and the estimated thickness of the basalt at the various coast exposures. All coast sections are well exposed, but those on the west coast north of Aesha Head are not readily accessible. The following are characteristic, easily accessible sections:

The Koam, west shore of Hamna Voe. Here three flows of intensely weathered basalt with sandstone veins are seen. The lowest flow has an exposed thickness of 15 ft (4·6 m) and is scoriaceous throughout. Its upper surface is uneven, and its topmost 3 to 8 ft (0·9–2·4 m) contain sandstone-filled cavities and irregular sandstone veins. It is overlain by a waterlaid deposit, up to 8 ft (2·4 m) thick, which fills the depressions on the basalt surface, and is composed of irregular fragments of basalt slag set in a pale purple sandy matrix. The second flow has a fairly even base, which dips gently to the south-west. Its lower part is composed of sparsely vesicular basalt with a number of near-vertical cracks which are bounded by intensely red-stained zones up to 1 ft (30 cm) wide and locally contain thin veinlets of hematite. The upper 3 to 7 ft (0·9–2·1 m) of the flow is highly vesicular, but virtually devoid of sandstone veins and cavity-fillings. A thin bed of basaltic tuff, containing basalt and sandstone clasts up to 1 cm in diameter, separates the second and third flows at the southern end of the exposure. The third flow is over 25 ft (7·5 m) thick and composed almost entirely of non-vesicular medium-grained dolerite.

Aesha Bight. The readily accessible shore exposures between Hirdie Geo and Aesha Head are formed of four flows of basalt ranging from 12 to 20 ft (3·6–6 m) in thickness. All but the highest have highly scoriaceous tops. That of the second flow is 6 to 7 ft (1·8–2·1 m) thick, and its cavities and vesicles are filled with hematite-stained fine-grained sediment which forms up to 30 per cent of its total volume. The third flow is 15 to 20 ft (4·5–6 m) thick, vesicular throughout, and has a highly scoriaceous top, 3 to 10 ft (1–3 m) thick. Many vesicles are highly elongated and a proportion are filled with an outer zone of red and white banded agate and a core of baryte. The largest agate-baryte amygdales are 4 in (10 cm) in diameter. A little violet-blue fluorspar is present in some amygdales. Successive flows are not separated by thin beds of sediment, but at the north end of the exposure the highest flow is overlain by up to 15 ft (4·5 m) of coarse basaltic breccia with a sandy matrix.

Culla Voe. The basalt and dolerite exposed on the shores of Culla Voe and on the north-west shore of West Voe probably form a single thick flow with a central portion, about 50 ft (15 m) thick, of relatively coarse-grained spheroidally weathering dolerite, passing up into fine-grained basalt with columnar jointing. In some places the flow has a 30 ft (9·1 m) thick scoriaceous top, but elsewhere its upper part is sparsely amygdaloidal. On the east shore of Culla Voe the basalt is overlain by basaltic tuff and tuffaceous sandstone.

Housa Voe. Along the north shore of Housa Voe two basalt flows are exposed. The upper part of the higher flow is scoriaceous and contains elongated inclusions of purple sandstone as well as large sandstone veins. The vesicles in the highest 6 ft (1·8 m) of the lava are empty, but in the 3 to 4 ft (0·9–1·2 m) thick zone beneath they are filled with calcite. The lower part of the flow contains a number of large irregular sandstone enclaves, which appear to be blocks of previously consolidated sediment caught up in the lava.

Kirk Sand. A small outcrop of highly amygdaloidal basalt underlies the rhyolite exposed just west of Kirk Sand. From this outcrop Heddle (1878, pp. 115–6) has recorded druses filled with calcite, baryte and fluorspar, which form pale violet and dark purple cubes, as well as chalcedony and saponite. Heddle also recorded fair specimens of cockscomb-baryte in druses, red heulandite in minute crystals coating some druses, and a single crystal of white stilbite. He recorded psilomelane coated with wad in veins at the east end of the exposure.

Scarvi Taing. The basalt and dolerite forming Scarvi Taing between 300 and 1100 yd (275–1000 m) SW of the west end of Kirk Sand is both overlain and underlain by sandstone and tuffaceous sandstone (p. 161). There is also an intercalation of sandstone, up to 12 ft (3·6 m) thick, between successive lava flows.

The basalt flows of Papa Stour are intercalated with tuff and sediment in the south-south-eastern part of the island. The sedimentary intercalations become progressively thicker and the basalts progressively thinner in a south-easterly direction.

PETROGRAPHY

The basalts and dolerites of Papa Stour have a considerable range in grain size and texture. The coarsest varieties from Culla Voe (S 30944, Plate XVII, fig. 4) and from Gorsendi Geo, west of Scarvi Taing (S 30964) are dolerites with randomly orientated euhedral to subhedral laths of zoned, largely albitized plagioclase ranging in length from 1·5 to 0·35 mm, and subophitic plates of colourless augite, which forms up to 30 per cent of the total volume of the rock. Accessory minerals include skeletal grains of iron ore and irregular interstitial patches of chlorite. In the specimens from Culla Voe the pyroxenes are unaltered, but in the Gorsendi Geo dolerite the subophitic plates of clinopyroxene are completely replaced by an aggregate of chlorite and carbonate. The latter also contains abundant pseudomorphs, probably after olivine, of bowlingite rimmed by iron ore.

Most of the compact centres of lava flows are porphyritic holocrystalline basalts with scattered plagioclase phenocrysts ranging in diameter from 1·6 to 0·7 mm, set in a groundmass of generally seriate laths of sericitized or kaolinized plagioclase, which range in size from 0·6 mm × 0·2 mm to 0·12 mm × 0·02 mm and may be randomly orientated or, more rarely, have a fluidal or variolitic texture. These are enclosed in a matrix which forms up to 35 per cent of the total volume. In the case of the freshest specimens from the west shore of Culla Voe (S 30946, 30948) this matrix consists largely of small subhedral to euhedral grains of augite, which normally do not exceed 0·03 mm in diameter. In one specimen (S 30946) some plagioclase phenocrysts are sieved with pyroxene grains, while the remainder are completely clear or have their outer rim only sieved with pyroxene. In this specimen there are also a small number of plates of green biotite and some pseudomorphs in bowlingite and calcite, possibly after olivine (see also S 30948).

In the vast majority of specimens examined both the feldspars and ferro-magnesian minerals are altered. The feldspars are albitized and patchily sericitized and kaolinized. Shapeless patches composed of carbonate and chlorite pseudomorph the interstitial ferromagnesian minerals. In many instances the carbonate and chlorite have partially replaced the adjoining feldspars. Iron ores are partially altered to hematite which forms a fine dusting throughout the specimens. Some thin sections contain small interstitial patches of secondary quartz (S 30699) and patches of calcite associated with epidote (S 30935).

In the fine-grained amygdaloidal upper parts of the flows stumpy plagioclase laths and rare plagioclase phenocrysts are set in an almost completely opaque hematite-stained matrix (S 30936, 30938), which in some instances contains small patches of chlorite. The scoriaceous tops of some flows (S 30947) consist entirely of opaque hematite-stained devitrified tachylite that contains quartz xenoliths with partially resorbed margins. The vesicles are most commonly filled with coarsely crystalline calcite, which in some instances has an outer rim of deep green chlorite. Amygdales of chalcedony or chalcedony together with banded calcite, baryte and zeolite are also common. On the north-east shore of Culla Voe small vesicles filled with potash-feldspar have been recorded.

LOWER TUFFS AND SANDSTONES

SOUTH-EAST PAPA STOUR

Housa Voe. Along the shores of Housa Voe a considerable thickness of sandstone and tuffaceous sandstone separates the basalt lavas from the over-lying rhyolite. Both the upper and lower junctions of the sediment are seen to be highly undulating erosion surfaces and it is probable that within 500 yd (450 m) SW of Housa Voe the base of the rhyolite transgresses the entire thickness of the sediment and rests directly on various horizons of the basalt series (Plate XVIII).

On the north-west shore of Housa Voe the basalt is overlain by the following sequence of sediments:

	ft	m
Alternating cosets of tuffaceous sandstone and purple silty sandstone	16	4·9
Tuffaceous sandstone, becoming coarser upwards ..	9	2·7
Sandstone, planar-bedded, banded grey and purple, with thin silty partings. Junction with basalts irregular, locally highly inclined .. .. 15 to 20		4·6–6·0

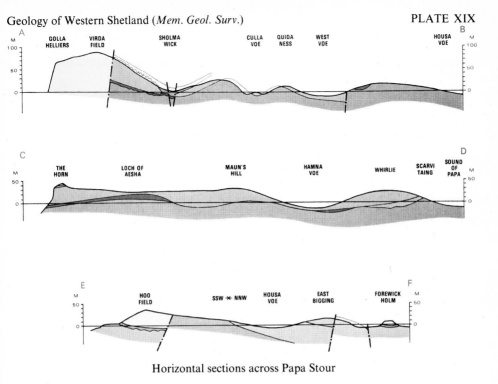

Horizontal sections across Papa Stour

The sandstone forming the lower half of this section contains only a few thin bands full of rhyolite debris. It is generally planar-bedded, but has a few ripple-marked surfaces and a number of sets up to 3 in (7·5 cm) thick, with convolute lamination. In the overlying tuffaceous sandstone, rhyolite clasts up to 5 in (12·7 cm) in diameter are set in a matrix of alternate laminae of sandstone and fine rhyolitic grit. The topmost 16 ft (4·9 m) of the section consist of alternate cosets up to 4 ft (1·2 m) thick, of:

(a) Purple and grey colour-laminated sandstone and sandy siltstone containing some sets up to 6 in (15 cm) thick with small-scale disturbed bedding, closely resembling the possible bioturbation structures in the Melby Formation at Djubabery (p. 147), and

(b) thin bands and lenses of tuffaceous sandstone with rhyolite clasts in a flaggy fine-grained sandstone. The tuffaceous rocks locally fill small irregular channels in the sandstone.

Sandstone and tuffaceous sandstone is exposed at intervals along the west and south shores of Housa Voe. In the tuffaceous beds rhyolite clasts, up to 2 in (5 cm) in size, predominate, but there are also scattered fragments of sediment and rare clasts of basalt. There is also one exposure of a 6-ft (1·8-m) bed of fine-grained irregularly laminated, disturbed, possibly bioturbated silty sandstone. The presence of these disturbed sediments supports the proposed correlation of the Papa Stour beds with the sediments underlying the Ness of Melby Rhyolite (p. 147).

The junction of the sandstone with the overlying rhyolite is exposed at Lambar Banks, 200 yd (180 m) NW of Brei Holm, where it is erosive and highly undulating, 8 ft (2·4 m) of sediments having been cut out within a horizontal distance of 20 yd (18 m). Locally the inclination of the erosion surface reaches 60°.

Scarvi Taing. At least 82 ft (25 m) of sediment, consisting of about 70 ft (21 m) soft purplish red micaceous sandstone and over 12 ft (3·6 m) of grey micaceous sandstone, overlies the basalt along the south-west shore of Scarvi Taing. This sediment thins out completely northwards within a distance of 500 yd (450 m) and thins westwards along the coast to less than 10 ft (3 m).

WEST PAPA STOUR

On the west coast of Papa Stour, between Hirdie Geo and Akers Geo, the basalt, which locally has a brecciated top, is overlain by rhyolitic agglomerate interbedded with tuff and tuffaceous sandstone. As is shown on Plates XVIII and XIX the thickness of this deposit, which is locally over 40 ft (12 m), is highly variable, due partly to its irregularly eroded top and partly to its undulating floor which caused marked variations in the original thickness of the tuff. The eroded surface at the top of the tuff is well seen in the cliff section on the north shore of Hirdie Geo (Plate XX, A), where the base of the rhyolite transgresses eastward across 30 ft (9 m) of tuff on to the underlying basalt. An example of the variation in original thickness of the tuff occurs at the north shore of Aesha Bight (Plate XX, B) where the west-south-westward thinning from 35 to 25 ft (10·7–7·5 m) of the entire bed within a distance of 20 yd (18 m) corresponds closely with a similar extent of thinning in individual beds.

The stack at the west end of Hirdie Geo (Plate XX, A) is composed of at least 40 ft (12 m) of evenly bedded tuff which contains a number of large isolated blocks of rhyolite up to 3 ft (1 m) in diameter. The most accessible section of the Lower Tuff is on the north shore of Aesha Bight, close to Aesha Head, where bedded agglomerate and coarse tuff with a sandy matrix is interbedded with layers, up to 18 in (45 cm) thick, of reddish brown, locally cross-bedded sandstone, virtually free of igneous detritus. The coarse tuffaceous layers contain subangular rhyolite blocks averaging 9 in (23 cm) in diameter, but attaining a maximum of 2 ft (60 cm), set in a matrix of sand and fine rhyolite detritus. The tuff or sandstone overlying the large rhyolite blocks is arched upwards and strongly attentuated over the top of the blocks. The tuffaceous sequence is here overlain by 3 to 5 ft (0·9–1·5 m) of fine-grained sandstone with small scattered grains of rhyolite. North of Aesha Head rhyolite blocks become less common and the tuff is mainly fine-grained, with a sandy matrix, and averages 10 ft (3 m) in thickness.

There is no rhyolitic tuff between the basalt or basaltic tuff and the overlying rhyolite exposed on the shores of Culla Voe in the north of the island. On the shores of Sholma Wick, however, the Lower Tuff may be present, but appears to be directly overlain by agglomerate of the Upper Tuff Group (Fig. 20, p. 165).

In all exposures nearly all the coarser clasts are composed of rhyolite, whereas large fragments of basalt are very rare. Both the tuff and tuffaceous sandstone have a matrix of quartz and feldspar grains and sedimentary structures which indicate that they were deposited by water. It is likely that the larger rhyolite clasts were ejected by a volcanic explosion and have since either remained where they fell or have been transported for only short distances by water. The fine-grained rhyolite detritus is without doubt retransported and partially sorted by water, but the fact that the tuff is highly compressed above and below large blocks suggests that the smaller clasts were relatively uncompacted at the time of deposition.

RHYOLITES AND INTER-RHYOLITIC TUFF
FIELD RELATIONSHIPS

The evidence for the existence of two separate flows of rhyolite is confined to the north-west and north-east shores of the island, where there are a number of exposures of a thick and variable deposit of rhyolitic tuff and agglomerate which is both overlain and underlain by rhyolite. It is, however, not known whether the tuff exposures on the north-west and north-east shores of the island are part of the same bed, or if this tuff originally extended southwards over the area of the entire island. The mapping of a lower and upper rhyolite sheet, as

---

EXPLANATION OF PLATE XX

A. North shore of Hirdie Geo, south-west shore of Papa Stour [132 606]. Lower Rhyolite resting on eroded top of rhyolitic tuff at right of picture. Foreshore rocks at left of picture are amygdaloidal basalt. Islands and cliffs in background are rhyolite. (D 920).

B. South side of Aesha Head, on west shore of Papa Stour. [148 611]. Lower Rhyolite on rhyolitic tuff which, in turn, rests on irregular surface of basalt rubble. (D 922).

C. North shore of Papa Stour, 98 yd (90 m) SE of Cribbie [157 624]. Rhyolite with large spherulites (lithophyses) and vertical banding, which is normal to the dip of the flow. (D 928).

A

B

C

attempted in Plate XVIII, must therefore be regarded as tentative, and it is possible that in the central and southern parts of the island both rhyolite sheets, not separated by an intervening tuff, are present. The following account is based on the assumption that there are two distinct flows and that their outcrops are as shown in Plate XVIII.

Lower Rhyolite

As is shown in Plate XIX the Lower Rhyolite rests on varying horizons of the Lower Tuff, tuffaceous sediment and basalt, indicating that the rhyolite was laid down on a surface which had been or was being strongly eroded. Though this land-surface had many minor irregularities its major feature appears to have been a broad north-north-east trending depression which crossed the present central part of the island. Indications from the sedimentary structures and the large-scale distribution of the volcanic rocks would suggest that the depression was formed by streams flowing from north-east to south-west.

As over the greater part of the area either the top or the base of the flow is not exposed, it is not possible to estimate its range in thickness. At Fogla Skerry, in the extreme west, the Lower Rhyolite forms unbroken cliffs over 100 ft (30 m) high, near Wilma Skerry and Calsgeo Taing close to the west and east shores of Hamna Voe it is at least 70 ft (21 m) thick, and on the north-east shore, close to Ram's Geo, its estimated minimum thickness is 150 ft (46 m). Between Shaabergs and Sholma Wick, on the north coast of the island, however, the rhyolite has been deeply eroded and at Sholma Wick the Inter-rhyolitic Tuff (T2) appears to rest directly on the Lower Tuff (T1) (p. 165). The original (i.e. pre-folding) base of the flow varied from almost level, as at the west coast of the island between Aesha Head and Akers Geo, to highly undulating, as at Lambar Banks near Brei Holm (p. 161). At Aesha Head the basal 10 to 15 ft (3–4·6 m) of the rhyolite are autobrecciated, but in most other exposures of the base of the flow there is no sign of brecciation.

The Lower Rhyolite forms orange-red cliffs with marked columnar jointing. It is commonly banded, the banding being emphasized by the presence of alternate orange and purple laminae. Though the strike and dip of the banding is in many cases consistent over several hundred metres, it bears no relation to the dip of the rhyolite sheet. In some areas, as on the north-west and north-east shores of Hamna Voe and the shore between Cribbie and Skaabergs, the inclination of the banding ranges from 70° to vertical (Plate XX, C). In places (e.g. Cribbie and north-west shore of Hamna Voe) the rhyolite contains closely packed spherical bodies, which have been termed lithophyses by Finlay (1930, p. 680). These range in diameter from less than $\frac{1}{8}$ to $1\frac{1}{2}$ in (3–38 mm). In some sections (e.g. Hirdie Geo) these enlarged spherulites are confined to the basal part of the flow, and in others (e.g. south coast of Fogla Skerry) there are a number of discrete spherulitic bands scattered throughout the flow. The rhyolite is almost everywhere sparsely porphyritic with euhedral feldspar phenocrysts up to $\frac{3}{8}$ in (10 mm) in size.

The upper surface of the Lower Rhyolite is well exposed in several sections on the north-west and north-east shores of the island. On the south shore of the Geo of Bordie the top of the rhyolite is undulating and deeply weathered to a depth of 15 ft (4·6 m). The weathered rhyolite is soft, and has a pale greenish colour, which contrasts strongly with that of its orange-red feldspar phenocrysts.

M

It has a number of near-vertical fissures filled with greenish, slightly indurated sandstone. On the north shore of the Geo of Bordie the upper surface of the rhyolite has been eroded into a number of roughly north-east trending ridges and depressions (Plate I), with the rhyolite highly weathered in the upper parts of the ridges only. Along the north shore of the island, between Shaabergs and Sholma Wick, the rhyolite, which is relatively unweathered, has an extremely irregular surface with some deep near-vertical breccia-filled clefts, and some partially or wholly detached blocks. Approximately 100 yd (90 m) W of Sholma Wick the erosion surface dips steeply eastwards, cutting out nearly 100 ft (30 m) of rhyolite, and on the north-west side of Sholma Wick the Lower Rhyolite appears to have been completely eroded away (p. 165).

Along the north-east coast of Papa Stour, at Ram's Geo and Doun Hellier, the top of the flow has not been affected by erosion and the original structures of its upper part have been preserved. Here the topmost 10 to 20 ft (3–6 m) of the rhyolite consist of a jumbled mass of greenish-weathered rhyolite blocks and pillow-like ovoids up to 1 ft 6 in (45 cm) in diameter, with the spaces filled with either rhyolitic detritus or a greenish sandy sediment. This rubbly zone is underlain by a 10 ft (3 m) thick zone composed of large pillow-like masses of rhyolite which are up to 6 ft (1·8 m) long and elongated in a north-west to south-east direction. These 'pillows' characteristically have a ropy surface. Below them the rhyolite is massive, purplish-weathering and sparsely porphyritic.

Inter-rhyolitic Tuff and Agglomerate

The tuff and agglomerate which separates the two rhyolite flows gives rise to spectacular cliff sections at both the Geo of Bordie and Sholma Wick (Plate I and Fig. 20). Along the north-west shore of the former locality the irregular upper surface of the Lower Rhyolite is buried under at least 30 ft (9·0 m) of bedded rhyolitic tuff and agglomerate. The bedding planes within the basal 10 to 15 ft (3–4·5 m) of this deposit are sub-parallel to the contours of the irregular rhyolite pavement, with a marked attenuation of individual beds above the ridges. This type of bedding suggests that a high proportion of the rhyolite detritus was deposited directly from the air. The higher part of the sequence is composed of beds of rhyolitic lapilli-tuff and agglomerate with a sandy matrix and scattered blocks of rhyolite, locally up to 3 ft (0·9 m) in size, interbedded with thinner sets of tuffaceous sandstone. The highest sets of this deposit are almost entirely planar-bedded.

The highly irregular rhyolite surface between Shaabergs and Sholma Wick is overlain by agglomerate with irregular bedding and some very large rhyolite blocks, which also contains some thin beds of flaggy sandstone. The agglomerate is coarsest in the eastern part of the section, where it is banked up against the steep erosion surface cutting the rhyolite (p. 163). In the cliff forming the west shore of Sholma Wick (Fig. 20) the coarse agglomerate rests, with an irregular base, on up to 45 ft (13·7 m) of interbedded tuffaceous sandstone and calcareous sandstone with rhyolite clasts not usually exceeding 1 in (2·4 cm) in size. This tuffaceous series overlies two small patches of basalt exposed at the south end of Sholma Wick and it seems likely that it is the stratigraphical equivalent of the Lower Tuff (p. 161). As the Lower and Inter-rhyolitic tuffs cannot be distinguished by lithological character alone, it is, however, not possible to determine with certainty the true stratigraphic position of this deposit.

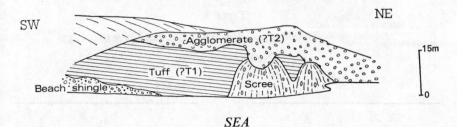

SEA

FIG. 20.   *Cliffs on north-west coast of Sholma Wick, Papa Stour, showing bedded tuff (?T1) overlain by agglomerate (?T2)*

Between Sholma Wick and Lamba Ness the tuff is on average 20 ft (6 m) thick and is composed of dark greenish lapilli-tuff irregularly interbanded with sandstone.

The tuff exposed on the north-east coast at Doun Hellier and Ram's Geo ranges in thickness from 12 to 18 ft (3·6–5·5 m). It is evenly bedded and composed of thin beds of fine-grained tuffaceous sandstone alternating with beds up to 18 in (0·45 m) thick, of agglomerate with rhyolite clasts up to 6 in (15 cm) (and, exceptionally, 2 ft (60 cm)) in size.

Upper Rhyolite

The upper flow of rhyolite is strongly banded and, in its eastern outcrop, sparsely porphyritic. It lacks the zones packed with spherical lithophyses, which are characteristic of the lower flow. If the interpretation of the structure shown in Plates XVIII and XIX is correct it must be at least 300 ft (90 cm) thick at Virda Field (north-west corner of the island) and possibly even thicker at Hoo Field, in the north-east of the island.

PETROGRAPHY

Rhyolites

The relatively constant thickness of the Papa Stour rhyolites and their probable wide extent well beyond the limits of the island are features more characteristic of incandescent ash-flow deposits than of acid lavas which consolidated *in situ* from a molten state. Incandescent ash flow or *nuée ardente* deposits give rise to welded vitric tuffs and sillars (non-welded vitric tuffs), both of which are generally called ignimbrite (Marshall 1935; Gilbert 1938, p. 1833). Ignimbrites are recognized in thin section by the presence of glass shards and pumice fragments which are flattened and welded to varying degrees, depending on their position within the flow and the temperature of the flow, together with varying proportions of crystal clasts (usually broken and with embayed margins), lithic clasts and interstitial vitric dust (Rast 1962, pp. 97–8). In many of the older ignimbrites the original vitroclastic texture is, however, to some extent obliterated by recrystallization during devitrification, which has led to the formation of crystalline silica and sanidine and to the development of spherulitic and axiolitic textures (Enlows 1955) and of branching plumes of sanidine and tridymite crystallites.

The Papa Stour rhyolites are completely devitrified and, if vitroclastic textures were ever present within either of the two flows, they have been

TABLE 3

Analyses of Papa Stour rhyolites

(No localities given)

NORMS

	1	2		1	2
SiO_2	71·23	69·12	Q	28·81	19·56
TiO_2	0·61	n.d.	C	0·00	0·00
Al_2O_3	11·08	14·55	or	51·83	60·10
Fe_2O_3	4·18	1·70	ab	8·17	10·75
FeO	0·59	0·14	an	0·00	3·96
MnO	0·01	n.d.	ac	6·22	0·00
(Co.Ni)O	Nil	n.d.	di	0·00	2·72
MgO	Nil	0·52	hy	0·00	0·03
CaO	0·05	1·57	ol	0·00	0·00
BaO	Nil	n.d.	mt	0·17	0·45
Na_2O	1·80	1·27	hm	1·91	1·39
K_2O	8·77	10·17	ilm	1·16	0·00
$H_2O > 105°$	1·22	0·67	ap	0·09	0·12
$H_2O < 105°$		0·12	pr	0·43	0·12
P_2O_5	0·12	0·05	Others	1·42	0·79
CO_2	0·12	n.d.			
FeS_2	0·43	0·12	Total	100·21	100·00
Total	100·21	100·00	Q	32·44	21·64
			or	58·36	66·47
			ab	9·20	11·89
			Total	100·00	100·00
			or	86·39	80·34
			ab	13·61	14·36
			an	0·00	5·30
			Total	100·00	100·00
			ab	100·00	73·05
			an	0·00	26·95
			Total	100·00	100·00

1. Analyst T. C. Day (Finlay 1930, p. 693)
2. Analyst R. R. Tatlock (Finlay 1930, p. 693)
n.d. = not determined

obliterated. The feldspar and quartz phenocrysts within the flows are almost invariably unbroken, and there is no noticeable variation of texture within the vertical profile of either flow. The only recognizable glass shards and pumice fragments occur in the tuff bands immediately above the lower flow (S 30955). As vitroclastic textures are reasonably well preserved in the ignimbrites of both the Melby Formation (p. 151) and the Clousta Volcanic Rocks (p. 96) it would be surprising if they had been completely obliterated at Papa Stour. Until further textural data are forthcoming, it must thus be assumed that the Papa Stour Rhyolites are not ignimbrites, but may be true devitrified acid lavas.

Finlay (1930, pp. 679–81) has stated that two rock types occur; a compact, highly porphyritic, felsitic type in the south and south-west, and a more vitreous type with platy jointing, 'flow-banding', and spherulitic and 'lithophysal' structures in the north and north-west. This distinction is not true, as spherulitic and strongly banded rhyolites are present in both flows, both in the north and south of the island, and considerable areas with closely packed 'lithophyses' occur in the lower flow at both Cribbie in the north-west and Hamna Voe in the south. Abundant feldspar phenocrysts are confined to the lower flow and plagioclase laths occur only in the eastern part of the island. Porphyritic quartz crystals are present in significant numbers only in the upper flow.

There is a great range in the texture of the groundmass of the rhyolite, and all the main textural types are present in both flows throughout the island. Basically the groundmass consists of minute simple or branching laths or microlites of potash feldspar (?sanidine) in a micropoikilitic base of quartz. The size of individual crystals varies greatly and the texture ranges from cryptocrystalline to poikilitic or micropegmatitic with distinct feldspar laths. The feldspars are commonly stained orange by hematite and the micropoikilitic aggregate contains minute grains or needles of iron ore (magnetite) which in many cases form a fibrous or branching network. Interspersed with this aggregate are irregular patches or veinlets of quartz which form either fine-grained mosaics or aggregates of larger irregular crystals which have a shadowy extinction and areas full of minute inclusions (S 30698).

The arrangement of the laths or microlites within the quartz base gives rise to a variety of textures. Where the laths are randomly orientated or arranged in plumose branching aggregates, the quartz base commonly forms patches in which the extinction direction is uniform, thus producing a mosaic effect under crossed nicols (S 30937). Very commonly the laths or microlites are radially grouped into spherulites (or 'radiolites' according to Bryan's (1965, pp. 20–3) classification) which in most cases range in diameter from 0·8 to 0·3 mm but are as small as 0·08 mm in parts of the upper flow (Plate XVII, fig. 5). Depending on the closeness of the packing, the spherulites vary in outline from round (S 30933) to polygonal (S 30952), diamond shaped (S 30930) or square. In some bands of tightly packed spherulites (S 30930) individual spherulites are incompletely developed and have, in thin section, an hour-glass structure, with two bundles of microlite sheaves with axes at right angles to the banding. Individual microlites normally extend from the centre to the periphery of the spherulite. They vary in shape from thin needles to distinct laths with square ends. The outer ends of the laths are euhedral and the laths are slightly uneven in length, forming a finely micropegmatitic intergrowth with the quartz patches adjoining the spherulites (S 30931, Plate XVII, fig. 6). The fine needles or plumes of iron ore tend to be concentrated near the periphery of the spherulites, and in the case of closely packed spherulites they produce an overall polygonal pattern (S 30952).

Spherulites normally occur in closely packed clusters or irregular bands separated by interstitial areas of coarse or mosaic quartz. The volume percentage of the quartz areas within the rhyolite is very variable, and the banded rhyolites are commonly formed of alternate irregular quartz-rich zones with loosely scattered spherulites and quartz-poor zones with tightly packed spherulites.

Individual bands are 1 to 1·5 mm thick. In all cases examined the banded structures are of secondary origin, and as in many cases the inclination of the banding is at a high angle to the inclination of the rhyolite sheet, it seems unlikely to be connected with true flow banding.

At Cribbie in the north-west of the island and on the north-west shore of Hamna Voe the banded rhyolite contains closely spaced spherical bodies which range in diameter from 1·5 mm to well over 20 mm (Plate XX, C). These enlarged spherulites have both a radial and a concentric internal structure and have been called lithophyses by Finlay (1930, p. 680). Lithophyses are, by definition, hollow spherical bodies, and as the large complex spherulites of Papa Stour rarely have true cavities the term is not strictly applicable. They could be termed 'solid spherulites' according to the classification of Bryan (1965, table i). They normally have an irregularly shaped core composed of either an ochre-stained micropoikilitic feldspar-quartz aggregate (S 30700) occasionally with fairly large feldspar laths, or a quartz mosaic, or an aggregate of both (S 30701). The core may be the original gas cavity which was later filled by rhyolitic glass. It is encased in an inner mantle formed of very thin radiating feldspar microlites which are clear of iron staining, but have an outer sub-marginal rim, up to 0·1 mm thick, with a high concentration of magnetite dust or with fine branching magnetite needles (S 30700). This radiating fibrous zone commonly occupies the greater part of the spherulite and is surrounded by an outer shell of a cryptocrystalline feldspar-quartz aggregate in which the feldspar microlites may be branching and plumose. The interstices between adjacent shells are filled by irregular patches of quartz. The various shells are not necessarily concentric, and in some instances (S 30704) the fibrous material from the inner shell extends outward beyond the spherulite to form irregular patches outside the spheres or to form a direct connection with the fibrous zone of an adjacent sphere. The 'lithophyses' from Cribbie differ from those from Hamna Voe, in that they generally have a very large and more coarsely poikilitic inner zone, a narrow central fibrous zone, and small spherulites (i.e. radiolites) developed in the outer zone. Scattered 'lithophyses' in the rhyolite along the west coast of the island have a similar structure.

The Lower Rhyolite sheet is highly feldspar-phyric throughout, and over the greater part of the island only potash feldspar phenocrysts are present. The latter are generally euhedral, unbroken, and normally range in diameter from 3 to 0·4 mm, the smaller phenocrysts being more or less equidimensional in outline (Plate XVII, fig. 8). The feldspar is normally Carlsbad-twinned, but some specimens have a closely spaced polysynthetic cross-twinning (S 30962, 30959). Most phenocrysts are slightly cloudy and patchily ochre-stained. Plagioclase (sodic oligoclase) laths which are patchily kaolinized are present in the rhyolite along the east shore of the island where they range in diameter from 1 to 0·15 mm (S 30942) and, in some instances, form up to 30 per cent of the total volume of the rock.

Apart from iron ore (magnetite, ilmenite, leucoxene, hematite) the only mafic minerals recorded in the Papa Stour rhyolites are isolated small plates of strongly pleochroic green biotite (S 30959) and, near the top of the flow, rare euhedral crystals of zircon (S 30962).

Specimens from the top of the lower flow are amygdaloidal with the original vesicles elongated possibly parallel to the direction of flow, and filled with mosaic quartz (S 30924). Cavities close to the top of the flow are in places filled with a fine vitreous tuff, which contains, in addition to unflattened glass shards and pumice fragments, abundant small angular grains of quartz, small flakes of mica as well as subhedral to euhedral grains of zircon, tourmaline and apatite.

Tuff

The agglomerate and tuff between the two flows of rhyolite in the north-western part of the island almost invariably has a sandy matrix. Along the north-east coast, however, the matrix of the tuff is generally fine-grained, locally argillaceous. In some instances it is a non-welded vitric tuff. An example of the latter, from Ram's Geo, north-east of Hoo Field (S 30955), consists of angular clasts of cryptocrystalline rhyolite and rounded fragments of spherulitic rhyolite as well as a few clasts of andesite, set in a matrix containing unflattened shards of glass and pumice with rounded vesicles. In other specimens (S 30956) the fine-grained matrix consists of structureless argillaceous material with clasts of rhyolite-glass with perlitic fractures. This is cut by irregular veins of carbonate and bright green celadonite.

STRUCTURE

Folds. Though the volcanic series of Papa Stour has a number of very gentle flexures its overall disposition is virtually horizontal, as is shown by the fact that the base of the Lower Rhyolite is exposed at intervals all round the shore of the island (Plate XVIII). Along the south-west shore of the island all formations dip gently to the west-south-west, but this dip flattens out a short distance from the shore. On the north coast west of Culla Voe, the tuffs and lavas are seen at intervals to dip at 20°–25° to the west-north-west but along the greater part of this shore they are again almost horizontal. In the eastern part of the island there is a consistent slight southerly dip which varies from south-west in the north to south-south-east in the south.

Faults. Papa Stour is cut by a large number of faults with relatively small throw and a great range of directional trend. The faults with the largest displacement have a north-westerly trend and a downthrow to the north-east, which effectively cancels out the south-westerly or southerly dip of the formations. The most important of the north-west trending faults is that extending from Akers Geo towards Dutch Loch. This has a possible maximum throw of over 200 ft (60 m) at Akers Geo. Due to the very variable thickness of the various groups it is difficult to calculate the throw of any of the faults.

In addition to the faults with discernible displacement there are a number of wide zones which contain numerous sub-parallel near-vertical crush planes, which in the rhyolite and Lower Tuff form zones of weakness along which the geos and extensive caves of the west coast have been excavated.

Mineral veins carrying mainly baryte have been emplaced along a number of these crush belts in the area just north-west of Hamna Voe and on the west coast of Vidra Field.

GEOLOGY OF FOREWICK HOLM

The small island of Forewick Holm off the south-east coast of Papa Stour (Plate XVIII) has not been visited by the author. According to the mapping by S. Buchan, it consists of a flow of porphyritic rhyolite, of which possibly 150 ft (45 m) may be exposed, underlain by at least 80 ft (24 m) of red and purple tuff with subordinate sandstone.

The tuffaceous sediment appears to be cut by two north-north-west trending faults the more easterly of which separates it from rhyolite, which forms the eastern extremity of the Holm. The two formations, which dip at 30° to west-north-west, are probably the equivalents of the tuffaceous sandstone and Lower Rhyolite, which form the East Bigging peninsula of Papa Stour.

References

BRYAN, W. H. 1965. Spherulites and Allied Structures. Part V. *Proc. R. Soc. Queensland*, **76**, 15–25.

ENLOWS, H. E. 1955. Welded tuffs of Chiricahua National Monument, Arizona. *Bull. geol. Soc. Am.*, **66**, 1215–46.

FINLAY, T. M. 1930. The Old Red Sandstone of Shetland. Part II: North-western Area. *Trans. R. Soc. Edinb.*, **56**, 671–94.

FLINN, D., MILLER, J. A., EVANS, A. L. and PRINGLE, I. R. 1968. On the age of the sediments and contemporaneous volcanic rocks of western Shetland. *Scott. Jnl Geol.*, **4**, 10–19.

GEIKIE, A. 1879. On the Old Red Sandstone of Western Europe. *Trans. R. Soc. Edinb.*, **28**, 345–452.

GILBERT, C. M. 1938. Welded tuff in eastern California. *Bull. geol. Soc. Am.*, **49**, 1829–62.

HEDDLE, M. F. 1878. *The County Geognosy and Mineralogy of Scotland, Orkney and Shetland*. Truro.

MARSHALL, P. 1935. Acid rocks of the Taupo-Rotorua district. *Trans. R. Soc. N.Z.*, **64**, 323–66.

PEACH, B. N. and HORNE, J. 1884. The Old Red Volcanic Rocks of Shetland. *Trans. R. Soc. Edinb.*, **32**, 359–88.

RAST, N. 1962. Textural evidence for the origin of Ignimbrites. *Lpool Manchr geol. Jnl*, **3**, 97–108.

SUMM. PROG. 1935. *Mem. geol. Surv. Gt Br. Summ. Prog. for* 1934.

Chapter 12

OLD RED SANDSTONE OF FOULA

INTRODUCTION

FOULA is a pear-shaped island, $3\frac{1}{2}$ miles (5·6 km) long from north to south and $2\frac{1}{2}$ miles (4 km) wide from west to east (Fig. 21). It is situated in the Atlantic Ocean 14 miles (22 km) WSW of Wats Ness in the Walls Peninsula and has a prominent skyline, which, when viewed from the mainland, presents three steep north-facing escarpments with relatively gentle south-facing dip slopes. The most northerly escarpment forms Soberlie Hill, 721 ft (220 m) high; the central one has three peaks which are, from west to east, The Kame (1220 ft, 372 m), The Sneug (1373 ft, 406 m) and Hamnafjeld (1126 ft, 343 m). To the south of these the island is traversed by a west-north-west trending glacially moulded valley, known as The Daal. South of this lies The Noup (803 ft, 245 m), a hill which has a less prominent north-facing escarpment than the others.

The greater part of Foula is formed of relatively soft sandstone with subordinate shale and siltstone bands. This sedimentary series appears to be of Middle or Upper Old Red Sandstone age and has a total exposed thickness of about 6000 ft (1800 m). The sandstone forms prominent vertical cliffs all along the west coast of the island. These have a maximum height of 1220 ft (372 m) at The Kame, which forms one of the highest sea-cliffs in the British Isles.

The Old Red Sandstone sediments are bounded in the east by a $\frac{1}{2}$-mile (800-m) wide strip of metamorphic rocks and the two formations are separated from each other by a slightly curving north-north-west trending fault. At its outcrop on the north coast this fault is inclined at 60° to the west, but at its southern coastal outcrop it is inclined at only 35° to the south-west.

Owing to its unique position and prominent topography Foula attracted many of the early geological investigators who visited the northern isles. There are accounts of the topography and geology of Foula by Jameson (1798), Hibbert (1822), Nicol (1844) and Gibson (1877). The first comprehensive geological account of Foula is by Heddle (1878, pp. 124–30). More recent accounts dealing with the geology of Foula are by Peach and Horne (*in* Tudor 1883, p. 395), Finlay (1926, pp. 564–5), and Wilson (*in Summ. Prog. geol. Surv.* 1934, pp. 71–3).

FIELD RELATIONSHIPS

The Old Red Sandstone of Foula consists of approximately 6000 ft (1800 m) of relatively soft, predominantly medium-grained, grey to buff sandstone, with small isolated pebbles of quartz and granite and with a number of partly or predominantly argillaceous or silty horizons. Many of the sandstones are cross-bedded with a consistent dip of foresets towards the east to south-east sector. The structure is simple, consisting of a very open, southward plunging syncline. Close to the fault which separates the sediments from the metamorphic rocks to the east, the dip of the strata is locally much steeper and the strike is roughly parallel to the fault.

171

Though the sediments of Foula display a remarkable uniformity throughout their thickness, it is possible to recognize the following nine more or less distinct members (Fig. 21):

	ft	m
9. *Noup Sandstone*. Yellow, medium-grained, cross-bedded sandstone, in sets up to 6 ft (1·8 m) thick, with dips of foresets predominantly to east-north-east. Scattered small rounded quartzite pebbles are present throughout, with some lenses containing pebbles up to 2½ in (6 cm) in diameter	800+	240+
8. *Daal Flaggy Sandstone*. Cross-bedded sandstone with scattered pebbles interbedded with yellow-weathering flaggy sandstone and a few beds, up to 18 in (45 cm) thick, of greenish grey sandy siltstone. Disseminated indeterminate plant scraps occur in silty and flaggy beds at Biggings (south coast) and Surpeidles (south-east coast). Dip of foresets is mainly to east	500	150
7. *Wester Hoevdi Sandstone*. Predominantly massive medium-grained cross-bedded sandstone, with very thick sets and few flaggy partings. Scattered small rounded pebbles of quartzite, quartz and granite ..	1600	490
6. *Kame Banded Beds*. Buff, greenish, purplish and brownish evenly bedded, mainly flaggy sandstones with fairly thick shaly and silty partings	520	160
5. *Kame Argillaceous Beds*. Grey and purplish silty shale and siltstone with subordinate sandstone ribs. Indeterminate plant fragments in siltstone and shale are found at North Bank, 1770 yd (1620 m) NE of The Kame, in Blobers Burn 300 yd (275 m) WNW of Harrier, and in a roadside quarry, 250 yd (230 m) ENE of Harrier ..	120	35
4. *Soberlie Hill Sandstone*. Sandstone, medium-grained, ochre-yellow-weathering, in cross-bedded sets up to 2 ft (60 cm) thick with small scattered pebbles up to 3 in (8 cm) in diameter throughout. Some gritty beds with abundant rounded pebbles are present near the base. Dip of foresets is predominantly to east-south-east. Flaggy and silty partings are very rare	1400	400
3. *East Hoevdi Beds*. Alternate thick cosets of (a) medium-grained cross-bedded sandstone ranging in thickness from 10 to 80 ft (3–25 m) and (b) thinly interbedded and interlaminated grey and purplish flaggy sandstones, siltstones and shales up to 60 ft (18 m) thick. Contorted and slumped bedding is common in the thicker cross-bedded sets. Dips of foresets to SE–ESE sector. Sand-filled desiccation polygons are present in mudstones; and mudstone chips occur in some of the intercalated sandstone sets. Base of group is predominantly sandy..	450	140
2. *Ness Beds*. Sets of thinly bedded buff cross-stratified sandstone up to 4 ft (1·2 m) thick, commonly with pellets and chips of purple mudstone and some red shale partings alternating with sets of grey and purple mudstone, silty mudstone and siltstone 4 in to 4 ft (10 cm–1·2 m) thick, with sand-filled desiccation polygons. Sandstone fillings of cracks are commonly distorted, and irregular		

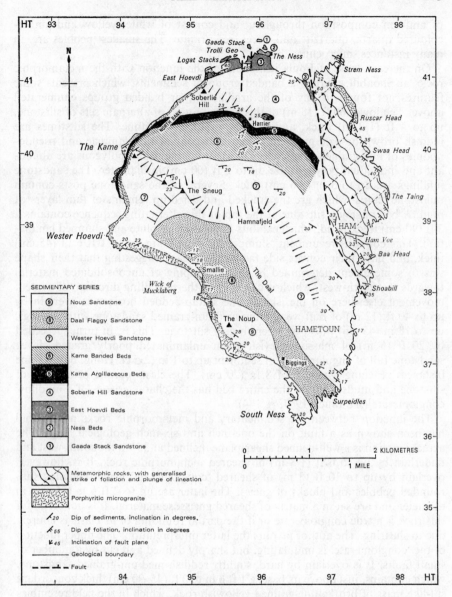

FIG. 21. *Geological sketch-map of Foula*

	ft	m
convolute pseudonodules of sandstone are present in the adjoining mudstone (Fig. 22)	100	30
1. *Gaada Stack Sandstone*. Predominantly massive, yellowish-buff-weathering sandstone	250+	75+

These nine groups have been mapped along the west coast and through the centre of the island. Though small isolated rounded pebbles are present in the cross-bedded sandstones throughout the sequence lenses of extraformational conglomerate with abundant large pebbles are absent. The pebbles appear to be of uniform composition throughout and consist of white, yellow and amber-coloured quartz, quartzite and, rarer, red granite. The smallest pebbles are, in many instances, microcline.

On the east coast of Foula just south of the junction with the metamorphic rocks at Shoabill, there is a banded group of sediments, which presents some features not found in any of the argillaceous or banded groups enumerated above. The lower 60 ft (18 m) of this group consist of alternate sets of siltstone, up to 4 ft (1·2 m) thick, and medium-grained sandstone. The siltstones are reddish brown or pale green in colour and contain thin laminae and pseudo-nodules of fine-grained sandstone. Desiccation cracks and polygons are abundant and individual polygons are up to 2 ft (60 cm) in diameter. The sandstone infillings are up to 3 in (7·6 cm) thick. Several of the sandstone posts contain mudstone 'chips' which are the cracked and dried-up remains of thin layers of mud. A bed of sandy siltstone at the top of this alternating sequence contains a 3-ft (90-cm) thick band, which exhibits intensely convolute and slumped lamination. This contains recumbent 'slump balls' which are up to 1 ft 6 in (45 cm) thick, and have their convex side facing south-east, suggesting that their shape was to some extent determined by lateral slipping of unconsolidated material towards the north-west, which is contrary to the prevailing direction of current movement elsewhere on the island. The slump-bedded horizon is overlain by up to 40 ft (12 m) of buff-weathering medium-grained sandstone with partings, up to 18 in (45 cm) thick, of greenish grey siltstone. This is, in turn, succeeded by 20 ft (6 m) of massive greyish green unlaminated poorly graded clayey sandstone full of angular clasts of sediment up to 1 in (2·5 cm) long, but in some instances reaching a length of 8 in (20 cm). The clasts consist principally of siltstone and mudstone and the entire bed has the characteristics of a mudflow conglomerate (*see* Bluck 1967, p. 144).

The junction between the sedimentary and metamorphic rocks at Shoabill has been shown as a fault on the one-inch and six-inch geological maps. The actual junction is a well-defined shear-plane inclined at 35° to the south-west and underlain by up to 6 ft (1·8 m) of sheared metamorphic rock. It is, however, overlain by up to 10 ft (3 m) of sheared 'conglomerate' which contains sub-rounded pebbles and blocks of gneiss. The latter are up to 2 ft 6 in (75 cm) in diameter and are set in a matrix of sheared gneissose material. It is not certain if this rock is a true conglomerate or if the partial rounding of its clasts is entirely due to shearing. The author favours the latter interpretation. The upper junction of the 'conglomerate' is undulating, but sharply defined and cut by a number of small faults. It is overlain by hard, slightly reddish medium-grained sandstone, which contains, just above its base, a 1 ft 6 in to 3 ft (45–90 cm) thick concordant sill-like mass of brittle, fine-grained yellowish rock, which in the field resembles a felsite, but is in thin section seen to be an intensely sheared sandstone. The junction between metamorphic rocks and sediments in this area is without doubt

---

EXPLANATION OF PLATE XXI

A. Wilson's Noup, Northmaven [302 716]. Brecciated basalt in granite and grano-diorite. (D 1345).

B. Wilson's Noup, Northmaven [302 716]. Brecciated partly permeated basalt cut by a stream of granodioritic material full of elongate, variably assimilated basaltic enclaves. (D 1347).

Geology of Western Shetland (*Mem. Geol. Surv.*)   PLATE XXI

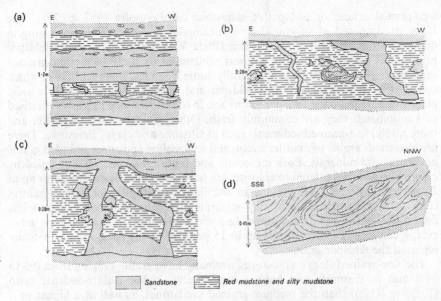

FIG. 22.   *Sedimentary structures in the Old Red Sandstone sediments of Foula*

(a) North shore, opposite The Brough [961 416]. Buff sandstone interbedded with red mudstone, with sand filled desiccation cracks in mudstone and mudstone chips in sandstone.
(b) and (c) Locality as in (a). Distorted sandstone 'dykes' and slump-balls in red silty shale.
(d) East shore, South Ness, 270 yd (250 m) SSW of Shoabill [974 379]. Convolute bedding in sandstone.

a faulted one, but if the 'conglomerate' described above is of sedimentary origin and is not a true fault breccia as the author believes the present junction must be close to the original unconformity between the two groups. If that were the case the subsequent fault movement would here appear to have been more or less along the unconformable junction.

CONDITIONS OF DEPOSITION

The greater part of the Foula succession appears to be composed of fluvial sandstones laid down by relatively fast-flowing, possibly braided, rivers. The lower part of the sequence was probably deposited by currents from the west to north-west sector, and the upper part by currents from the west or south-west. The presence of red and purple siltstones and mudstones with sand-filled desiccation cracks in the lower half of the sequence, well seen on the north and east coasts of the island, suggests that there were some periods with a quieter fluvial regime, possibly dominated by meandering rivers with flood plains on which top stratum deposits were preserved. The uniform composition of the sandstones (p. 176) and their scattered pebbles (p. 174) throughout the sequence suggests that the source area remained much the same during the entire period.

PETROGRAPHY

The medium- to coarse-grained pebbly sandstones of Foula are generally bi-modal with scattered well-rounded pebbles, 3 to 1 cm in diameter, set in a well-graded arkose or feldspathic sandstone (cf. Pettijohn 1957, p. 291). The latter is composed of subrounded to subangular grains commonly ranging in diameter from about 0·75 to 0·1 mm (Plate VIII, fig. 6). The quartz-feldspar ratio varies from 50:50 in the coarsest sandstones to 70:30 in the fine-grained sandstones. Quartz grains are generally more rounded than feldspars, which consist of both untwinned potash feldspar and rarer microcline, as well as sodic plagioclase. Though the feldspar grains are, in some instances, patchily sericitised or kaolinitized, they are commonly fresh. Other grains consist of felsite, and more rarely, fine-grained sediments such as siltstone and clayey limestone. There are also small grains of apatite, zircon and tourmaline and small rounded grains of opaque ore minerals. Both muscovite and biotite, in some cases mantled by chlorite or partially altered to chlorite, are present in all sections and form up to 20 per cent of the total volume in the flaggy partings. The detrital matrix composed of an aggregate of clay minerals and chlorite normally forms less than 5 per cent of the total volume. In some sandstones of all grain sizes, however, a calcareous cement forms up to 15 per cent of the rock and has patchily replaced the feldspar grains.

The fine-grained flaggy sandstones normally have grains ranging from 0·5 to 0·08 mm in diameter and have a considerably higher quartz-feldspar ratio (70:30 to 80:20) than the medium-grained sandstones, as well as a higher proportion of partially chloritized mica flakes. The dark grey ungraded sandstone with scattered pebbles of sediment, exposed on the shore south of Shoabill, close to the junction with the metamorphic rocks (p. 174), is unique in Foula. It consists of widely scattered subangular to subrounded grains of quartz and feldspar (quartz-feldspar ratio 9:1) together with subrounded fragments of siltstone and clayey limestone which range in size from over 5 to 0·03 mm, as well as some grains of felsite and tourmaline. The matrix forms over 25 per cent of total volume of the rock and is composed of a matted mass of mica flakes set in an indeterminate brownish aggregate. This highly ungraded immature sediment contrasts strongly with the generally well-graded sandstones of Foula.

STRUCTURE

The sedimentary rocks of Foula form an open syncline plunging at 20° to 30° to the south. The eastern limb of this syncline steepens appreciably close to the eastern boundary fault and the strike of the beds in a 109 yd (100 m) wide zone bounding the junction is roughly parallel to this fault. In the exposures along the north shore where the fault is inclined at 70° to W10°S, the adjoining sediments are intensely shattered and cut by a number of small faults which trend N60°W and range in inclination from 60° to 85° to the south-west, as well as by several faults which are sub-parallel to the main fault.

Apart from the faults associated with the eastern boundary fault the Foula sediments are relatively free from major faulting. Two north–south trending faults with small throws and westward inclinations cut the north coast at Trolli Geo and a small geo 150 yd (135 m) farther east. These have small downthrows to the west and both appear to die out southwards.

The sediments throughout the island are cut by several systems of closely spaced faults with very small displacements. These are well seen on the large expanses of bare bedding planes such as those forming the south-east side of

Logat Stacks. On the latter three intersecting sets of minor faults and joints are developed. These are:

1. Strike faults, trending E25°S, spaced 10 to 15 yd (9–14 m) apart with small downthrows (1 to 3 ft, 0·3–0·9 m) to the south-south-west.
2. Very small faults trending E30°N with very small downslip to south-south-east.
3. Small faults trending N22°E, about 25 yd (23 m) apart with small displacement to east-south-east.

Intense jointing and possible small-scale faulting is also well seen on the shores of South Ness, where the dominant trend is N10°W. On the hill slope north-east of the Wick of Mucklaberg the intersection of the south-south-west dipping bedding planes in the Daal Flaggy Sandstone, which are parallel to the steep undercut hillside, with west-north-west trending joints provides conditions ideal for landslides. The Smallie, a deep cleft over 250 yd (230 m) long and 6 ft (1·8 m) wide and as much as 100 ft (30 m) deep (Heddle 1878, p. 128) is the most prominent of a series of joints which have been opened by the down-dip slipping of the sandstone to the south.

REFERENCES

BLUCK, B. J. 1967. Deposition of some Upper Old Red Sandstone conglomerates in the Clyde area; A study of the significance of bedding. *Scott. Jnl Geol.*, 3, 139–67.

FINLAY, T. M. 1926. The Old Red Sandstone of Shetland. Part I: South-eastern Area. *Trans. R. Soc. Edinb.*, 54, 553–72.

GIBSON, G. A. 1877. On the Physical Geology and Geological Structure of Foula. *Rep. Br. Ass. Advmt. Sci.*, 90.

HEDDLE, M. F. 1878. *The County Geognosy and Mineralogy of Scotland, Orkney and Shetland.* Truro.

HIBBERT, S. 1822. *A Description of the Shetland Islands.* Edinburgh.

JAMESON, R. 1798. *An Outline of the Mineralogy of the Shetland Islands, and the Island of Arran.* Edinburgh.

NICOL, J. 1844. *Guide to the Geology of Scotland.* Edinburgh.

PETTIJOHN F. J. 1957. *Sedimentary Rocks.* New York.

TUDOR, J. R. 1883. *The Orkneys and Shetland: Their Past and Present State.* London.

SUMM. PROG. 1934. *Mem. geol. Surv. Gt Br. Summ. Prog. for 1933.*

Chapter 13

THE NORTHMAVEN–MUCKLE ROE
PLUTONIC COMPLEX

INTRODUCTION

UNFOLIATED igneous rocks occupy most of the area west of the Haggrister and Walls Boundary faults from Muckle Roe to the northern margin of the One-inch Sheet (Figs. 5 and 23). They form the southern part of the intrusive North-maven–Muckle Roe complex, of presumed Old Red Sandstone age, which extends from Swarbacks Minn to the Beorgs of Uyea near the north coast of Mainland. The total area over which rocks of this complex crop out is at least 50 sq. miles (>130 km²) and of this total about one-quarter, or 14½ sq. miles (37·5 km²) lies in the Western Shetland One-inch Sheet under description. The rocks occurring in this smaller portion are representative of the whole complex in rock types and probably also in the nature and sequence of the components; however, they include no member of the aegirine-riebeckite minor intrusive group which is a feature of the complex north of Ronas Voe, but on the other hand include two small outcrops of ultrabasic rock which is unrepresented to the north but is similar to material in the Sandsting Complex (p. 216). The descriptions which follow are based almost entirely on the six-inch maps and collections made during the field survey in 1931–33 and can be regarded as no more than preliminary to a more comprehensive study of the whole complex. Unlike the Walls district the Muckle Roe–Northmaven area has not been the subject of revision survey in recent years.

Age and sequence of events. The rocks of the complex comprise a wide range of plutonic types from ultrabasic to aplo-granophyric in composition and a suite of dyke rocks ranging in composition from olivine-dolerite to very acid felsite or rhyolite. They are not foliated but are traversed by many crush zones which are thought to be related to the later movements along the Walls Boundary Fault. They are intrusive into regionally metamorphosed country rock which is largely basic hornblendic gneiss with important mica-schist, quartzo-feldspathic granulite, and siliceous granulite components (Chapter 3). The complex consists essentially of granite, diorite and gabbro and is considered here to belong to one period, probably a prolonged one, of magmatic activity during the Devonian period. It should be recalled however that in the early description by Hibbert (1820; 1822) and the later by Peach and Horne (1879; 1884) the gabbro-diorite portion was considered a pre-Old Red Sandstone intrusion. In the Muckle Roe–Gunnister Voe area there is no geological evidence of an upper limit of age other than that of faulting and shattering along the Haggrister and Walls Boundary faults. In One-inch Geological Sheet Northern Shetland, however, xenoliths of rock which is considered by the writer to be andesitic lava comparable with the Eshaness volcanic rocks are enclosed in the Ronas Hill granite at Colla Firth (Summ. Prog. 1933, p. 78). It is therefore probable, or at least possible, that the complex is later than the Old Red Sandstone volcanic series of Eshaness. The age of rhyolite lava from Eshaness has been given by Flinn and others (1968) as 373 ± 2 m.y. from Rb-Sr whole-rock determinations, while the age of the Ronas Hill granite determined by the Rb-Sr method on biotite separated from the rock is reported as 358 ± 8 m.y. (Miller and Flinn 1966), a value practically the same as that determined recently by Snelling for the Sandsting granite (p. 211).

While the age of the granite component of the plutonic part of the complex can thus be regarded as defined within narrow limits by these isotopic determinations on analogous granites there is geological evidence within the area under description that the magmatic activity continued over a long period. From field data the complex can be separated into three major time groups: (i) a stage of early minor intrusion, (ii) a period of plutonic intrusion which can probably be further subdivided into three stages and (iii) a period of late minor intrusion. The plutonic period comprised a stage of basic intrusion, one of acid intrusion and hybridization, and one of late granophyre intrusion; further, at some late stage or stages within the period hydrothermal activity was widespread, and is manifested by retrogressive changes in the thermally altered country rocks (p. 33), uralitization of the basic plutonic rocks (pp. 195–6), and quartzification of retrograde hornfels.

Later in time than the three-stage sequence directly referable to the complex, but perhaps also a final phase of the magmatic activity, there was an episode of mineralization leading to scapolitization and zeolitization. This episode has been shown by Mykura and Young (1969) to be closely associated in time with the crushing and faulting which affect the Sandsting Granite and adjoining Old Red Sandstone sediments. In the Muckle Roe–Northmaven area the same time association of scapolitization and faulting is observed. Moreover there is evidence that the scapolitization is later than the quartzification of hornfels and that a still later phase of zeolitization resulted in replacement of scapolite by analcime.

EARLY HYPABYSSAL INTRUSIONS

Intrusive rocks earlier than the main gabbro-diorite-granite complex have been identified with certainty in only one case. A specimen collected from the small mass of gneiss enclosed in diorite, in western Egilsay, shows a 2 cm vein of contaminated granodiorite separating country rock which has different aspects on the two walls of the vein. On one side the rock is a foliated banded pelitic gneiss (S 44282A), on the other the rock is black, fine-grained and structureless (S 44282). The black structureless rock is a thermally altered porphyrite which next to the vein also shows accession of quartz. Irregularities of the gneiss and the porphyrite contacts correspond on the opposing sides of the vein as if the latter had made its way along a surface of discontinuity. The conclusion is clear that the porphyrite represents a minor intrusion in the gneiss earlier than the diorite-granite complex in which both are enclosed. Petrographically the porphyrite consists of phenocrysts of zoned plagioclase, centrally $\sim An_{55}$, varying seriately from 1 mm downwards in length, and microporphyritic pseudomorphs of a ferromagnesian mineral in a base of plagioclase ($\sim An_{30}$), green hornblende, olive-brown biotite, minor ore granules and epidote grains, and interstitial quartz. The mineral constituents of the base are closely intercrystallized and encroach on the faces and along the terminations of the phenocrysts; they have a fresh, recrystallized aspect and their form varies from xenoblastic granular to idioblastic prismatic, up to 0·05 mm long. The

N

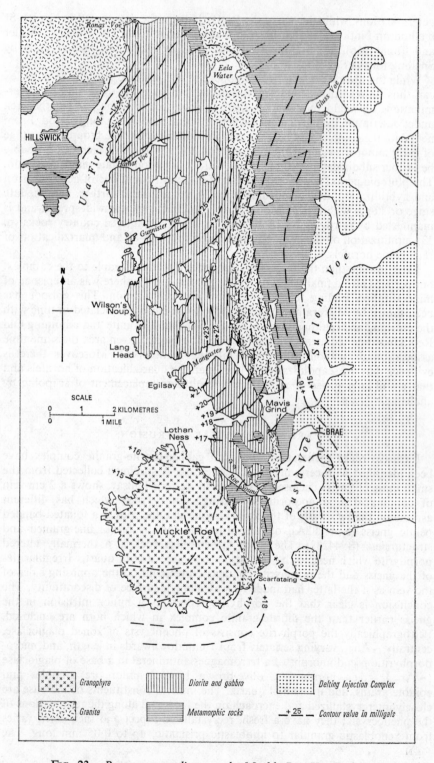

FIG. 23.   *Bouguer anomalies over the Muckle Roe–Ura Firth area*

ferromagnesian pseudomorphs are composed of microgranular aggregates of green hornblende which may represent pyroxene; larger ragged prisms of hornblende may have a different parent. At the margin against the vein plagioclase and quartz become more coarse and abundant and enclose aggregates and small crystals of the cafemic minerals (S 44284, Plate XXII, fig. 1).

On the north face of the Ward of Runafirth [343 676] (Fig. 5), a basalt or dolerite, in which the plagioclase prisms range from 0·3 to 3 mm in length, shows some evidence of contact alteration by the gabbro-diorite. It may be an early dyke but its field relations are not known. The pyroxene is ophitic to the plagioclase ($\sim An_{70}$) and though of a faint pink colour is monoclinic (S 55658); some small grains of squat octagonal shape may be of orthopyroxene. The mesostasis is a cryptocrystalline green and brown aggregate, with accessory iron ore, which seems to be mainly amphibole but in part is composed of compact clusters of brown microcrystalline scales of biotite which suggest thermal reconstruction.

In the eastern coastal area of Muckle Roe there is some evidence of intrusion of dolerite prior to the emplacement of the granite which forms sheets and veins in the gabbro-diorite but definite conclusions are difficult to reach partly because of the prevalence of mineral change associated with crushing close to the Walls Boundary Fault-zone, and partly because of scapolitization penecontemporaneous with the crushing. The difficulty of distinguishing and classifying basic rocks in exposures along this coast—hornblende-schist, gneiss, or granulite, gabbro-diorite, or doleritic minor intrusion—has already been mentioned (p. 29). The least doubtful field evidence is found in large masses of gneiss enclosed in granite in the Scarfataing Burn [337 639] (p. 30). The mass which is exposed 250 to 300 yd (230–270 m) upstream from the burn mouth shows good foliation in dark grey and green gneiss intruded by thin wedging sheets of granite (Fig. 5) which are nearly conformable with the foliation and contain streaks of gneissic rock. The gneiss includes thin bands of dark grey massive structureless basic rock which is ophitic dolerite (S 45131). This rock is composed of prismatic labradorite, An_{65} but less calcic in outer zones, colourless augite and less abundant green hornblende, both ophitically related to the plagioclase; plates of leucoxenized ilmenite are abundant and microcrystalline to cryptocrystalline pale green aggregate, locally associated with epidote, is interstitial and occupies areas 2 to 3 mm across resembling irregular vesicles. Both augite and green hornblende are extensively replaced by turbid brownish or greenish grey streaky uralite. Neither quartz nor apatite is seen. The rock shows no textural recrystallization other than that usual in uralitization, but the finely divided green interstitial material, and similar aggregate occupying narrow, impersistent shear fractures, appear to consist entirely of amphibole; this occurrence of amphibole instead of chlorite is taken to indicate low temperature hydrous recrystallization and may be connected with the intrusion of the granite sheets. Similar relations of metadolerite and hornblende-rich gneiss xenolithic in granite were observed in outcrops between the road and the coast about 250 yd (230 m) N of Scarfataing crofts [342 642] (Fig. 5).

A dark grey, fine-grained rock sprinkled with white phenocrysts, up to 3 mm in length, collected from the north-east corner of Roedale Water [315 736], ½ mile (800 m) SW of Gunnister Voe, is doubtfully included among the early hypabyssal intrusions; its field relations are uncertain. The rock is composed essentially of particoloured brown and green hornblende intersertal or subophitic to calcic labradorite (An_{65-70}) prisms 0·1 mm long; fibrous or short acicular green amphibole of lower refractive index than the hornblende, scarce chlorite, and granules of epidote and sphene are interstitial; there is no iron ore (S 55269). The phenocrysts are idiomorphic but are very thoroughly altered to saussuritic epidote and white mica; relics of fresh feldspar indicate the original composition as >An_{75}. The crystallization of the white mica is unusually coarse, and some plates run the length of the phenocryst and enclose the epidotic aggregates. The alteration of the porphyritic feldspar occurred prior to inclusion in the magma which the groundmass represents since the latter locally invades

---

EXPLANATION OF PLATE XXII

PHOTOMICROGRAPHS OF ROCKS OF THE NORTHMAVEN–MUCKLE ROE PLUTONIC COMPLEX

FIG. 1.   Slice No. S 44284. Magnification × 25. Plane polarized light. Early basic dyke, thermally altered. The plagioclase (An 50%) phenocrysts are recrystallized at their margins to interlock with the recrystallized base consisting of microgranular andesine and small idioblastic prisms of hornblende and biotite; small ferromagnesian phenocrysts are recrystallized to compact aggregates of interfering microprisms of hornblende. West side of Egilsay [316 695].

FIG. 2.   Slice No. S 30016A. Magnification × 14·5. Plane polarized light. Bytownite-peridotite. Numerous small crystals of olivine are enclosed poikilitically in large plates of bytownite and clinopyroxene. The relative proportions of these minerals are variable; orthopyroxene and reddish brown hornblende and biotite are variable minor constituents. South-east shore of Glussdale Water [332 734].

FIG. 3.   Slice No. S 30017A. Magnification × 11·5. Plane polarized light. Gabbro. Large plates of augite enclose prisms of calcic labradorite ophitically; brown hornblende margins augite and also forms smaller ophitic plates. A late crystallization of deep brown biotite is moulded on plagioclase. West shore of Glussdale Water [332 734].

FIG. 4.   Slice No. S 55647. Magnification × 10. Plane polarized light. Hornblende-gabbro (bojite). Large partly uralitized plates of brown and green hornblende enclose zoned plagioclase (An 65–45%) prisms ophitically. Deep brown biotite is moulded on plagioclase. Cliva Hill, 150 yd (130 m) SSE of Mavis Grind [340 682].

FIG. 5.   Slice No. S 30023. Magnification × 10·5. Plane polarized light. Hornblende-augite-quartz-diorite. Prisms of augite are subophitic to plagioclase; brown hornblende is moulded on plagioclase and idiomorphic against quartz. Plagioclase prisms are zoned from acid labradorite core to oligoclase margin. South shore of Gunnister Voe [315 742].

FIG. 6.   Slice No. S 55665. Magnification × 12. Plane polarized light. Biotite-hornblende-quartz-diorite. Tables of zoned plagioclase (An 55–15%) interfere with hypidiomorphic brown and green hornblende which show good crystal forms against quartz (top centre). Peninsula 195 yd (180 m) WSW of Nibon [304 730].

FIG. 7.   Slice No. S 45034. Magnification × 14. Plane polarized light. Scapolitized gabbro. On the left, gabbro of unaltered plagioclase, hornblende and augite resembles that of Figs. 3 and 4; on the right, plagioclase is completely, hornblende partially replaced by xenomorphic, coarsely crystalline scapolite. East coast of Muckle Roe, 435 yd (400 m) NNE of Scarfa Taing [334 639].

FIG. 8.   Slice No. S 29505. Magnification × 14·5. Crossed polarisers. Granophyre. Muckle Roe, west side of Roda Geo [314 674].

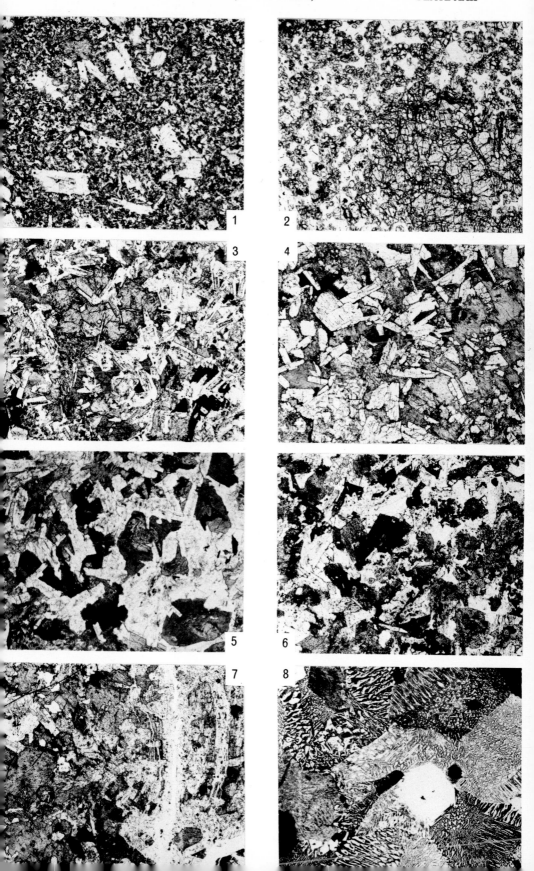

the phenocrysts and incorporates the mica and epidote alteration products. The rock is unlike any other collected from the complex. From the mineral composition it can be classified as a hornblende-basalt; in the complete absence of pyroxene and iron ore it is a very unusual petrographical type and no corresponding rock is described by Johannsen (1937) under his families 2312 and 3312.

Other early igneous rocks which occur in the main gabbro-diorite area north of Mavis Grind (Fig. 5) but are not separable in the field as geological units are described in the section dealing with the plutonic complex (see pp. 184–5).

THE PLUTONIC COMPLEX

FIELD RELATIONS

Ultrabasic rocks

Ultrabasic members of the plutonic complex have been found in only two localities, at the south-east corner of Glussdale Water [333 732] and on the east bank of the northern of the Moora Waters [328 729]. Both localities are close to the junction of the Gunnister road with the main road and are easily accessible. The outcrop at Glussdale Water shows black ultrabasic rock while that at Moora Waters is represented by large boulders, the nearest rock in place at both localities being diorite. Both outcrops are in drift and the nature of the contact of the dioritic and ultrabasic rock is not known. The rock at Glussdale Water has a greenish black aphanitic base in which glisten ophimottled cleavage surfaces up to 0·5 cm across. The Moora Waters rock is similar but blacker; it has been chemically analysed and described briefly by P. A. Sabine who has classed it as an enstatite-harrisite (Guppy and Sabine 1956, p. 35). The petrography of rocks from the two exposures is described on pp. 193–4; they are composed of orthopyroxene, clinopyroxene, calcic plagioclase, olivine, and minor hornblende and biotite.

Diorite and Gabbro

The basic and intermediate rocks of the complex can be considered only together since it is impossible to separate them in the field. On the published one-inch Geological Sheet areas in which the rock has been noted as gabbroic or doleritic have been outlined and are coloured as gabbro. The outlines on the map are however only lines necessary in the production of a colour-printed sheet and do not indicate bodies possessing a definite form within or showing contact relations to the dioritic portion of the complex. No contact relations between gabbroic and dioritic rocks have in fact been noted during the survey, and as recorded by Finlay (1930, p. 685) 'in the more basic modifications, rocks of different composition are often confusedly intermingled, gabbros passing upwards or laterally within a small area into dioritic or even more acid types'. He stated also that gabbro is more common in the south of the Northmaven area and along the eastern margin from Mavis Grind northwards (op. cit., p. 690) and that 'in a north-west direction one finds a gradual increase in acidity and occasionally the rock is a granodiorite'. Finlay regarded the gabbro-diorite-granite as an intrusion of sheet form inclining gently westwards, showing a transition from granite above to gabbro at the base of the sheet and extending from near the north coast to Muckle Roe, the granite reappearing again in Sandsting from beneath the Walls Syncline. In the Muckle Roe–southern Northmaven area, however, mapping has shown that areas of gabbro are distributed through the diorite as far west as the entrances to Gunnister and Mangaster voes, Turvalds Head (Fig. 5) and the western end of Roe Sound. In contrast to Finlay's observation that there is a gradual increase of acidity westwards it will be seen from the geological map that granite veining and acidification of the diorite is as common in the east as in the west. A phenomenon unremarked hitherto is the presence of areas where inclusions of black basaltic rock are numerous in the diorite and in places give the rock the aspect of a breccia (Plate XXI). The petrography of these rocks (pp. 198–9) suggests that some represent early volcanic or sub-volcanic rocks which have been thermally altered by the diorite. Unfortunately only a few specimens were collected during the field survey and it is not possible to make a statement of the dominant rock-type. They are mentioned in this section dealing with the gabbro-diorite rather than in the previous section on early intrusions because there is no evidence that they were intruded into gneiss country rock as minor intrusions. Detailed study of the breccia-like areas and of the xenoliths is necessary to show whether gneiss enters into or is absent from their composition. In the latter event the xenolithic material must represent early volcanic or sub-volcanic material which has been disrupted prior to or concurrently with intrusion of the plutonic rock.

In the following paragraphs the distribution of the basic rocks and their associations is dealt with by areas.

Gunnister Voe–Mangaster Voe. In this district low rocky hills and knobs separated by peat-filled hollows are composed mainly of dioritic rocks. A speckled medium-grained type, ranging in general colour from grey to dark grey as the proportion of mafic minerals—mainly hornblende—to feldspar increases, is dominant but wide variation in grain and texture and of composition occurs. Thus the rock may possess a homogeneous aspect corresponding with uniform distribution of dark and pale minerals or may show regular mottling by feldspathic clusters or irregular, impersistent feldspathic veining. Homogeneous members differ in aspect owing to difference in nature or morphology of the dark mineral; the presence of much augite is associated with a more sparkling appearance and a cuboidal fracture while the occurrence of hornblende in longer prisms, usually in the more feldspathic varieties, produces a felted as distinct from the usual granular texture. Variations from diorite to more basic and more acid types are common, and are found in any part of the district. Along the eastern limit of the diorite dark greenish grey, medium to coarse-grained gabbroic rocks are common south of Glussdale Water to the head of the Mangaster Voe, plagioclase-phyric types being frequent in the complex of granitic and basic rock between Innbanks [337 700] and Scora Water [337 717]. In the central area coarse gabbro occurs around the Gill of Mangaster [323 708] and medium-grained feldspathic gabbro at intervals northwards to Gunnister Voe [315 473], for example on the north-east of Brei Water [320 713], on the south-west of Brei Water of Nibon [315 713] and east of the southern loch of Moora Waters. Along the west coast coarse gabbroic types appear south of Lang Head [305 703], on the east side of the Isle of Nibon and astride North Sound [306 743] at the mouth of Gunnister Voe.

Varieties of granodioritic composition occur throughout the diorite. Locally, as at Noons Vird [317 739] south-west of Gunnister, such rocks occupy outcrops large enough to be mapped as bodies of granodiorite which, however, have no well-defined demarcation from the diorite; in places they are contiguous with red granite of the type which forms most of the larger acid masses shown within the diorite on the One-inch Geological Map. They show irregular variation in mineral composition and contain basic xenoliths more or less digested. As well as in these larger occurrences variably acid types are numerous as bodies of small extent and irregular form, and occur also as networks of veins. These appear at intervals throughout the diorite outcrop and are particularly well exposed on Wilson's Noup [301 718], on the west coast (Fig. 23), where their wide variation in mineral composition from aplite to granodiorite coupled with varying degrees of assimilation of fine-grained basic material is clearly seen.

In addition to this range of varieties, which because of their continuous transition from ultrabasic to acid exemplify a *serial* type of variation, another type of variation which may be referred to as *hiatal* is found. The hiatal type is characterized in this area by an abundance of xenoliths of a very fine-grained basic igneous rock in a matrix of diorite or granodiorite of normal medium grain. On the evidence of the specimens collected during the survey of the area most of the xenoliths make sharp contact with the enclosing rock and generally no change in either rock is visible across the contact. There is usually, however, some small part of the contact along which reaction occurs involving felds-pathization and coarsening of the xenolith though its original margin remains visible; from these parts feldspathic material may penetrate into the xenolith as irregular, impersistent apophyses with diffuse margins. The hiatal xenoliths are in places so numerous, as at Wilson's Noup [301 718] (Plate XXI, A) and south of Nibon between Middis Vird and the coast [305 728], that the rock has the appearance of a breccia and the sharp margins of the fine-grained basic individuals, in some cases separated by only a few millimetres of acid material, confirm the appropriateness of the designation. To the eye the xenoliths look like basic microdiorite or feldspathic basalt and some are aphanitic. Usually they contain microporphyritic thin laths of feldspar up to 2 mm in length. Petrographical examination of a number of xenoliths shows that some have been thermally recrystallized but that most retain their original, in some cases fluidal, igneous texture. The matrix to the xenoliths may be diorite or grano-diorite and usually contains ghosts of fine-grained basic and even ultrabasic material or shows diffuse variation in proportion of dark to light-coloured minerals. These characteristics indicate assimilative rather than hybridization activity. The narrow veins and strings separating parts of xenoliths or pen-etrating into them are generally of feldspathic or quartzo-feldspathic composi-tion and their composition and manner of occurrence suggest that fluids rich in silica and alkali have been active over a long period and represent also the final stages of magmatic action. An unusual kind of xenolith occurs a short distance north-east of Roedale Water [315 736]. Described from its field relations as diorite marginal to the granite, it is a hornblende-porphyrite showing opaque white tables of altered calcic labradorite up to 4 mm long, and is perhaps an early dyke rock (p. 181).

The 'serial' and 'hiatal' styles of variation appear to reflect two geological processes. The serial is essentially a magmatic process of differentiation leading to the formation of ultrabasic, basic and intermediate rocks and late quartzo-feldspathic residuals while the hiatal represents an early though cognate hypabyssal or volcanic phase perhaps representative of the undifferentiated magma. There is no evidence in this area of the structural relations of the ultra-basic, basic, and intermediate components of the complex and no directional structure has been noted. The more acid components are clearly of later formation whether their formation has been by the action of alkali-silica liquors on rocks consolidated in place or by the consolidation of intrusive granitic magma. *Mangaster Voe–Roe Sound.* This area comprises the Islesburgh peninsula, lying north-west of Mavis Grind, the islands of Egilsay, The Hogg [313 697], and Black Skerry [319 692], and the Busta peninsula between Mavis Grind and Roe Sound (Fig. 5). The same irregular distribution of patches of gabbroic rock in medium-grained speckled and fine-grained grey diorite is found as in the country north of Mangaster Voe but the gabbroic patches seem more numerous; this, however, may be occasioned by better exposure in the hilly topography of this area, which is almost free of peat cover. Here also as in the northern district, ground composed of basic and intermediate rock, in places cut by individual dykes or sheets of granite or aplite, alternates with ground in which a network of acid veins penetrates the diorite with local production of variable granodioritic material.

The most common rock of the area is diorite of medium grain which shows an equal and homogeneous distribution of pale and dark minerals, but variation to dark rocks speckled, or occasionally clotted by feldspar and to more feldspar-rich types speckled by black mineral is usual. Variation in grain appears to be less common but fine-grained varieties occur, in some cases feldspar-phyric, which on their macroscopic appearance could be classed as fine-grained dolerite or basic microdiorite. The relation of these finer-grained members to the medium-grained diorite is not known, but in one case, on the shore 300 yd (274 m) NW of Mavis Grind, the fine-grained rock is noted as forming a basic band in diorite. Breccia-type concentration of very fine-grained inclusions in diorite or granodiorite, such as have been described above from Wilson's Noup, occurs only rarely in this area, for example on the western coast south of Turvalds Head, but the basic inclusions there are, in some cases at least, of metamorphic rock (pp. 27–28). The most common type of gabbroic rock is a medium-grained, greenish grey, homogeneous gabbro, but varieties with specks of pale feldspar and scattered scales of biotite are almost as plentiful. Varieties with poikilitic feldspars up to fully 1 cm across are found, for example, on the knoll 500 yd (450 m) NW of Busta House and on the summit of Cliva Hill [341 682] above Mavis Grind. The most accessible localities for examination of the gabbro and diorite, and their variation lie along the coast east and west of Mavis Grind, along the rocky ascent of the main road north from the Grind, and in the quarry bordering a stretch of this road on the Ell Wick coast. Coarse, greenish grey gabbro with plates of augite up to 0·5 cm across, ophitically enclosing plagioclase, and sparkling medium-grained doleritic gabbro with lathy plagioclase both occur in the road-cutting. In the small quarry west of the road gabbro, with primary brown hornblende subordinate to pyroxene, forms the main rock and contains veins and patches of coarse feldspar-hornblende-pegmatoid (S 43535) which carries both thomsonite and analcime as interstitial filling to albitized plagioclase tablets. Diorite composes the Mavis Grind isthmus and is cut by granite dykes on the eastern coast while vein networks of granite in dioritic and gabbroic rock are to be seen on both sides of the head of Mangaster Voe and in the large quarry [342 683] which borders the road along Ell Wick. The main material in this quarry is grey fine-grained diorite cut by irregular veins of pink granite; in the centre of the quarry a coarse gabbroic rock overlies massive red granite and both the basic and the acid rock appear to be cut off by the grey diorite. If this relation is a real one there must be two periods of granite intrusion. Proximity to the Mangaster Voe Fault is reflected in the broken and crushed condition of the rock in this quarry; lines of dislocation strike NNE–SSW and ESE–WSW, and along the north-north-easterly lines ribs and bands of white or pinkish scapolite are present.

Muckle Roe. Two comparatively small areas of diorite and more basic rock occur on Muckle Roe (Fig. 5). They are separated by an outcrop of gneiss which extends from Roe Sound southwards to Kilka Water and from the margin of the Muckle Roe granophyre eastwards to the bridge over Roe Sound. That an irregular contact of gneiss and igneous rock lies close to the existing surface is shown by the appearance of outcrops of gneiss within the diorite, and of diorite within the gneiss in the Stabaness–Burn area [330 664] and by the embayed contact of the two formations around the north end of Kilka Water. Over the northern area the igneous rock is mainly medium-grained grey diorite cut by granitic veins and traversed by granitic vein networks between Lee Skerries [323 670] and Otter Ayre and on Roe Sound north of Burn [333 662]; with these networks coarser-grained types of diorite with small feldspathic clots and granodioritic varieties are associated. The association of basic gabbroic types usual farther north is repeated in the occurrence of fine-grained gabbro on the higher ground west of Stabaness croft [328 664].

The southern area of dioritic and gabbroic rock on Muckle Roe extends from Kilka Water almost as far south as the Burn of Scarfataing. It is separated from similar but generally more gabbroic types along the Busta Voe coast by a strip of gneiss of irregularly changing width. The contact between diorite and gneiss is seen only in a low knoll west of Orwick Water. At this exposure the igneous rock is fine-grained dolerite, in part porphyritic, and has a chilled appearance against hardened hornblende-rich gneiss; the contact curves from E–W to NE–SW to E–W again on the low surface but its inclination is not observable. The contact of the diorite with granite or granophyre on the west and south of its outcrop is conjectural owing to poor exposure under drift. In the most southerly exposure, south-west of Orwick Water, the diorite has granodioritic variations which produce a directional structure striking south by east. Combined with the similar directions of elongation of the gneiss enclaves in the granite along the Burn of Scarfataing this directional structure in the diorite may indicate a tongued type of contact of the diorite and granite rather than the blunt contact shown on the map.

Along the Busta Voe coast gabbro is found on Scarfataing [340 637] and in the cliffs under Lubba and Southpund [343 654] crofts. Diorite is most in evidence along the stretch between Greentaing [343 645] and a fault 600 yd (549 m) to the south but here also gabbroic and granodioritic variation is common. The more basic rocks enclose masses of basic gneiss and the assemblage is riddled by granitic sheets and dykes. Superimposed on this mixture of rocks are the effects of crushing along the Walls Boundary Fault which obscure the original relations of the formations and which by convergent retrograde changes in the mineralogy of the basic components make field determination of their igneous or metamorphic origin uncertain or impossible. Still further complication is caused locally by scapolitization, concurrent with or later than the faulting, which results in the production of medium-grained speckled rocks without foliation, resembling diorite.

Structure of the diorite-gabbro complex. No clear evidence of the form of this intrusive mass has been found in the region under description. On the east it is bounded by granite which may be of dyke-form (p. 190) and by the Walls Boundary Fault. On the west it is cut off by the ocean and by the granophyre of Muckle Roe. At the south end there are indications of a steep orientation striking south-south-east in a variable, probably hybridized diorite. On the north it is continuous with the diorite of Northmaven, the form of which has not yet been determined. Internal evidence of structure such as banding of rock types and mineral orientation is practically non-existent; though variation in grain and composition is everywhere apparent in only one instance has a banded type of relation between the fine-grained and the usual granular diorite been noted. The distribution of gabbroic facies throughout the diorite seems to have no such regularity as would indicate any layered or zonal arrangement. Finlay (1930) suggested that the diorite-gabbro of Northmaven forms the lower, non-uniform component of a great granite-diorite-gabbro sheet inclining gently and becoming more acid westwards, and he noted the common occurrence of gabbros in the southern part of Northmaven, that is in the area under description here. The widespread occurrence of gabbro from east to west of this area, however, gives no support to the hypothesis of a basic layer reaching deeper levels westwards. The only regularity shown by the mapping is that of a general increase in outcrops of gabbro to the south—in Busta peninsula and in the east of Muckle Roe. If the diorite-gabbro mass is in fact the lower part of a sheet which becomes more basic in its lower levels the evidence of greater abundance of gabbro in the south would indicate a northward inclination of the sheet. It has been pointed out in an earlier chapter (pp. 25–30) that, so far, there is no clear evidence whether the gneiss of the Skipadock [342 691] and Busta–Muckle Roe areas overlies the igneous rock as a roof with pendants or underlies it with upward projections from an uneven floor. If the gneiss forms a roof the general slope of the surface between the two formations must rise northwards from Muckle Roe and must be faulted down by the Mangaster Voe Fault to allow reappearance of the gneiss at Skipadock, on the assumption that the superpositional relation is the same there as in the Busta peninsula. On Finlay's hypothesis, however, the surface between gneiss and igneous rock will fall northwards and consequently the gneiss at Skipadock will lie on the upthrow side of the Mangaster Voe Fault. Steep north-easterly dips in the gneiss close to the fault suggest that the gneiss is downthrown on the Skipadock side, if the fault is a normal one, thereby favouring the gneiss-roof alternative. The occurrence of a massive enclave of gneiss in granite at Djubi Dale [336 743], 3 miles (4·8 km) N of Skipadock also indicates that gneiss overlies the igneous rock. The nature of the Mangaster Voe Fault is not known. From the general picture of Bouguer anomaly (Fig. 23) in this gravitationally confused area no reliable conclusions can be drawn whether downthrow is to south-west or to north-east. It is possible, too, that the superposition relations of gneiss and igneous rock are not the same at Skipadock as around Busta. At Skipadock the gneiss may be roof and in the Busta peninsula floor. In such circumstances the direction of throw of the fault is not critical to the problem.

In the Busta peninsula the gravity map (Fig. 23) shows low anomaly of saddle-form over the area where gneiss crops out most extensively. Here also the gravity picture is a confused one since the gravity changes in an E–W direction resulting largely from low values over granophyre in Muckle Roe and high values over pyroxenite south of Brae. The coincidence of the saddle-form low anomaly over the gneiss outcrops appears, however, to indicate that the underlying rock is more probably a basement of gneiss than basic igneous rock of batholithic depth. The consistent increase in gravity northwards from this area implies increasing thickness of the basic rock in that direction, with maximum thickness occurring between Gunnister Voe and Hamar Voe, just north of the area under description.

The tentative conclusions reached on consideration of the possibilities and uncertainties is that the gabbro-diorite has the form of a sheet which thins southward and that it may be a lopolith with its feeder conduit lying between Gunnister and Hamar voes. The occurrence of amphibolite and garnet-magnetite rocks of Clothister type among the gneiss outcrops south of Bays Water [334 668] suggests that the lopolith may have been intruded along a folded thrust dislocation like the one postulated (p. 22) between the two metamorphic formations on the east side of the Busta–Haggrister Fault.

Granites and Granophyre

As shown on the geological map and Fig. 23 there are three main bodies of granitic rock in the area. The most northerly of these is here termed the Eastern Granite. It consists of coarse red granite and extends from the north edge of the Sheet south to the east end of Roe Sound. A small mass of coarse red granite crops out in the south-east part of Muckle Roe around Scarfataing. The largest granitic body is the Muckle Roe Granophyre which occupies the greater part of that island. Numerous smaller bodies of granite appear within the gabbro-diorite, and these are of considerable size north of Mangaster Voe. From their irregular and tongued junctions with the diorite it appears probable that they are, in part at least, of sheet form. In this district however it is obvious in the field that granite spreads from its main N–S trunk into the diorite as a complex stockwork of thick and thin veins and sheets. Away from this complex area the smaller granitic bodies appear mainly in the form of dykes, of short length in relation to their thickness, and in small oval outcrops which may be of stock or lens form. In many cases the outcrops are mapped as irregularly embayed, as for example to the south of Gunnister Voe, and it seems probable that the acid rock has penetrated the diorite as a sheet-dyke complex. This probability is strengthened by the common occurrence, close to masses mapped as red granite, of areas in which granitic and aplitic veins of random direction are abundant; such areas are indicated on the geological map by short line ornament in red. Dyke-form aplites are common as individual intrusions in places where granitic networks have not been observed, for example on the north side of outer Mangaster Voe and in the centre of the Islesburgh peninsula, and this independent mode of occurrence may indicate them as the last phase of granitic intrusion. Pegmatites have been noted only rarely and only in association with granitic vein networks; they appear to be merely local coarse-grained facies of inconsiderable magnitude. The field characters and lithology of the granitic bodies are described below in the order: the Scarfataing Granite, the Eastern Granite, the Muckle Roe Granophyre, the smaller bodies and veins.

Scarfataing Granite. This granite is intrusive into and encloses large enclaves of the gneiss. From the elongated shape and generally vertical dip of the enclaves and from dentate contact of granite with the larger enclaves it would appear that the granite was intruded vertically as sheets. Exposures in Scarfataing Burn show that in the broad outcrop 300 yd (275 m) up from the river mouth (Fig. 5) the gneiss and granitic sheets wedging into the gneiss have a moderately high dip to the north-west, while in the next outcrop upstream red granite is interposed between two vertical strips of gneiss. In these stream exposures and in the outcrops between the stream and Scarfataing crofts the granite contains streaks of gneiss relics and the xenoliths of gneiss are locally penetrated by granitic laminae. Both granites and gneiss have a sheared or crushed appearance in some outcrops. The enclave of gneiss farthest upstream shows sheared granite-ribbed gneiss cut by an unsheared clean red leucogranite. Along the coast between Scarfataing and Pobies Geo [337 633] several exposures of gneiss appear as vertical or steep bands within the granite which also contains inclusions of gneiss elongate in a vertical direction so that they simulate dykes. West of Pobies Geo the granitic rocks are greatly crushed, at least seven dislocations with a north to north-westerly trend being present between Pobies Geo and the burn at Knowe; granophyre becomes identifiable with certainty on the coast about 100 yd (91 m) E of the mouth of this burn. Thus the contact relations between the Scarfataing Granite mass and the granophyre are uncertain, but since the granophyre is in general unaffected by shear while the granite is so affected along with the gneiss which it penetrates, it is probable that the granophyre is younger and that either its eastern boundary was guided by the existence of a marked zone of crushing or that its imminent intrusion produced a zone of crushing in the granite-gneiss cover. The contact relations between the Scarfataing Granite and the gabbro-diorite complex are unknown.

From the observations described above it is suggested (i) that the Scarfataing Granite is in the main early in the magmatic history and was intruded in sheet form under physical conditions which allowed some assimilation of the gneiss country rock to take place, and (ii) that at a much later period, separated from the first phase by an interval during which crush lines were formed, a further intrusion of granitic material was emplaced. It is of interest to recall here the occurrence of layers of ophitic dolerite in the gneiss enclaves (p. 30) as further evidence of early intrusive units probably even earlier than the Scarfataing granite.

Eastern Granite. This intrusion extends along a 9 mile (14·5 km) long outcrop from the eastern end of Roe Sound northwards to Ronas Voe. Its width is 1 mile (1·6 km) near Eela Water, in One-inch Geological Sheet Northern Shetland, but varies down to 500 ft (150 m) in the stretch between Busta Voe and Ell Wick [344 680]. The form of the outcrop is thus that of a dyke, and the intrusion was so described by Finlay (1930). Its eastern margin is concealed over most of its length by superficial deposits but exposures are sufficiently numerous and close to show that the boundary is everywhere against metamorphic rocks and runs north–south in a course which is only slightly undulating. In the area under description the contact with the metamorphic rocks is a steep faulted one well exposed on the west side of the Bight of Haggrister [346 700]. It has been suggested on an earlier page (p. 22) that the eastern limit of the intrusion was controlled by the existence of early dislocations along the line of the Walls Boundary Fault-zone.

On its western margin, however, the granite presents contact relations entirely different from those of a steep dyke-form intrusion. Everywhere on this side it is in contact with the gabbro-diorite but the contacts take the form of a plexus of large and small sheets, dykes, and veins of granite which can be mapped only in a general way, so that in places diorite appears on the map to be surrounded by granite, in other places granite by diorite. This complex is well exposed in the knolls and cuttings along the road between the head of Mangaster Voe and Glussdale Water where the interweaving of dark diorite and bright pink or red granite in a rocky topography produces a vivid scenic effect. The basic rock in contact with the granite is in places gabbro or coarse dolerite which may be of the porphyritic variety and locally is granitized to granodiorite. It is clear that the granite was intruded into the consolidated basic rock with only limited reaction and interchange of material.

Thus while the outcrop of the granite is in general that of a dyke-form body the complex relations on its western margin show that it was intruded under conditions which permitted easy lateral penetration of the consolidated gabbro-diorite. The ease of penetration may have been due partly to incipient cracking as the basic mass cooled, partly to shivering on reopening of an ancient locus of faulting along which the granite later ascended.

Throughout its extent from Busta to the north margin of the One-inch Sheet the granite maintains the aspect of a pink to red leucogranite composed of a base of coarsely crystalline feldspar in which are set blebs, up to 3 mm across, or less regular and larger aggregates of quartz. Ferromagnesian mineral is consistently present but in small proportion. It is largely biotite and usually forms interstitial aggregates of grain much finer than the quartz and feldspar; this character and the occasional laminar form of the aggregate suggest its genesis as xenolithic rather than primary crystallization. Some specimens of the rock are slightly cavernous; the small vacuoles may be druses but in many cases seem to be due to disappearance of loose interstitial aggregate. Slight variation from the type is shown by a specimen, from the roadside above Innbanks [338 700], which contains porphyritic feldspar of very pale pink colour in a base which is of finer grain than normal. Pink aplite with small porphyritic feldspar tables and fine-grained biotite-granite with irregularly distributed coarse feldspathic aggregates, the whole having a hybrid aspect, have been noted as 'bands' in the coarse red granite on the west side of the Loch of Haggrister [337 705].

Muckle Roe Granophyre. This granophyre forms a mass which, though its western and southern boundaries are cut off by the sea, seems to have been of roughly circular outline. As now exposed it extends for 3 miles (5 km) in a north–south direction. As already stated (p. 190) the nature of its contact against the Scarfataing Granite is obscured by a broad zone of crush lines between Pobies Geo and Knowe, and the location of the boundary on this 400 yd (360 m) stretch of coast has not been determined. No contact is visible along the eastern boundary which runs from the coast on a general but slightly undulating S–N to SSE–NNW course towards Kilka Water and can be located to within 120 yd (110 m) by outcrops of granophyre and diorite to the west of Orwick Water and by outcrops of granophyre and hornblendic gneiss west of the north end of Kilka Water. West of this loch the trend of the boundary swings from north-north-west to north-west to reach the sea in the notch of the coast which forms the root of Lothan Ness. Along this stretch the country rock to the north-east is at first hornblendic gneiss, within which lie bodies of granite and of diorite, followed by gabbro-diorite south of Stabaness, and at the north-west end the country rock is diorite invaded and locally granodioritized by a granitic

network of veins. Throughout its outcrop the granophyre maintains a distinctive aspect conferred partly by the almost aphanitic nature of the pink micropegmatite base, partly by a characteristic close ragged jointing which on the
hilly slopes in the interior of the island is responsible for the weathering of the
rock into immense screes. The rock is of pink colour locally tinged with yellow
and shows numerous blebs of clear quartz in a stony base in which individual
crystal outlines of feldspar are only rarely seen but in which feldspar cleavage
surfaces up to 1 cm long are usual and impart a ragged fracture to the rock.
Drusy cavities containing more and less well terminated crystals of quartz are
common and black ferromagnesian mineral is a minor constituent but always
present in shapeless clots. Rarely, as in a specimen collected in the south-west
of the island, 400 yd (360 m) W of Gilsa Water [302 633], the base of the rock
is locally microgranitic (S 28909). The occurrence of small masses of hornfelsed
banded hornblende-rich gneiss in the granophyre has already been mentioned
(pp. 30–31).

The roughly circular form of the granophyre, though the existing outcrop is
only partial, and its cross-cutting relations to the gneiss and gabbro-diorite, and
also to the later granitic vein network in the latter suggest that the mass is of
stock form and of late date in the plutonic history of the district. Its intrusion
was earlier than the dykes of felsite, porphyrite, and basalt or dolerite by which
it is cut. The shape of the low gravity anomaly over the granophyre (Fig. 23)
though incomplete is consistent with its occurrence as a stock-form body of
roughly circular section and low density.

Smaller Acid Bodies and Veins. Small acid bodies and veins are intrusive into
both the diorite-gabbro and the gneiss. Macroscopically they include three
main types: coarse-grained granite, feldspar-phyric fine-grained granite, and
aplite. The coarse granite contains less or more abundant clots of dark minerals
and closely resembles the rock of the Eastern Granite. It occurs as small boss-
like bodies in the gneiss of the Busta peninsula and as thick forked intrusions
in the gneiss at Skipadock. Throughout the gabbro-diorite it forms masses of
any size from 500 yd (450 m) downwards in length, in places of oval outline, in
places irregularly and repeatedly embayed, as for example in the central part
of Mangaster Voe. Though these masses are numerous towards the head of that
voe as if spatially related to the Eastern Granite they are equally as large and
common south-west of Gunnister. Even as small bodies they appear to make
outcrops less regular than dykes. In places they appear where no granitic vein
network has been mapped in the gabbro-diorite, which suggests that they
represent a distinct phase in the evolution of the complex. Finlay regarded the
larger granites within the gabbro-diorites as forming two broad dykes running
respectively north-west from the eastern end of Roe Sound and north-north-
west across the Islesburgh peninsula to Gunnister Voe and he noted that 'on
higher ground they are in places impersistent, failing to reach the surface
altogether or represented by a medley of smaller dykes and narrow, fine-grained
or felsitic veins ramifying through the diorite' (Finlay 1930, pp. 691–2). From
their distribution, shape of outcrops, and variability of direction of their length
as shown on the Geological Survey six-inch maps it appears more probable
that these coarse-grained granites have penetrated at random as sheets and dykes
and combinations of sheets and dykes, and that no persistent direction of extension from a focus exists.

The fine-grained and aplitic rocks have the form of dykes or veins and no definite trend can be discerned. Exposures along the east coast of Muckle Roe and around Mavis Grind provide examples of the coarse-grained and the feldspar-phyric types cutting diorite. Aplite dykes or veins are common on the high ground [333 693] west of Islesburgh croft.

PETROGRAPHY

Ultrabasic Rocks

The ultrabasic rocks of Glussdale Water and Moora Waters show cleaved poikilitic crystals, up to 5 mm across, in a dense black microgranular base. Fracture surfaces have a greasy lustre like that of fractured serpentinite. In thin section the rocks are seen to be composed of orthopyroxene and clinopyroxene, plagioclase, olivine and minor hornblende, biotite, iron ore and serpentine; the relative proportions of the plagioclase and the two pyroxenes vary within the area of a thin section though olivine is uniformly distributed. In mineral composition the rock therefore varies from that of a troctolite-dolerite resembling harrisite, through olivine-dolerite to lherzolite. The plagioclase forms large irregular plates without noticeable zoning (S 30016) or large and smaller prismatic crystals slightly zoned with optically negative core and positive envelope (S 29994). In powder from the analysed specimen (S 33683) the highest refractive index observed was $\gamma = 1 \cdot 576$ and the lowest $\beta = 1 \cdot 565$ corresponding to variation from bytownite (An_{80}) to calcic labradorite (An_{65}). The large plates are irregularly interlocked with both pyroxenes but smaller prisms are subophitic. The pyroxenes generally are in the form of irregular plates but orthopyroxene locally has good prismatic shape in its smaller crystals and may be enclosed along with the small crystals of olivine within plates of clinopyroxene (S 29994). It shows a very faint pleochroism and in powder from the analysed specimen has $\gamma = 1 \cdot 697$–8 corresponding to $En_{72}Fs_{28}$; the optic axial angle is about 75° (by Mallard's method) and the sign negative. Olivine is uniformly present as almost fresh to considerably altered, small (0·2 to 0·4 mm) crystals singly or in clusters enclosed in the plagioclase, pyroxenes, hornblende, and biotite (Plate XXII, fig. 2). The optic axial angle $2V\alpha$ 90°− and refractive index $\beta = 1 \cdot 691$ indicate the composition $Fo_{83}Fa_{17}$. Cracks radiate in plagioclase from its slightly serpentinized crystals (S 33683). The minor constituents hornblende and biotite form grains of irregular shape interstitial to and intergrown with plagioclase and pyroxenes; they appear to be primary crystallizations and local poikilitic growth enclosing plagioclase and pyroxene indicates that they are the last products of an aqueous cafemic residuum. They have the same dichroism α pale yellow, $\beta = \gamma$ reddish orange, the colour being deeper in biotite. Iron ore is abundant as small octahedral grains in plagioclase and pyroxenes and as rods, streaks, and chains of granules in serpentinized olivine. The serpentine is usually pale brown and isotropic but locally passes into interstitial packs of curved flakes of a pale brown material possessing moderately high birefringence; marginal between olivine and plagioclase however the colour is persistently apple-green.

The chemical composition and norm of the Moora Waters rock (S 33683) are given in Table 4 where the composition is compared with those of some other Scottish ultrabasic rocks of similar composition; distinction from the hornblende-peridotite of Lugar, which is associated with titaniferous alkaline rocks, is apparent in the difference in titania and in alkalies relative to alumina. In view of the heterogeneous distribution of the mineral constituents it is relevant to note that all the rocks with which the Moora Waters rock is compared in the table are in the field transitional types or portions of variable bodies.

Gabbro and Diorite

Corresponding to the difficulty of separating gabbro and diorite in the field, the collection of sliced rocks from the complex shows continuous variation from true

TABLE 4

Chemical analyses of enstatite-harrisite from Northmaven and comparable Scottish rocks

	1	A	B	C	D
SiO_2	40·58	40·82	40·35	43·09	42·52
Al_2O_3	7·57	10·66	3·75	7·51	8·12
Fe_2O_3	3·18	1·80	3·53	$FeO\frac{1}{2}$ 0·52	2·41
FeO	9·01	8·92	9·86	7·14	8·72
MgO	27·50	28·08	25·69	33·48	27·42
CaO	4·51	6·11	4·64	5·75	6·27
Na_2O	0·76	0·58	3·14	$NaO\frac{1}{2}$ 0·65	0·72
K_2O	0·15	0·21	0·80	$KO\frac{1}{2}$ 0·12	0·23
$H_2O > 105°C$	5·74	2·00	5·28	1·18	1·90
$H_2O < 105°C$	0·33	0·16	0·83	—	0·17
TiO_2	0·39	0·16	2·12	0·16	0·49
P_2O_5	0·04	—	0·25	nil	0·04
MnO	0·20	0·19	0·20	0·05	0·18
CO_2	0·09	0·00?	tr.		0·56
FeS_2	0·11	S 0·02			
Fe_7S_8	tr.				
Cr_2O_3	0·26	0·25		0·39	0·20
BaO	0·02		⎫		
SrO	0·01 (s)		⎬ 0·06		
Cl	—	—	⎭		
	100·45	100·12	100·56	100·04	99·95

(s)   Spectroscopic determination

Norm of Lab. No. 1065

or	0·89	ol	48·01
ab	6·29	mg	4·64
an	17·07	il	0·76
di	3·72	em	0·38
hy	12·79		
			94·75

Symbol 4.1.4.1.2: *Custerose*

1. Enstatite-harrisite. NE shore of Moora Waters, Northmaven, Shetland. One-inch 128 Scot., Six-inch 24 SE. Lab. No. 1065. Anal. C. O. Harvey; spect. det. H. K. Whalley. Guppy and Sabine 1956, pp. 35–6.
A. Harrisite, Tertiary intrusion. Roadside near Dornabac Bridge, Rhum. The total includes (Ni, CO) O 0.11, CuO 0.05. Guppy and Thomas 1931, pp. 97–8.
B. Hornblende-peridotite, Lugar sill. Glenmuir Water, Ayrshire. Tyrrell 1916, pp. 113–4.
C. Feldspathic peridotite. An Garbh-choire, Skye. Weedon 1965, p. 57.
D. Ultrabasic dyke No. 1, 4 ft 4 in from edge. Ben Cleat, Skye. Gibb 1968, p. 628.

gabbro to hornblende-biotite-diorite with minor quartz and potassium feldspar, which with increase of the latter components becomes a mesocratic granodiorite. No chemical analysis of any of the basic or intermediate rocks of this area has been made. For the purpose of petrographical description the rocks of the complex are grouped as gabbro, dolerite, gabbro-diorite, and diorite.

Gabbro. The gabbros include feldspar-phyric and non-porphyritic types. The former, which are less common, show stout prisms of plagioclase in a base of plagioclase laths

0·3 to 1 mm ophitically related to colourless augite which forms shapeless grains up to 1 mm across. In the freshest specimen (S 55657) the augite is locally replaced by and grades marginally into amphibole of a neutral brownish grey tint; microcrystalline pale green amphibole felt forms a patchy cement and occupies larger pockets resembling vesicular fillings. Titaniferous magnetite is accessory. The feldspar phenocrysts are usually aggregates of large stout broadly twinned prisms which have a composition of about 85 per cent An, by the simultaneous extinction method, are optically negative, and have refractive index β close to 1·575; they may show a narrow marginal, slightly less calcic zone (S 55656) and in this rock the larger crystals are seriate to the ground-mass laths. The latter are continuously zoned from centre at least as calcic as An 70 per cent to margins in which extinction angles of about 20° indicate their composition as andesine with about 35 per cent An. The augite is colourless or faintly brown, shows an isogyre with normal curvature for augite, and has $\beta = 1·693$. No olivine has been seen but some small, 0·3 mm, round aggregates of bright green microcrystalline amphi-bole or chlorite may represent early crystallizations of olivine (S 55644). The place of augite may be taken by hornblende which may retain the ophitic texture (S 55656) or may form subhedral prismatic grains interfering with the lathy plagioclase (S 55670). In this rock the plagioclase is traversed by many chlorite-filled cracks and contains granular epidote and small groups of actinolitic hornblende like that of the interstitial filling. This small group of porphyritic rocks might be classed as bytownite-phyric dolerite in view of the small grain size of the groundmass but since the rocks show seriate transition from phenocryst to groundmass plagioclase and local change of grain size in the groundmass they are retained in the gabbro group. There is no indication in the field that these relatively fine-grained feldspar-phyric gabbros are earlier or later than the non-porphyritic gabbros but since most occur close to the margin of the complex, they may represent an early partly chilled phase of the gabbro.

The non-porphyritic gabbros are in general coarse-grained rocks (Plate XXII, fig. 3) in which plates of colourless augite up to 5 mm across ophitically enclose prisms of calcic plagioclase which is zoned at the margins but centrally has an average com-position of An_{70} as indicated by the high refractive index and by an optic axial angle $\sim 90°$. Brown hornblende varies from minor (S 53593) to major (S 30017) proportions in relation to pyroxene. It may occur only as marginal portions of the pyroxene plates and is present also as large shapeless crystals ophitically enclosing the plagioclase. No fresh olivine has been observed but several specimens (S 29992, 30006, 30017) contain small (0·2 to 0·4 mm) aggregates of colourless microcrystalline amphibole and finely granular ore which are pseudomorphs of a mineral on which the pyroxene and plagio-clase are moulded and which may have been olivine. These pseudomorphs occur singly and are not poikilitically enclosed by pyroxene as in the ultrabasic rocks. Both augite and hornblende are replaced in variable degree by uralitic amphibole and the pseudo-morphs of ?olivine tend to act as centres for the growth of the pale green fibrous and acicular amphibole which forms a constant cement to the main minerals. Biotite of a deep brown colour occasionally forms short micropoikilitic plates among this infilling but more usually occurs in accessory proportion as irregular flakes marginal to or moulded on the main minerals and often enclosing large grains of black ore. The only other accessory mineral is magnetite. In distinction to the extensive alteration of pri-mary brown hornblende and pyroxene to secondary uralite or amphibole felt the feldspar of the gabbros is not altered. A variety of gabbro containing orthopyroxene has been collected from the lenticular basic body exposed between outcrops of gneiss on the coast of Busta Voe, 200 yd (183 m) ESE of Northpund, Muckle Roe [343 655]. Of three specimens representing this mass two are gabbro composed of large shapeless crystals of colourless augite and brown and green hornblende, both pyroxene and amphibole being locally uralitized (S 45133-4). These crystals ophitically enclose plagioclase prisms of very varying size and brown biotite, locally chloritized, also shows ophitic relations to the plagioclase. The third specimen is of finer grain and is composed mainly of stout, zoned prisms of labradorite enclosed subophitically by equant grains

O

of augite and less abundant green hornblende. The augite is pale green or pink in colour and encloses small grains of orthopyroxene (S 45035). Fresh dark brown biotite forms long plates which enclose both plagioclase and augite poikilitically. The genetic relationship of the rock to the coarse gabbroic specimens is clear but whether it is a local variant, an enclave, or a penecontemporaneous minor intrusion in the gabbro is not known.

Most of the gabbroic rocks in the collection are more appropriately classed as *hornblende-gabbro* or *bojite* but some less coarse rocks are hornblende-dolerites. The coarser rocks retain the texture of the gabbros but hornblende is as abundant (S 29415, 55642, 55651, 55653) as the pyroxene or considerably more abundant (S 55647–9) (Plate XXII, fig. 4). In those rocks hornblende includes both the primary brown variety and the neutral tinted and pale green type which is not uralitic in habit but appears to replace pyroxene and, in some cases, brown hornblende by reaction concurrent with crystallization of the rock. The plagioclase is still calcic with average composition 65 to 75 per cent An and the optic sign is variable. Brown biotite is usually present in flakes irregularly moulded on ore and plagioclase and enclosed in the interstitial fibrous amphibole felt. Deuteric alteration causes patchy replacement of feldspar by microcrystalline zeolite, opacization of biotite, and formation of chlorite and pyrite in the uralite.

The hornblende-gabbros are transitional to diorite through rocks which are texturally similar, but the strongly zoned plagioclase of which shows cores of bytownite or labradorite, in many cases turbid or sericitized, enveloped in more acid material marginally of oligoclase composition (S 44324). A little quartz (S 35706) or alkali-feldspar (S 55643) and analcime (S 55645) may be present interstitially. In these rocks biotite is usually a minor primary constituent and may occur in plates enclosing plagioclase and uralitized pyroxene (S 44324). The iron ore is titaniferous and small stout prisms of apatite are accessory. Another type of transition involves the appearance of interstitial quartz as an essential minor constituent in a rock in which both colourless augite and brown hornblende are coarsely ophitic to labradorite laths in a base of stout plagioclase prisms zoned externally to andesine-oligoclase composition (S 55664). The term *gabbro-diorite* is appropriate to those rocks though Johannsen considers that this term should not be used (1937, vol. III, p. 226).

The basic portions of the complex also include fine-grained varieties which resemble the gabbros but petrographically are to be classed as *dolerite* or *basalt*. The mass shown as gabbro on the one-inch sheet $\frac{1}{4}$ to $\frac{1}{2}$ mile (400–800 m) south by east of Roe Bridge [343 656] affords examples of both gabbro or coarse dolerite and basalt though no distinction has been drawn in the field. Specimens from this mass have been described on an earlier page (p. 195) as a variety of gabbro containing orthopyroxene. Farther south of the Muckle Roe coast a similar rock with micropoikilitic biotite in ophitic dolerite of varying grain-size occurs in the stretch mapped as diorite (S 55169). Basalts occur also on the east coast at Mavis Grind (S 55750, 55659), as a basic band (S 55652) in diorite on the north shore of Mangaster Voe 350 yd (320 m) WNW of Mavis Grind, and on the northern margin of the Skipadock gneiss area where the 'diorite' in contact with the gneiss is an ophitic basalt in which colourless cores of augite persist within the dominant green and uralitic hornblendes while deep red brown biotite with sagenite appears as an interstitial mineral (S 55165). In these rocks the plagioclase is a labradorite (An_{65}) or bytownite (An_{75}) but is commonly much altered so that a basic sericitized core is surrounded by a strongly zoned mantle, and may be locally so altered that the laths within fresh augite plates are argillized (S 55169). Transition to the porphyritic gabbro type is seen in a rock from the east coast at Mavis Grind (S 55650). It is noteworthy that these basaltic variations are petrographically similar to those which occur within xenoliths of gneiss in granite in the Scarfataing area, and have already been described among the early hypabyssal intrusions (p. 181). These petrographical similarities and the occurrence of the basaltic and porphyritic dolerite varieties at the southern and south-eastern extremities of the gabbro-diorite complex suggest that in this direction early units of the basic phase of intrusion have entered the country gneiss as basaltic sills and that while in the south they remained as layers in the gneiss which formed the country rock to granitic intrusion, further north they were incorporated as bands in the main irruption of basic magma.

Diorite. The diorites, which occupy by far the greatest exposed area of the complex, are in general medium-grained, black and white speckled rocks which in mass give the impression of fairly homogeneous composition. Comparison of specimens from the large area of exposed rock when juxtaposed in a collection or examined under the microscope shows wide variation in grain and texture and in the relative proportions of mafic and felsic components. These variations do not have any relation to their position in the complex.

The essential minerals of the diorites are plagioclase, hornblende, biotite, and, in some specimens, pyroxene. Quartz and potassium feldspar may be entirely absent or present as accessory or minor constituents; potassium feldspar is much less commonly present than quartz and is a minor essential in only one specimen (S 55662) in which microcline is concentrated with quartz in felsic pools which produce a pink mottling in the hand specimen. The main accessory mineral is black iron ore which is probably titaniferous and shows leucoxenic coatings on many grains. Apatite, sphene, and zircon are present in notable amount only in quartz-bearing varieties. Though leucoxene is common in most specimens, crystals of clear sphene, irregular in shape (S 55663, 53582) or skeletally idiomorphic (S 53666), have been observed only in association with interstitial quartz. Zircon, always sparse, has a broken and cracked appearance as though it were relict (S 30022, 55662).

The variations in mineral composition and in texture are of petrogenetic interest. Augite is present in a minor proportion of specimens and is a colourless type with close salite striation. Only in one specimen is it present in its original primary state of crystallization. In this rock (S 30023, Plate XXII, fig. 5) it forms small shapeless crystals moulded ophitically on plagioclase and stout hypidiomorphic prisms of pinkish colour which in many cases are enveloped in brown hornblende, itself ophitically related to the plagioclase. In the other specimens pyroxene forms more and less turbid cores in, but not sharply separable from, hornblende (S 29412, 30022, 55646, 55690); in one case the pyroxenic core retains ophitic structure (S 44322). Hornblende is the principal mafic mineral in all specimens (Plate XXII, fig. 6) and occurs in several forms; as ophitic or subophitic plates of brown, green or particolour (S 55655, 55690), but in most specimens as hypidiomorphic or ragged, green and brown prisms which interfere with plagioclase and usually also with one another in shapeless groups. Amphibole is abundant also in aggregates of pale green, locally bluish green blades and as fibrous uralite replacing both pyroxene and brown or green primary hornblende (S 29412, 30020, 30023). The aggregates of the bladed or subprismatic pale green and bluish green amphibole may be intimately admixed with small scales of biotite, large and small grains of epidote and iron ore, and, less commonly, pyroxene (S 30022, 53582, 55646, 55665–6); these feldspar-free clots clearly represent recrystallized ultrabasic material. The earliest precipitation of amphibole is the brown variety but this was crystallized peneconcurrently with the green form which continues the ophitic crystallization of both pyroxene and brown amphibole. There is evidence that the brown variety crystallized also comparatively late in the consolidation of some rocks since in these it has interfering relations to anhedral oligoclase and may be in part anhedral, in part euhedral against interstitial quartz (S 55669). Biotite, which is absent in only one specimen (S 55646), also is variable in type. It may be brown or green, may show fine sagenite structure (S 29998, 55669) in fresh crystals, or occur as coarse flakes interleaved with leucoxenic streaks (S 55666–7). It may be fresh or chloritized and enclose epidote (S 30020, 55662–3, 55665). In part it appears to be an early crystallization enclosed in the primary brown or green hornblende; in part it is, like brown hornblende, a late crystallization interfering with interstitial oligoclase and quartz (S 55669), and in some cases becoming poikilophitic (S 29998, 30008). Large ragged flakes which are spattered with epidote and streaked by leucoxene show growth of small flakes orientated with the basal plane normal to the base of the large crystal (S 55666); no colour difference exists. Plagioclase, the principal mineral constituent of the diorites, likewise shows great variety of composition, textural relations, and alteration. It is always zoned and in most specimens the crystals have a core of andesine-labradorite, An 40 to 55 per cent, which is surrounded by a zoned mantle with the An content decreasing marginally in the range 25 to 15 per cent as indicated by the refractive index relative to quartz. The core has usually a prismatic shape and is slightly zoned across the prism whereas the mantle may be hypidiomorphic prismatic, idiomorphic against quartz or quite allotriomorphic by growth against contiguous feldspars. Higher anorthite content in the core is in some cases indicated by saussuritization; in a few fresh augite-bearing types An contents of 75 to 65 per cent can be deduced from the optic sign, shape of the isogyres, and symmetrical extinction angles. Such calcic plagioclases are ophitically enclosed in pyroxene or amphibole but even within these minerals they show outward normal zoning to less calcic composition and corrosion of the pinacoids with discontinuous deposition of plagioclase of markedly lower refractive index (S 30023, 55690).

Xenolithic and hybrid rocks. Xenoliths occurring in the Wilson's Noup–Lang Head area in concentrations so great as locally to produce the appearance of breccia are black fine-grained rocks of basaltic appearance. They are composed essentially of a base of granoblastic oligoclase and green hornblende with minor green biotite; quartz is present in some specimens (S 55674, 55693) but it may have been introduced since it forms comparatively large grains enclosing small plagioclase and hornblende crystals; iron ore coated by sphene, apatite and epidote is accessory. Most are microporphyritic with prismatic plagioclase crystals up to 2 mm in length and aggregates of granular, colourless to bluish green hornblende of equant shape, about 1 mm across, which probably represent original pyroxene; these aggregates and the large plagioclase prisms are locally cumulophyric (S 55674). The plagioclase phenocrysts consist of a calcic core mantled by zoned feldspar which is intercrystallized anhedrally with the minerals of the groundmass and at the margin has a composition in the An 15 to 20 per cent range. The core is usually turbid and altered by epidotization, sericitization, or analcitization but can be shown to be at least as calcic as An 55 per cent (S 55693) and in one case probably An 75 to 80 per cent (S 55676). The oligoclase grains of the groundmass usually show central lathy relics of andesine-labradorite composition; the transition may be continuous or discontinuous and in the latter case alteration of the core to analcime (S 55674) can emphasize the discontinuity. A parallel arrangement of subprismatic groundmass and porphyritic plagioclase indicates the existence of fluidal structure in one xenolith (S 55693), and to the former presence of this structure the local roughly parallel orientation of large andesine-oligoclase ovoid crystals in the coarsely recrystallized portion of another xenolith may be due also (S 55692). While most of the xenoliths show sharp but intercrystallized contacts with the more acid rock which encloses or cements them, interaction both mechanical and chemical is shown by the occurrence of slivers of xenolith, and diffusion of quartz and potassium feldspar into reconstructed xenolith. The latter process is particularly well observed in those specimens (S 55692–3) which show fluxional structure under the microscope and in hand-specimen a platy structure, to which the fluxional structure is parallel. The macroscopic platiness corresponds with a parallel lenticular arrangement of granoblastic xenolithic material. The latter contains microaugen of microcline, some coarser tonalitic rock with potassium feldspar locally interstitial to or perforating the plagioclase, and granodioritic rock composed of granular quartz and microcline, turbid hypidiomorphic plagioclase, and hypidiomorphic and idiomorphic hornblende and minor biotite. Allanite appears in association with the dark minerals of the granodioritic portions. The composition and the microscopic and macroscopic structures in combination with the areal concentration of the xenoliths suggest that they represent a formation of basic andesitic or basaltic lavas broken up, perhaps by the forces leading to irruption of the gabbro-diorite intrusion, and incorporated into the latter by penetration of its more mobile acid fractions. It is relevant to draw attention to the frequent occurrence of concentrations of fine-grained basaltic xenoliths in the gabbro-diorite in its northern outcrop between Hamar Voe and Ronas Voe (One-inch Geological Sheet Northern Shetland). Though these concentrations are not of the 'breccia' type characteristic of the Wilson's Noup–Lang Head area they also indicate the existence of a fine-grained basaltic formation prior to intrusion of the main gabbro-diorite.

Other xenoliths occurring sporadically in the area north of Mangaster Voe are of the same thermally altered type as those described in the earlier part of the preceding paragraph. Some are porphyritic fine- to very fine-grained basic rocks (S 55268, 55679), in which compact aggregates of granular hornblende coated by biotite represent pyroxene phenocrysts and altered prisms of calcic plagioclase are mantled by andesine-oligoclase. Others are non-porphyritic (S 55678). None shows fluidal structure but in one a radial grouping of the larger plagioclase laths gives a vague suggestion of variolitic structure (S 55677). Petrographically they may represent basic lava or fine-grained hypabyssal types. Their contacts with the host rock, generally quartz-diorite but in one case granite (S 55677) of the eastern mass, show very restricted marginal intermingling with penetration of quartz into spaces and channels in the xenolith close to the contact. In one specimen, however, there has been extensive potassium metasomatism leading to the formation of shapeless plates of microcline, 3 mm across, which isolate and enclose turbid plagioclase prisms and granular and idioblastic hornblende and biotite of the xenolith (S 29993).

The *hybrid rocks* fall into two groups on the basis of their field occurrence: (1) heterogeneous hybrid rocks of strictly local development usually in association with xenoliths, and (ii) hybrid rocks which form more acid facies of the main diorite mass. The heterogeneous hybrids are exemplified by the acid material cementing the xenoliths of the Wilson's Noup–Lang Head area already described (S 55674–6, 55678–9). They range from tonalitic to granitic in mineral composition and characteristically show uneven proportions of potassium feldspar to plagioclase and of mafic to quartz-feldspar constituents. In some the difference in habit of the hornblende and biotite derived from xenolithic material and that of these minerals of the invading rock is clear, but the distinction can be obscured by idiomorphic recrystallization of the mafic minerals against the invasive quartz (S 55267). Late idiomorphic recrystallization of the hornblende and biotite is seen also in a fine-grained type which appears to be composed entirely of a leucocratic granular quartz and microcline base enclosing small plagioclase laths and grains of hornblende and biotite of xenolithic origin (S 30005). Rarely the granodiorite may show indications of fluxional structure by orientated elongation of mafic streaks or of basic xenoliths (S 55267); one of the xenoliths in this rock shows a curved lineation of its plagioclase laths.

The most common hybrid type occurring as a facies of the main diorite mass is a rock in which potassium feldspar is present as specks, irregularly distributed interstitial matter, or patches (segregations), up to about 1 to 2 cm across but occasionally 5 cm (S 30011). These rocks are composed of turbid, sericitized or epidotized prisms of plagioclase with which hypidiomorphic hornblende and biotite interfere, interstitial granular quartz and potassium feldspar and accessory iron ore, sphene, apatite and epidote. The plagioclase prisms usually show a calcic core mantled by a more sodic zone of oligoclase in both the more dioritic and the quartzo-feldspathic portions of the rock but in the latter the plagioclase is partially clarified of the alteration products and seems less calcic (An_{35-15}) than the original was. A noteworthy feature of the leucocratic patches is the common predominance in them of potassium feldspar over quartz (S 29411, 29416, 55696), and idiomorphic crystallization of hornblende adjoining that feldspar. Unlike hornblende, biotite is only rarely recrystallized in quartzose areas in fresh brown books (S 53569). In the sections just referred to and also in S 29506 it is extensively replaced by chlorite and iron ore or epidote, and it is possible that potassium has been withdrawn from the dioritic rock to add to the concentration in the potassic patches. The variability of this group is illustrated by specimens S 55691, 55694–5 in addition to those already cited.

Less common acid facies of the diorite are tonalitic in composition and differ from the usual dioritic types by the presence of interstitial quartz as an essential mineral and by the greatly diminished proportion of dark to light components. They consist mainly of zoned plagioclase prisms with which ragged prismatic prisms of green horn-blende interfere. Small amounts of biotite may be present in clots of small scales or as flakes. Quartz is abundant interstitially. Of the accessory minerals apatite and iron ore rimmed or interlayered by sphene are abundant, zircon and potassium feldspar scarce. The plagioclase may show epidotized cores (S 30009, 43768) with zoned oligoclase-albite mantles or it may be mainly oligoclase with still more sodic margins (S 55673). Hornblende characteristically is idiomorphic against quartz but is otherwise hypidio-morphic and has a close polysynthetic twinning. The mafic constituents tend to be aggregated in shapeless groups. A more leucocratic specimen (S 29999) contains green biotite as a major constituent; accessory microcline occurs interstitially and as a honey-comb pattern of replacement in plagioclase.

The larger granite masses

Scarfataing Granite. The single specimen available of the *Scarfataing Granite* is com-posed of mutually interfering coarse-grained prismatic albite, granular perthitic potas-sium feldspar, and quartz (S 55680), with a minor amount of matrix composed of the same minerals, ragged aggregates of oxidized and chloritized biotite, and grains of magnetite and epidote. The rock contains no quartz-feldspar intergrowth but has a characteristic enclosure of tiny idiomorphic prisms of turbid plagioclase in the perthite. Similar turbid plagioclase occurs as subhedral grains interstitially and in small aggre-gates enclosed in large perthite crystals. This rock occurs within the broad shear zone on the east of the Muckle Roe Granophyre and all the minerals in it are conspicuously deformed.

Eastern Granite. The coarse *Eastern Granite* consists of large irregular plates of perthitic feldspar and large grains of quartz which in places interlock with the feldspar in a coarse intergrowth without pattern. Prismatic alkalic plagioclase (An 5–10%) is sub-ordinate and usually has a stout prismatic habit, interlocking with the perthite and also crystallized in smaller prisms enclosed in quartz. It may contain irregular patches of potash feldspar (S 46601) or show an antiperthitic pattern (S 29419). The close associa-tion of two phases of perthitic feldspar is also shown by crystals composed of a micro-perthite core, a sericitized sodic plagioclase shell, and a mantle of microperthite opti-cally continuous with the core (S 29997). This rock contains patches of finer-grained albite rock in which uniformly distributed olive-green biotite is a minor constituent.

The accessory minerals of the granite, mainly chlorite after biotite and epidote, form clots in which grains of magnetite, ilmenite, and sphene are common. Zircon is abun-dant in one specimen (S 46601). New biotite and sphene sometimes crystallize in the mafic aggregates (S 29997) and alteration of biotite at an early stage in the crystalliza-tion of the rock is suggested by the manner in which quartz and feldspar grow through the mafic aggregates.

Quartz is always strained and dusty, in patches very dusty owing to an unusual abundance of very minute inclusions (S 28913).

The Muckle Roe Granophyre. The granophyre of Muckle Roe consists of micropegma-tite crystallizations which, usually spreading each from a nucleus of idiomorphic quartz or feldspar, interfere with one another (S 29505, Plate XXII, fig. 8) or coalesce with the quartz of interstitial coarse-grained pools (S 28909, 55689). Thus quartz always seems to exceed feldspar. The feldspar is a very turbid orthoclase or microperthite; only rarely is albite twinning seen and then it affects prismatic crystals which appear to be surrounded by the micropegmatitic growth rather than to form a nuclear part of it (S 44320). The mafic aggregates, meagrely represented in the thin sections though conspicuous in hand specimen, are composed mainly of iron ore grains and scraps of chlorite. The only variation from this type of rock is one in which albite occurs in equant prisms along with granular quartz and potash feldspar as aplitic patches within the granophyre (S 28905).

The smaller acid bodies and veins

Specimens of red granite from the smaller acid bodies are of finer grain than the typical rock of the Eastern Granite. They are composed of perthitic feldspar in tabular or shapeless crystals, prisms of albite (An_{5-10}), and granular quartz. Both perthitic feldspar and albite may in some cases have marginal micropegmatitic intergrowths with quartz (S 44331), but in others no micropegmatite may occur (S 29410). Biotite forming stout prisms moulded on plagioclase is only slightly chloritized and oxidized and is more abundant than in the Eastern Granite.

The feldspar-phyric acid rocks occurring as small bodies in gneiss or diorite consist essentially of large crystals or perthitic potassium feldspar, prisms of plagioclase in some specimens, and large grains of quartz in a groundmass of quartz, perthite or microcline, and minor albite (An_{5-10}) with accessory biotite, chlorite, ore and epidote in small clots or dispersed aggregates. The textures show that these rocks are not direct consolidations from a magma. The large perthitic feldspars show post-crystallization effects such as recrystallization to a patchwork of smaller grains and invasion by quartz (S 55681), rounded outlines of large cores mantled by similar potassium-rich feldspar or by micropegmatite (S 55685), granulitized margins grading into the quartz-feldspar base (S 55687–8), replacement by quartz with formation of micropegmatite (S 55684). Composite aggregates may show zonal alteration which passing without interruption through contiguous crystals is clearly later than their consolidation (S 55686). Some large irregular xenocrysts of highly strained (S 55685) or fractured (S 55684) quartz are mantled by micropegmatite. The ratio of perthite to plagioclase changes so abruptly that the rock is clearly of mixed origin (S 29410, 55686). Mafic minerals are usually present in only small amount and include biotite, chlorite, epidote and iron ore. They tend to form clots of small grains which appear to be derived from earlier rock (S 55687–8) but biotite also occurs in thick flakes as a primary constituent (S 29410, 55684–6). Hornblende is either not present or is only accessory in these rocks. It is however an important constituent of two small quartz-syenite bodies which cut the diorite in Egilsay. In these rocks (S 55682–3) hornblende forms primary hypidiomorphic crystals which interlock with feldspar and occurs also as fibrous uralite, with cores of augite, which form clots of prismatic grains incorporated from inclusions of more basic rock. Scarce accessory minerals in the small granitic bodies include sphene, epidote, and rare zircon and allanite. Sphene in some cases forms large crystals, partly idiomorphic, partly fractured or irregular (S 55684, 55687) which appear to be relict from incorporated material. Rarely such small bodies have a granodioritic composition. A specimen from Green Ward [326 740] is composed of large and small prismatic crystals of sericitized and epidotized plagioclase, $\sim An_{10}$, in a granular base of quartz and microcline with groups of raggedly terminated prisms of olive brown biotite as a minor constituent. The plagioclase often shows a zoned alteration and some crystals are mantled by microcline (S 29991).

The aplitic veins are composed of allotriomorphic feldspar and quartz with little or only accessory biotite, chlorite, and epidote. The feldspar is in some cases perthite and minor microcline and albite, An_{5-10} (S 55271), in others microcline and oligoclase, An_{10-15} (S 55272), or a cryptoperthite with a moiré pattern of extinction which suggests potash-oligoclase (S 55273). Quartz in all cases is strained and in some appears to replace perthite marginally. Perthite may be granulitized in pockets or along short lines which are also the loci of chloritization and oxidation of ore (S 55273).

LATE HYDROTHERMAL ACTIVITY

FIELD RELATIONSHIPS

Late hydrothermal activity affecting the gabbro-diorite mass and associated gneiss along the line of the Walls Boundary Fault-zone and minor parallel crush lines has resulted in scapolitization and analcimization of the rocks and in the case of the gneiss is of later date than the late phase of silicification to which reference has already been made (p. 37).

The scapolitized rocks are exposed on the coast of Muckle Roe to north and south of the mouth of the Burn of Scarfataing, in the quarry and exposures along the main road bordering Ell Wick, and on the high ground of Hurda Field between Mangaster and Sullom voes. In the Scarfataing exposures scapolitization affects both gneiss and basic igneous rocks to produce a mottled or speckly medium-grained material resembling black and white diorite so that recognition of scapolitized hornblendic schist, which is the main component of the gneiss in this area, from scapolitized dolerite or gabbro-diorite is possible with certainty only when banded or foliated structure is visible.

In the quarry at Ell Wick and on Hurda Field scapolite occurs in veins in sheared dark rock. The veins may be thick and composed of massive aphanitic pinkish scapolite as seen in the quarry (Plate XXIII) or they may form a zone of narrow veins, from 3 or 4 in (8 or 10 cm) to a tenth of an inch (3 mm) thick, separated by crushed dark rock, as observed on Hurda Field. At both localities the country-rock traversed by the veins is diorite.

PETROGRAPHY

Specimens from the Scarfataing coast include both basic igneous rock (S 45034) and gneiss (S 28904. 45029–31). The former is a coarse ophitic biotite-hornblende-gabbro. Most of the plagioclase and much of all varieties of amphibole have been replaced by coarse shapeless grains of scapolite (Plate XXII, fig. 7), while very scarce and local tourmaline, dichroic from bluish green to colourless, has replaced pale green amphibole. Biotite appears to have been replaced in part leaving a powder of sphene in the replacing scapolite but much has recrystallized in groups of small flakes which persist enclosed in scapolite. Locally aggregates of calcite and idioblastic chlorite replacing hornblende and biotite may represent earlier alteration products which were stable during recrystallization. Grains of sphene are abundant within biotite and hornblende and large round grains of apatite are local in the scapolite. Scapolitization appears to have been aided by an earlier fracturing, mainly without shearing, of the rock.

The scapolitized gneiss of the Scarfataing area includes a variety of banded rock-types, and the nature of the pre-scapolitization rock may be uncertain. A calcite-chlorite-biotite-scapolite rock (S 28904) shows foliation in respect of a roughly parallel banding of biotite-rich and calcite-chlorite-rich bands. Original reddish brown biotite is commonly margined by green biotite, and similar green biotite in groups of unorientated small flakes is enclosed in scapolite; in part large grains of the reddish brown

EXPLANATION OF PLATE XXIII

A. Roadside quarry, south of Mavis Grind, Northmaven [342 682]. Scapolite vein in diorite-granite complex. (D 1344).
B. East shore of Cow Head, Vementry Island [309 607]. Straight clean-cut junction between Vementry granite (pale) and metamorphic rocks. (D 904).

biotite are irregularly corroded and embayed by scapolite without intervention of the green variety. Both reddish brown biotite and chlorite show strained cleavages indicative of shear stress prior to scapolitization. Other scapolitized rocks include biotite-hornblende-andesine-granulites which are banded by alternating dominance of the component minerals or by variation in grain-size (S 45029–31). The coarser layers of these rocks greatly resemble diorite and in the field this resemblance is enhanced by the resemblance of the granular scapolite to feldspar. The mineral layering, the grano-schistosity of plagioclase, and the hornfels texture of some laminae are sufficient to identify them with the contact-altered gneiss of the area.

The degree of scapolitization in the Scarfataing gneiss is very variable and its distribution is patchy though controlled at least in part by the foliation and grain size of the gneiss; the very fine-grained granulites are not scapolitized. Within coarse-grained bands transition from fresh gneiss to rock in which the plagioclase is almost completely replaced is abrupt though relics of plagioclase persist and in general the disposition of biotite and hornblende persist unchanged. Locally recrystallization of hornblende to sharp, notched prisms of a blue species has occurred (S 45030). The crystals, entirely enclosed within scapolite, vary in depth of colour from pale to deep greenish blue; the darkest variety has pleochroism X yellow, Y bluish green, Z indigo blue, refractive indices $\alpha < 1\cdot699$, $\beta > 1\cdot699$, $\gamma = 1\cdot712$, 2V (by Mallard's method) 30–35°. The paler varieties have lower refractive indices and higher optic axial angles. These properties are similar to those of ferrihastingsite in the Clothister skarn (Phemister, in press). The scapolite itself is usually turbid because of alteration to analcime but fresh material was found to have $\alpha = 1\cdot555$–6, corresponding with that of the sodic scapolite described from elsewhere in Shetland (Mykura and Young 1969). The alteration initially produces a turbid streakiness parallel to the c axis of the mineral and proceeds through a diffuse network until the scapolite is replaced by large patches of greatly cracked turbid isotropic zeolite in which optically continuous but ill-defined spots of scapolite persist. The refractive index of the zeolite is $1\cdot490$ and it is therefore referred to analcime. Most of it is isotropic, yielding no figure in convergent light, but some patches yield a good negative uniaxial figure which is thought to arise from unaltered scapolite not otherwise detectable.

The specimens of scapolitized rock from Ell Wick, the quarry south of Mavis Grind and Hurda Field are essentially of vein scapolite containing rock relics. They afford much fresher scapolite than the Scarfataing rocks and its refractive indices have been determined on several specimens as $\omega = 1\cdot555$ or $1\cdot556$, $\varepsilon = 1\cdot544$ or $1\cdot545$. There is evidence of several phases of crystallization of scapolite. For example fragments of biotitic rock in which large idioblastic scapolites have grown are embedded in finer grained scapolite felt (S 55701). Rock replaced by coarse granoblastic scapolite is cut by irregular channels of fine-grained lathy scapolite (S 53595) or embedded in finer grained scapolite (S 55698). Scapolite is locally replaced by analcime (S 55701) and the zeolite occurs also in narrow channels cutting scapolite and quartz (S 55697). Late crystallization of pale green prochlorite in shapeless ophiblastic plates enclosing prisms of scapolite is seen in several specimens (S 53604, 53635, 53636); the chlorite is almost isotropic, optically positive, with γ $1\cdot631$.

The rock material which is enclosed in the scapolite consists mainly of quartz, biotite aggregate, chlorite aggregate, rare plagioclase, calcite, and accessory black ore, sphene, apatite and zircon; a single sharply wedged crystal of yellowish colour, strong relief, probably biaxial negative, which is idioblastically intergrown with scapolite, may be axinite (S 53636). Quartz is present in all specimens generally as large shapeless or angular, strained or fractured grains which are corroded by scapolite. It occurs in association with prismatic acid plagioclase and interstitial biotite and chlorite in one specimen as fragments in a scapolitized granitic breccia (S 53699). In other rocks it is granular, unstrained, and cemented by pale green biotite aggregate and this material may represent a paraschist (S 53635). Brown biotite is commonly present as aggregates of microcrystalline and unorientated flakes with or without granular quartz and the texture of these groups as well as the sparse occurrence of linear trails of ore granules suggests derivation from a quartzified pelitic hornfels (S 53604). Biotite of a pale brown colour occurs also in very fine-grained almost cryptocrystalline aggregates in which prisms of pale green hornblende are locally enclosed; in some groups the minute scales are schistose and the schistosity is interrupted by idioblastic scapolite (S 55701). Schlieren of this type of biotite foliated with strained calcite and deformed chlorite and enveloping small prisms of pale green hornblende suggest derivation from a schistose rock (S 55698), but in this case and in the preceding also the rock is considered to have been decomposed sheared diorite; grains of plagioclase occur only rarely (S 55700). It seems probable that rocks of both igneous and pelitic origin and vein quartz also are represented in the relics within the scapolite.

EVOLUTION OF THE IGNEOUS COMPLEX

In reviewing the field and petrographical descriptions set out in the foregoing pages it becomes clear that the phenomena characteristic of this complex and cardinal to a hypothesis of its evolution are the following:

1. A complete transition, spatial and mineralogical, between ultrabasic, gabbro and diorite components;
2. the lack of any regularity in spatial arrangement of those components, relative to one another or to position in the massif;
3. the ubiquity of minor acid material in varying amount throughout the gabbro-diorite;
4. the separateness of major granitic members from the gabbro-diorite body;
5. the predominance of primary hornblende throughout the gabbro-diorite;
6. the predominance of secondary amphibole over biotite and chlorite in the interstitial component of all basic rocks plutonic or hypabyssal.

From the random distribution of gabbroic rock in the more voluminous dioritic mass and from the total absence of contacts between these components it is deduced that the magma at the time of emplacement was of heterogeneous composition. The ubiquitous conversion of pyroxene to amphibole contemporaneously with consolidation shows that the intrusive magma carried an important content of water which, in addition to promoting mineralogical change, would assist in maintaining fluidity during emplacement. The absence of compositional banding within the gabbroic components and their lack of any regular form against the diorite indicate that the process of emplacement was slow and tranquil, effectively under hydrostatic conditions. The diorite component is variable in respect of the content of pyroxene, the proportion of mafic to felsic constituents, and degree of conversion of calcic to more sodic plagioclase. This variability combined with the evidence of the existence of minor quantities of quartzo-feldspathic fluids, varying in amount but present throughout the gabbro-diorite body, suggests that during the emplacement continuous acidification of the basic rock was operating, that in effect there did not exist at depth a magma of diorite composition, but that two 'primary' magmas were available for intrusion—one of basic composition with ultrabasic differentiates and acid residues which were dispersed in the process of intrusion, the other of granitic composition which was tapped only in small amount during the irruption from the basic reservoir. This conclusion is supported by the separateness of the major gabbro-diorite and granite-granophyre masses and by the separateness of the late dyke swarms of acid and basic compositions (see Chapter 16). The diorite is therefore considered to be an essentially hybrid material produced from gabbro by reaction, during emplacement, with more hydrous and quartzo-feldspathic fluids derived partly from its own differentiation, partly from the adjacent reservoirs of granitic magma which were later intruded as dykes, sheets, the Eastern Granite mass and the Muckle Roe Granophyre stock.

This hypothesis affords also an explanation of the production of the recrystallized actinolitic amphibole which is so persistent as the interstitial material throughout the gabbro-diorite, since the close presence of acid magma in an immediately irruptible state would make available the heat and fluid required for the recrystallization. In conclusion attention is drawn to the many examples of partial replacement by mica, epidote, silica and by both calcic and sodic scapolite, which have been mentioned in the earlier pages. Some are only locally intense but all together indicate the activity of hydrothermal solutions in the intrusive and the country rock over a very long period of time extending to the last stages of dislocation on the Walls Boundary Fault-Zone.

SUMMARY

The sequence of the main events in the history of the unfoliated igneous rocks of Old Red Sandstone age in Muckle Roe and southern Northmaven has been tentatively deduced from the field and laboratory work as follows:

1. Intrusion of early basic dykes and sills into gneiss in Muckle Roe and Busta peninsula; local intrusion or extrusion of basaltic magma in the area north of Lang Head.
2. Intrusion of granite at Scarfataing as sheets into gneiss, with associated granitization.
3. Brecciation, possibly explosive, of the volcanic or subvolcanic rocks of the Lang Head zone and invasion of the breccia by quartzo-feldspathic liquor.
4a. Intrusion of basic magma carrying ultrabasic enclaves and containing an important proportion of hydrous component to form the main gabbro-diorite body; this intrusion caused thermal alteration of the gneiss country rock.
4b. Reaction within the gabbro-diorite intrusion to produce a variable dioritic rock and plexiform diorite-granodiorite-granite mélange; the formation of the mélanges may be due in an unknown degree to an initial phase of event 5.
5. Intrusion of coarse uniform granite as large and small dykes and sheets.
6. Intrusion of the Muckle Roe Granophyre stock.
7. Intrusion of a parallel swarm of felsite–acid porphyrite dykes originating at the locus of the Muckle Roe Granophyre.
8a. Intrusion of a parallel swarm of basic dykes.
8b. Intrusion of pyroxene-porphyrite dykes possibly from a focus in the Busta peninsula (N.B. for the sequence 7, 8a, 8b see Chapter 16).
9. Crushing and scapolitization associated with faulting along the Walls Boundary fault-zone, followed by local analcimization.

The main gabbro-diorite body may have the form of a lopolith, its main conduit lying in the area between Gunnister Voe and Hamar Voe (One-inch Sheet 130) approximately on the line of the Lang Head zone of earlier brecciation.

REFERENCES

FINLAY, T. M. 1930. The Old Red Sandstone of Shetland. Part II. North-western area. *Trans. R. Soc. Edinb.*, **56**, 671–94.

FLINN, D., MILLER, J. A., EVANS, A. L., and PRINGLE, I. R. 1968. On the age of the sediments and contemporaneous volcanic rocks of western Shetland. *Scott. Jnl Geol.*, **4**, 10–19.

GIBB, F. G. F. 1968. Flow Differentiation in the Xenolithic Ultrabasic Dykes of the Cuillins and the Strathaird Peninsula, Isle of Skye, Scotland. *Jnl Petrology*, **9**, 411–43.

GUPPY, E. M. and SABINE, P. A. 1956. Chemical Analyses of Igneous Rocks, Metamorphic Rocks and Minerals 1931–1954. *Mem. geol. Surv. Gt Br.*

—— and THOMAS, H. H. 1931. Chemical Analyses of Igneous Rocks, Metamorphic Rocks and Minerals. *Mem. geol. Surv. Gt Br.*

HIBBERT, S. 1819–20. Sketch of the Distribution of Rocks in Shetland. *Edinb. Phil. Jnl*, **1**, 269–314; **2**, 67–79, 224–42.

—— 1822. *A Description of the Shetland Islands.* Edinburgh.

JOHANNSEN, A. 1937. *A descriptive Petrography of the Ingneous Rocks.* Vol. III. Chicago.

MILLER, J. A. and FLINN, D. 1966. A Survey of Age Relations of Shetland Rocks. *Geol. Jnl*, **5**, 95–116.

MYKURA, W. and YOUNG, B. R. 1969. Sodic scapolite (dipyre) in the Shetland Islands. *Rep. No. 96/4, Inst. geol. Sci.*

PEACH, B. N. and HORNE, J. 1879. The Old Red Sandstone of Shetland. *Proc. R. Phys. Soc. Edinb.*, **5**, 80–7.

—— —— 1884. The Old Red Volcanic Rocks of Shetland. *Trans. R. Soc. Edinb.*, **32**, 359–88.

PHEMISTER, J. 1976. The Lunnister metamorphic rocks, Northmaven, Shetland. *Bull. geol. Surv. Gt. Br.*, in press.

SUMM. PROG. 1933. *Mem. geol. Surv. Gt Br. Summ. Prog. for 1932.*

TYRRELL, G. W. 1916. The Picrite-Teschenite Sill of Lugar (Ayrshire). *Q. Jnl geol. Soc. Lond.*, **72**, 84–131.

WEEDON, D. S. 1965. The layered ultrabasic rocks of Sgurr Dubh, Isle of Skye. *Scott. Jnl Geol.*, **1**, 41–68.

Chapter 14

VEMENTRY GRANITE

FIELD RELATIONSHIPS

GRANITE crops out in the north-eastern part of the Island of Vementry, where it occupies an area of a third of a square mile (0·85 km²) (Plate XXIV). Its outcrop may, however, continue northward under the Swarbacks Minn, possibly linking with the granophyre of Muckle Roe. The Vementry Granite, which forms Muckle Ward, 298 ft (91 m) OD, the highest hill on Vementry Island, consists of two lithological types, an outer coarse-grained pink leucocratic quartz-rich granite with a very low proportion of dark minerals, and an inner porphyritic granite composed of quartz and feldspar phenocrysts in a medium- to fine-grained rather darker matrix. The inner granite occupies an area of less than $\frac{1}{18}$ sq. mile (0·14 km²). Its boundary with the surrounding coarse-grained granite is not everywhere clearly defined, and in one area there is a transitional zone up to 100 yd (90 m) wide.

The granite is intruded into metamorphic rocks of the Vementry and Neeans groups (pp. 39–44) and the junction is everywhere clearly defined. It is best seen in the sea cliffs east of Cow Head, at the eastern end of the granite outcrop (Plate XXIIIB) where it is very straight, and inclined at 47° to SW with a number of apophyses up to 2 ft 6 in (0·75 m) thick passing into the overlying slightly indurated gneiss. The southern margin of the granite between Cow Head and Suthra Voe, though clearly defined, is not well enough exposed for an assessment of its inclination to be made. At the head of Suthra Voe, a 100 to 150 yd wide (90–135 m) tongue of granite extends for at least 300 yd (270 m) south-westward from the main granite mass. The western margin of the granite is bounded by a thin band of crush-breccia on the north shore of Suthra Voe, but farther north, close to the head of Northra Voe, it is unaffected by shearing and inclined very steeply to the south-west. Chilling of granite against gneiss is confined to a 1-ft (30-cm) wide zone, in which the normal coarse-grained granite becomes porphyritic, medium- to fine-grained. Effects of induration in the adjoining rocks are not marked (see p. 58).

Crush Belts and Jointing. The Vementry Granite and adjoining metamorphic rocks are traversed by a number of crush belts and the granite has strong joints. Many of these give rise to strong linear depressions inland and deeply cut geos along the coast. In the western part of the outcrop the mean trend of the major joints is N30°E to N40°E with inclinations of 60°–65° to the south-east. In the central part of the outcrop their trend is more variable but mainly between N20°E and N10°E. In the east the predominant joint direction is N35°W to N40°W, with a large number of closely spaced near-vertical joints developed north of Cow Head and in the Holms of Uyea Sound. North-easterly cross joints, are, however, also present in the Cow Head peninsula, while in the western part of the outcrop particularly on the west and south-west slopes of Muckle Ward and the east shore of Northra Voe there is a suite of closely spaced N40°W-trending minor joints which are steeply inclined to the north-east but do not form major topographic features.

PETROGRAPHY

Outer Granite

The outer granite (S 50139, 30737, 30734) is a characteristically pale pink coarse-grained rock, with prominent quartz and generally a lower proportion of dark ferromagnesian minerals than the typical Sandsting Granite. In thin section it is composed of up to 70 per cent feldspar and 30 per cent quartz.

Potash Feldspar. Approximately 70 per cent of the feldspar occurs as large anhedral crystals of microperthite which range in diameter from 5 mm to 1·2 mm with an average size of 1·8 × 1·6 mm. In the microperthite the exsolved sodic plagioclase most commonly takes the form of irregular branching, roughly parallel rods and ribs which in some instances form a reticulate network. The rods average 0·03 mm in width. Some microperthite has a replacement texture, consisting of irregular linked blebs of sodic plagioclase with closely spaced albite twinning in potash feldspar. Individual replacement blebs reach a maximum size of 0·15 × 0·1 mm.

Plagioclase is mid- to sodic-oligoclase and forms about 30 per cent of the total feldspar. The crystals are platy, euhedral to subhedral, with an average diameter of 0·8 mm and a maximum of 2 mm. The smaller plagioclase crystals are commonly totally enclosed in perthite.

Quartz forms rounded to subrounded crystals, which are usually grouped in clusters and range in diameter from 2·5 to 0·8 mm. There are no liquid inclusions as in the quartz of the Sandsting Granite. Margins of quartz crystals vary from straight to irregular and are rarely serrate. Locally a thin vein of secondary quartz, composed of minute acicular crystals whose axes are perpendicular to the length of the vein, is developed between the crystals of quartz and microperthite and more commonly between adjacent microperthites. The maximum widths of these veinlets is 0·6 mm and in some specimens incipient graphic texture is developed along the junction with microperthite.

Ferromagnesian minerals are usually subordinate, consisting mostly of irregular wispy crystals of strongly pleochroic biotite, locally associated with chloritic patches, and small clusters of subhedral grains of the ore minerals, ilmenite or pyrites (S 30712).

Accessory minerals are very rare. Rutile forms small euhedral crystals. Finlay (1930, p. 687) has recorded fluorite in the granite, but this has not been confirmed.

The Vementry Granite differs from the typical Sandsting biotite-granite in its relative lack of ferromagnesian minerals, the absence of apatite and sphene and the complete lack of liquid inclusions in the quartz. The presence of secondary quartz between adjacent microperthites has not been recorded in the Sandsting Granite. There is, however, a very wide range in composition and texture within the Sandsting Granite, and the Vementry Granite corresponds most closely to the coarse quartz-rich varieties from the eastern part of the Sandsting Complex.

Inner Porphyritic Granite

The inner granite of the Vementry Complex (S 30713) consists of feldspar and quartz phenocrysts set in a fine-grained matrix. There is a slightly higher proportion of dark minerals than in the outer granite. Phenocrysts form about 40 per cent of the total volume of the rock. Of these, 60 per cent are microperthite and untwinned orthoclase, the former reaching 4 mm in diameter; 20 per cent are quartz which is up to 1·7 mm in diameter and about 10 per cent are smaller phenocrysts, up to 0·9 mm long, of cloudy sodic-oligoclase. There are also some clusters of strongly pleochroic biotites, up to 1·3 mm long. The texture of the microperthites is more irregular than in the coarse-grained granite, and the exsolved phase consists of irregular rods which show little sign of orientation. Commonly plates of microperthite are clustered into groups about 5 mm in size, with adjacent crystals separated by thin veinlets of crystalline quartz.

The matrix is formed of roughly equidimensional grains of quartz (50% to 60% of volume), microperthite (30%) and plagioclase (about 10%), the grain size ranging from 0·3 to 0·15 mm. In some areas, however, quartz is interstitial, forming highly poikilitic patches up to 0·6 mm in diameter. Irregular scattered grains of opaque minerals (mainly ilmenite-leucoxene) are usually associated with biotite, and there are also isolated small euhedral crystals of allanite.

REFERENCE

FINLAY, T. M. 1930. The Old Red Sandstone of Shetland. Part II: North-western Area. *Trans. R. Soc. Edinb.*, **56,** 671–94.

Chapter 15

THE SANDSTING GRANITE-DIORITE
COMPLEX

INTRODUCTION

THE SANDSTING granite-diorite complex forms the south-eastern part of the Walls Peninsula (Plate XXV) and extends over a land area of approximately 12 sq. miles (31 km²). It is intruded into the sediments of the Walls Formation and in the western part of the outcrop the junction between the granite and over-lying sediments trends roughly east–west and is inclined at 40°–70° to the north. The north-eastern part of the complex, however, appears to form a series of near-vertical sill-like intrusions with a north-north-westerly trend, which thin and finger out in a northerly direction.

The complex consists of the following rock types arranged in probable order of intrusion:

1. Diorite, including melamicrodiorite, biotite- and hornblende-diorite, quartz-diorite and syenodiorite; gabbro which forms at least two small dyke-like masses within the diorite; and an ultrabasic rock, which has been located by the presence of blocks.
2. Granodiorite.
3. Coarse-grained biotite-granite grading locally into graphic granite.
4. Porphyritic microadamellite and fine-grained porphyritic granite.
5. Felsite and porphyritic microgranite forming dykes which cut both the intrusive complex and the adjoining sediments.

The members of the diorite group are very variable in grain size and colour index and are veined by leucocratic diorite and pegmatite. The veins are most abundant in the west. The more westerly diorite outcrops contain a number of enclaves of indurated sediment, which range from blocks a few feet (tens of centimetres) in diameter to large masses like that between Loch of Sotersta and Sand Water, which is over ½ mile (800 m) long. Melamicrodiorite forms rela-tively small near-vertical dyke-like masses within the diorite as well as in the granite and adjoining sediment. In a number of cases the melamicrodiorite is clearly earlier than the diorite as it is brecciated and veined by the latter, but there are also dykes cutting diorite and granite, indicating that intrusions of this type are among the earliest and the latest members of the complex.

The granitic rocks are almost everywhere demonstrably later than the diorites, which are commonly brecciated and veined by granite. The evidence regarding the relative ages of the various 'granitic' members of the complex is rather inconclusive, and it is possible that a single intrusive body may consist of several rock types. It is, however, considered that the coarse-grained granodiorite which crops out in a narrow strip extending from Keolki Field to Culswick is earlier than the other types (p. 221).

A feature of the coarse-grained biotite-granite forming the Wester Wick and Skelda Ness areas is the presence within the granite of near-vertical roughly north-north-west trending belts of intensely sheared and locally mylonitized rock. Sodic scapolite occurs as both a replacement mineral and as a vein mineral in these crush belts and as a vein mineral in the basic minor intrusions and in the sediment adjoining the granite. The scapolite appears to have been introduced by hydrothermal solutions along active shear belts, joints and other lines of weakness, shortly after the emplacement of the Sandsting Complex (Mykura and Young 1969).

Miller and Flinn (1966, pp. 107–9) produced a radiometric age date of 334 ± 13 m.y. for the Sandsting Granite. Potassium-argon age determinations by N. J. Snelling of granite and diorite from the complex are as follows: Biotite from granite, roadside ½ mile (800 m) NE of Culswick [279 457] % K 6·72, radiogenic 40 Ar 0·191 p.p.m., Age 360 ± 11 m.y.; hornblende from diorite, Hestinsetter Hill [291 457] % K 2·40, radiogenic 40 Ar 0·023 p.p.m., Age 369 ± 10 m.y. The Sandsting Complex may thus have been emplaced in late-Middle or early-Upper Devonian times.

The sediment overlying the granite-complex in the western part of the area is hornfelsed or highly indurated within a zone extending up to 1 mile (1·6 km) from the junction (Plate XII). Within this zone there is no evidence of the minor folds and lineations which are a feature of the less altered sediments to the north (p. 131). Instead, the baked sediments adjoining the folded rocks outside the aureole are merely shattered, suggesting that when the folding took place they were already hornfelsed by the granite intrusion. The sediment is in places cut by an anastomosing complex of granite sills and elsewhere by dyke swarms and isolated dykes of granite and felsite and by veins of scapolite. Several of these dykes and veins extend beyond the zone of induration into the intensely folded sediment where they are themselves deformed and have in some cases acquired the small-scale folding and regional lineation.

DIORITES AND ASSOCIATED ROCKS

FIELD RELATIONSHIPS

Diorite forms four large and several smaller outcrops within the complex (Plate XXV). The four main diorite masses are termed (from west to east) the (1) *Culswick*, (2) *Hestinsetter*, (3) *Scarvister–Loch of Arg–Skelda Voe* and (4) *Garderhouse* diorites. The first three of these may be part of an originally continuous sheet intruded into the Walls Sandstone. This sheet forms the first major intrusion of the complex (pp. 237–8) and contains numerous inclusions of sediment. It is now steeply inclined to the north in the Culswick and Hestinsetter areas, possibly arched into a north–south trending elongate dome in the Scarvister–Easter Skeld district, and steeply inclined to the east along the eastern shore of Skelda Voe. The thickness of the sheet cannot be estimated but it can be assumed that it is at least 1200 ft (360 m) thick in the west, and thins irregularly eastward and south-eastward to less than 600 ft (180 m) in the Easter Skeld and Scarvister areas. The sheet is intensely veined by granite close to its southern, presumably lower, margin along the shores of the Stead of Culswick, and throughout its width along the shores of Skelda Voe and at Scarvister.

Diorite

Culswick. In the western and southern parts of the outcrop, there is no evidence of flow-foliation or regular layering within the diorite. On the shores of the Stead of Culswick, the fine-grained dark diorites are commonly cut by coarser

P

more leucocratic diorite, with junctions ranging from sharp and angular to undulating. In some cases a gradual transition from fine- to coarse-grained diorite takes place within a few inches, but there are also some instances on this shore of two adjacent masses of virtually identical diorite separated from each other by a chilled margin in one of the diorites.

On the western shore of Keolki Field [253 454], between 500 and 700 yd (460 and 640 m) N of Culswick Broch, the oldest intrusive rock is medium-grained diorite which ranges in composition from quartz-biotite-diorite to dark hornblende-rich diorite, and in places contains blocks and fragments of indurated sediment. The diorite contains several near-vertical north-west trending bands of coarse-grained leucocratic diorite, with acicular hornblendes up to 1 in (2·6 cm) long, now largely replaced by biotite, set in a pale matrix of plagioclase.
Hestinsetter. The diorite forming Hestinsetter Hill and the ground extending southwards towards Lunga Water displays a marked differentiation into near-vertical layers with distinctive textural and mineral compositions. In the roadside quarries on the west slope of Hestinsetter Hill this banding trends E30°N and reflects a vertical alignment, not only of the constituent minerals of the diorite but also of the acid and granitic veins. Several thick vertical layers traversing the south slope of Hestinsetter Hill are composed of fine-grained diorite full of pale buff ovoids ranging in diameter up to 10 mm. These ovoids are composed of pyroxene-monzonite with a nucleus of sphene which may reach a length of 3 mm. Less clearly defined bands of diorite with pale ovoid patches containing large sphenes are exposed on the north slopes of the Knowes of Westerskeld, 200 to 400 yd (180–360 m) east of Lunga Water. Other members of the diorite suite exposed between Hestinsetter Hill and Housa Water are granodiorite, syenodiorite, quartz-diorite, and dark fine-grained diorite rich in hornblende and biotite. The relationships between these rock types are not clear and some may grade into each other.
Scarvister–Loch of Arg–Skelda Voe. The diorite forming the arcuate outcrop between Scarvister and Easter Skeld contains some of the most basic members of the complex and the rocks exposed 150 to 200 yd (135–180 m) NE of Loch of Arg are composed mainly of fine to medium-grained diorite (p. 216) the chemical composition of which borders on that of hornblende-gabbro (Guppy and Sabine 1956, pp. 14, 24). Finlay (1930, p. 689) has also recorded gabbro on the eastern shore of Skelda Voe, and augite-diorite at Tarasta, 550 yd (500 m) SSE of Loch of Arg. These records have, however, not been confirmed. The headland west of the Ayre of Swartagill, which forms the most southerly exposure of diorite on the east shore of Skelda Voe, consists of fine-grained meladiorite. The latter is sheared and contains a number of irregular veins or lenticles of very coarse-grained strongly ophitic gabbro, which has been amphibolitized and saussuritized. These veins trend west-north-west and range from 4 in (10 cm) to several feet (over 1 m) in thickness. Along the south shore of this peninsula the fine-grained meladiorite is veined by a pink hornblende-rich rock. Hybrid or metasomatized rocks are common among boulders on the Ayre of Swartagill, just east of the peninsula, the most widespread being a coarse-grained quartz-diorite with large porphyroblasts of potash feldspar.
Garderhouse. The 'diorite' cropping out on the west shore of Seli Voe south of Garderhouse is intensely shattered because of its close proximity to the Walls Boundary Fault. It is predominantly fine-grained and ranges in composition from quartz-diorite to ophitic gabbro with completely amphibolitized pyroxenes.

The impression gained from the field exposures is that this diorite forms the base of an irregular sheet inclined to the east. The outcrop is split into a number of discrete masses by thick granite veins.

Melamicrodiorite dykes

The range in grain size and composition of the diorite in the outcrops described above is considerable and its fine-grained basic members grade into hornblende-rich microdiorite. There are, in addition, a number of dyke-like bodies of very fine-grained almost basaltic-looking melamicrodiorite, the age of intrusion of which appears to range from pre-diorite to post-granite.

(a) *Pre-diorite intrusions*. The most accessible of these is a vertical sheet of melamicrodiorite exposed on the northern slope of Hestinsetter Hill [290 458] close to the northern margin of the Hestinsetter diorite (p. 212). It is approximately 180 ft (55 m) wide and has a trend and inclination parallel to the banding in the diorite. Along its southern margin there is an 18 ft (5·5 m) wide brecciated zone composed of angular blocks and fragments of melamicrodiorite embedded in a net-vein complex of diorite and quartz-diorite. These veins are, in turn, cut by veins of granitic material. It is not certain if the diorite of the vein-complex is derived directly from the flow-banded diorite bounding it to the south, but the presence of small fragments of melamicrodiorite in the latter suggests that the melamicrodiorite is older than the diorite.

A large mass of melamicrodiorite crops out a short distance above the east shore of the Stead of Culswick, 300 yd (275 m) S of the head of the bay (Plate XXV). Its shape and the character of its junction with the adjoining diorite are not easily determinable, as both are intensely net-veined by granite. The microdiorite is largely broken into angular blocks within a net-vein complex of granite, whereas the junctions between diorite and granite in this area are highly undulating, suggesting contact of two fluid magmas, and thus implying that the intrusion of melamicrodiorite preceded that of diorite.

(b) *Post-diorite dykes*. Two dyke-like bodies of melamicrodiorite, which appear to be younger than the surrounding diorite, crop out on the north-west shore of the Stead of Culswick (Plate XXV). Both are irregular in shape, with maximum widths of 25 and 7 ft (8 and 2 m) respectively. They have chilled margins against diorite and both contain belts of sheared rock lined with irregular anastomosing veinlets of sodic scapolite (Mykura and Young 1969, p. 3). The more westerly of the two dykes (Fig. 24(4)) is bounded by granite. It appears to have chilled margins against the granite, but is also cut by a number of granite veins.

The diorite cropping out on the shore west of Keolki Field, 570 yd (520 m) N10°W of Culswick Broch, is cut by a 14 in (35 cm) dyke of dark microdiorite which trends N10°W and is steeply inclined to the east. It has slightly undulating, strongly chilled margins and contains a number of irregular veinlets of a black almost completely hornblendic rock. Though clearly later than the diorite, the age-relation of this dyke to the granite is not known.

(c) *Post-granite dykes*. Dykes of hornblende-rich melamicrodiorite cut the shattered granite cropping out along the shore of Seli Voe north of Rea Wick, the granite sills interdigitated with sediment just south of Bixter Voe, and the indurated sediment around Bixter Voe. The dykes range in thickness up to 5 ft (1·5 m) and some are cut by thin irregular veinlets of sodic scapolite. Though

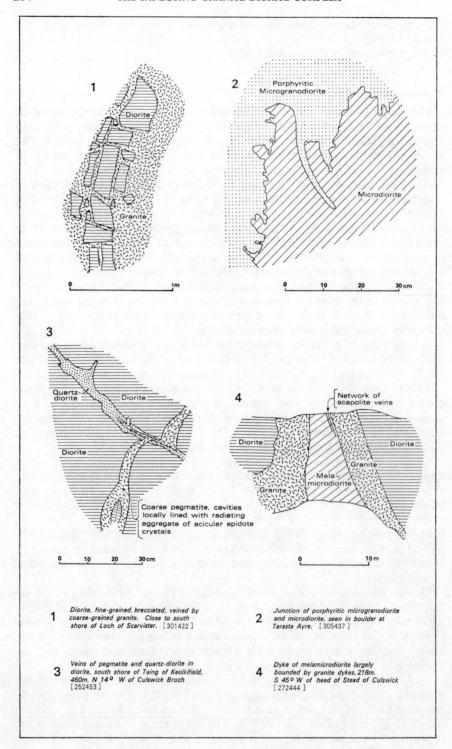

FIG. 24.   *Diagrams of igneous contacts, dykes and pegmatite veins in the Sandsting Complex*

these dykes are petrographically closely allied to the melamicrodiorites described above (p. 214), they are clearly the last igneous intrusions associated with the complex.

Acid Veins

The diorite of the western part of the Culswick area is cut by a network of generally leucocratic quartz-diorite veins, characterized in field exposures by the presence of small euhedral plagioclase phenocrysts. The quartz-diorite contains local concentrations of biotite. The veins are branching, irregular in outline, and do not normally exceed 1 ft (30 cm) in thickness.

The quartz-diorite veins are themselves cut by veins of pegmatite composed largely of crystals of pinkish white microperthite up to 2 in (5 cm) in size. The pegmatite veins contain small elongate cavities along their middle, which are filled or lined with radiating aggregates of acicular crystals, up to 1 in (2·5 cm) long of pale-green epidote. The veins vary greatly in thickness, with a recorded maximum of 18 in (45 cm). They are usually near-vertical and branching, but have no consistent trend. In some instances pegmatites are, for a short distance, intruded along the middle of a quartz-diorite vein (Fig. 24(3)).

Granite-diorite net-vein complexes

The areas of diorite intensely veined by granite are shown in Plate XXV. In some outcrops, as on the shores of the Stead of Culswick, granite forms up to 40 per cent of the total volume of rock, with individual veins ranging in thickness from a few inches to over 30 ft (9 m). Most commonly the diorite blocks enclosed in granite are angular and straight-sided (Fig. 24(1)) suggesting that the diorite had consolidated before the granite was intruded. In some instances, however, the contacts between diorite and granite are lobate or crenulate (Fig. 24(2)), which suggests that both granite and diorite magma were fluid at the point of contact. Diorite is not normally chilled at these junctions. At the Stead of Culswick there are several exposures of granite chilled to porphyritic microgranite within a 2 in to 1 ft (5–30 cm) wide zone along the more lobate margins. There are a number of instances at Shalders Taing [272 437] on the east shore of the Stead of Culswick of complete gradation from fine-grained granite to coarse-grained quartz-diorite which itself veins the finer-grained diorite.

Metasomatism at contacts between granite and diorite is uncommon in the Sandsting Complex, but good examples are seen on the shore south-west of Tarasta Ayre [304 436] 600 yd (550 m) SSE of Loch of Arg. Here the width of the zone of alteration at the junction of diorite and coarse-grained granite averages 1 ft (30 cm). It is characterized by the presence within the diorite of porphyroblasts of potash feldspar up to 1 in (2·4 cm) in size which increase in abundance as the contact is approached. Included masses of diorite within the granite are thoroughly altered to porphyritic granodiorite, and similar material forms veins within unaltered diorite, suggesting that the altered rock was locally mobilized.

The association of granite and basic rock in igneous complexes has been widely studied, and a number of observers (Bailey and McCallien 1956; Elwell 1958; Blake and others 1965) have shown that adjoining basic and acid magmas are commonly coeval.

Ultrabasic Rock

The only evidence for the presence of ultrabasic rock within the Sandsting Complex is found close to Stump Farm, 250 yd (230 m) E of Gossa Water, where there is a concentration of boulders of an olivine-rich rock (p. 220). There is little doubt that this represents a small outcrop which may be the surface expression of either an isolated pipe of ultrabasic rock that has penetrated through the diorite, or, less likely, a large block engulfed by the diorite magma. Small masses of similar ultrabasic rock, usually found only as detached surface blocks, occur in the Northmaven Complex near the northern margin of the present Sheet (pp. 183 and 193).

PETROGRAPHY

Diorite

The members of the diorite suite show a wide range in both composition and grain size, which does not appear to be related to either geographical distribution or order of emplacement. The rock types in the main dioritic masses appear to form a continuous series which ranges from granodiorite and syenodiorite through quartz-diorite (Plate XXVI, fig. 2) to biotite- and hornblende-diorite (Plate XXVI, fig. 4), the more basic varieties of which approach the composition of gabbro (p. 218; Guppy and Sabine 1956, p. 24). The finer-grained varieties are locally porphyritic. The granodiorites and syenodiorites are generally slightly coarser than the diorites.

Plagioclase. Plagioclase occurs in euhedral to subhedral crystals which range in length from 3·5 to 0·7 mm in the coarsest varieties, and from 0·8 to 0·12 mm in the finest. The crystals are almost invariably zoned, usually with a fairly large homogeneous central area and narrow outer rims. The composition of the core in the majority of specimens is sodic andesine but in some it is mid-oligoclase. Labradorite and andesine-labradorite have been recorded in two basic coarse-grained members of this suite (S 51543, 30035). The outer rims in most plagio-clase crystals are sodic-oligoclase, though in one case (S 30035) Guppy and Sabine (1956, p. 24) have recorded an outer rim of albite.

In many specimens the plagioclase is fresh throughout but the core of larger crystals is often altered to a cloudy aggregate which contains epidote and biotite.

Potash feldspar. Microcline occurs as an interstitial constituent in virtually all specimens. In the normal diorite it usually forms less than 5 per cent of total volume. In the syenodiorites potash feldspar forms between 15 and 20 per cent of total feldspar and in granodiorites up to 40 per cent. In the last two rock types the percentage of interstitial microcline is not much greater than in diorite, the additional potash feldspar taking the form of large, highly poikilitic pheno-crysts of microperthite or microcline-microperthite enclosing plagioclase. Individual crystals reach 3 mm in diameter and in one case (S 51547) crystal clusters up to 5 mm have been recorded.

Quartz. Quartz is a minor interstitial constituent in the diorites. In the syeno-diorites and granodiorites, the division between which is here drawn at 5 per cent quartz, quartz content ranges from 0 to 15 per cent by volume of total rock. In addition to interstitial quartz the granodiorite contains anhedral quartz crystals up to 0·5 × 0·3 mm in size, and in one case patches (possibly juvenile) up to 3 mm in diameter have been recorded.

Within the 'diorites' with less than 5 per cent potash feldspar, quartz is invariably interstitial. Quartz-diorite with up to 20 per cent of quartz (S 31141) is locally present.

Mafic minerals. The total content of mafic minerals within the diorite suite ranges from less than 20 to 45 per cent in the varieties approaching gabbro in composition, but in the melamicrodiorites it may exceed 50 per cent. Biotite and hornblende are usually moulded on to euhedral and subhedral feldspar crystals, but they are also found in fairly large clusters, which locally also contain apatite and epidote. The ratio of biotite to hornblende is very variable. In one mafic syenodiorite large hornblendes with small inclusions of biotite make up to 50 per cent of the total volume. The size of these hornblendes ranges up to 1·5 mm. The largest hornblendes are usually partially altered to biotite and an olive-green amphibole which is structureless or, less commonly, fibrous.

Pyroxenes are only rarely found in the diorites. Medium-grained diorite from the shore ¼ mile (400 m) NNW of Culswick Broch (S 31141) contains a number of subhedral or, more rarely, ophitic crystals, of pyroxene (?augite) which are partially altered to amphibole.

Epidote. Epidote is present in small quantities in many of the diorites. It occurs both in altered patches within and along the margins of plagioclase and as individual subhedral crystals, associated with the clusters of biotite and hornblende. It is abundant in certain restricted pockets, particularly in the vicinity of Culswick Broch, and appears to be a late hydrothermal replacement mineral, as it is abundant in the pegmatites cutting the diorites of this area. Epidote is also found as grains in melamicrodiorite.

Accessory minerals. Apatite and large sphenes are common. Allanite is a rare accessory in the granodiorites (S 51537) and syenodiorites (S 31143). Rutile inclusions are common in biotite and quartz.

Diorite with sphene-bearing ovoids

The pale spheres or ovoids with a large central sphene which are abundant in the banded diorite at and near Hestinsetter Hill (p. 212), are composed of coarse-grained pyroxene-monzonite. They normally range in diameter from 2 to 5 mm but may reach 20 mm (S 51528, Plate XXVI, fig. 8). The monzonite is composed of roughly equidimensional euhedral crystals of oligoclase set in large (up to 3 mm) poikilitic crystals of microcline. It contains numerous stumpy near-euhedral crystals of augite which are commonly about 0·2 mm in diameter, but may attain 0·5 mm. The augite is only locally altered to amphibole. Green hornblendes, some euhedral, are present only near the margins of the ovoids. Sphene is abundant, commonly forming euhedral crystals up to 3 mm long in the centre of the ovoid, but also occurring in smaller euhedral crystals throughout. Interstitial quartz forms approximately 5 per cent of the total volume of the rock. It is of interest that the rock forming these ovoids is the only member of the diorite suite which contains abundant unaltered pyroxenes, and completely fresh feldspars. This would suggest that they are cognate xenoliths which crystallized from a 'dry' magma, whereas the bulk of the diorite was formed from a magma rich in H_2O and other volatiles, probably due to saturation by volatiles released from the neighbouring acid magma (see Bailey 1958).

TABLE 5

Analyses of specimens from Sandsting Complex

	1	2	3	4	5	6
SiO_2	69·49	70·96	57·60	55·35	51·94	51·98
Al_2O_3	16·12	15·18	16·66	17·82	18·65	18·70
Fe_2O_3	0·83	1·69	2·60	2·28	1·49	1·62
FeO	1·63	—	4·34	3·86	4·04	3·95
MgO	0·90	0·70	2·99	3·58	5·86	5·94
CaO	1·94	1·52	5·27	6·60	8·96	8·88
Na_2O	4·14	4·32	3·97	4·28	3·18	3·20
K_2O	3·98	4·68	3·01	2·88	2·24	2·27
$H_2O > 105°$	0·65	0·66	1·05	0·76	1·20	0·86
$H_2O < 105°$	0·10	0·18	0·27	0·18	0·16	0·18
TiO_2	Trace	Nil	1·73	1·74	1·34	1·36
P_2O_5	0·09	Trace	0·59	0·66	0·68	0·66
MnO	Trace	—	0·11	0·13	0·10	0·07
CO_2	—	—	Trace	Trace	n.d.	—
SO_3	n.d.	n.d.	—	—	Trace	—
Cl	n.d.	n.d.	0·04	0·02	Trace	n.d.
S	n.d.	n.d.	0·05	0·03	—	n.d.
FeS_2	—	—	—	—	0·13	0·24
Cr_2O_3	n.d.	n.d.	—	—	n.d.	0·03
BaO	—	—	0·12	0·11	0·09	—
SrO	n.d.	n.d.	—	—	—	—
Rb_2O	n.d.	n.d.	0·01(s)	0·01(s)	—	—
Li_2O	n.d.	n.d.	—	—	—	—
TOTAL	99·87	100·00	100·41	100·29	100·06	99·94
Less O for Cl	—	—	0·01	—	—	—
	99·87	100·00	100·40	100·29	100·06	99·94
Sp. gr.	n.d.	n.d.	2·79	2·78	—	—

s = Spectrographic determination   n.d. = not determined

Gabbro

Gabbro is a very minor constituent of the Sandsting diorite, being confined to narrow dyke-like bands in the eastern part of the complex. The gabbro (S 51533) is coarse-grained and contains subhedral plates of sodic labradorite which are, in some instances, albitized to an irregular blotchy aggregate of chlorite, sodic plagioclase and muscovite. Deep green amphibole occurs along cracks and twin-planes, and there are small aggregates of epidote along some crystal margins. Individual feldspar plates are up to 10 mm long. They are enclosed in large ophitic crystals of largely amphibolitized pyroxene, which forms 30 per cent of the total volume. The pyroxene, a pale non-pleochroic augite, is commonly stained yellowish orange and peripherally altered to amphibole, which is deep green along the margins.

In the gabbro of the Garderhouse area (S 51543) the ophitic pyroxene is

TABLE 5—(*contd.*)

NORMS

	1	2	3	4	5	6
Q	23·88	23·62	8·61	2·46	0·00	0·00
C	1·68	0·24	0·00	0·00	0·00	0·00
or	23·52	27·66	17·79	17·02	13·24	13·42
ab	35·03	36·56	33·00	35·92	26·91	27·08
an	9·04	7·54	19·06	21·06	30·00	29·96
hl	0·00	0·00	0·11	0·05	0·00	0·00
di	0·00	0·00	2·61	5·89	8·00	7·80
hy	4·55	1·74	9·26	8·61	11·82	11·52
ol	0·00	0·00	0·00	0·00	2·23	2·38
mt	1·20	0·00	3·77	3·31	2·16	2·35
cm	0·00	0·00	0·00	0·00	0·00	0·04
hm	0·00	1·69	0·00	0·00	0·00	0·00
ilm	0·00	0·00	3·29	3·30	2·54	2·58
ap	0·21	0·00	1·37	1·53	1·58	1·53
pr	0·00	0·00	0·11	0·06	0·13	0·24
bao	0·00	0·00	0·12	0·11	0·09	0·00
Others	0·75	0·84	1·33	0·95	1·36	1·04
Total	99·87	99·89	100·41	100·29	100·06	99·94
Q	28·97	26·89	14·49	4·45	0·00	0·00
or	28·53	31·49	29·95	30·72	32·97	33·13
ab	42·50	41·62	55·56	64·83	67·03	66·87
Total	100·00	100·00	100·00	100·00	100·00	100·00
or	34·80	38·54	25·46	23·00	18·87	19·04
ab	51·83	50·95	47·25	48·54	38·36	38·44
an	13·37	10·51	27·29	28·46	42·77	42·52
Total	100·00	100·00	100·00	100·00	100·00	100·00
ab	79·49	82·90	63·39	63·04	47·28	47·48
an	20·51	17·10	36·61	36·94	52·72	52·52
Total	100·00	100·00	100·00	100·00	100·00	100·00

1. Granite, Skelda Ness, (Exact locality not known). Anal: W. H. Herdman (Finlay 1930, p. 693).
2. Granite, Gruting Voe, (Exact locality not known). Anal: R. R. Tatlock (Finlay 1930, p. 693).
3. Diorite, 400 yd (380 m) ESE of Wester Skeld. [299 438]. S 33678. Lab. No. 1066. Anal: G. A. Sergeant; spect. det. H. K. Whalley. (Guppy and Sabine 1956, pp. 14–15).
4. Diorite, ¼ mile (390 m) ENE of Lunga Water. [293 452]. S 33676. Lab. No. 1067. Anal: G. A. Sergeant, spect. det. H. K. Whalley. (Guppy and Sabine 1956, pp. 14–15).
5. Gabbro-diorite hybrid, 150 yd (140 m) NE of Loch of Arg, Sandsting. [304446]. S 33677. Lab. No. 1068. Anal: G. A. Sergeant. (Guppy and Sabine 1956, pp. 14–15).
6. 'Hornblende-gabbro', 150 yd (140 m) NE of Loch of Arg, Sandsting. [304446]. S 30035. Lab. No. 970. Anal: B. E. Dixon. (Guppy and Sabine 1956, p. 24).

completely uralitized. Primary hornblende, some showing polysynthetic twinning, is abundant. The feldspars are locally saussuritized. Epidote associated with calcite and chlorite is also present in veinlets.

Melamicrodiorite and other mafic intrusions

The petrographical characters of the larger intrusions of melamicrodiorite (Plate XXVI, fig. 5) grade into those of the fine-grained meladiorite but the thin dykes cutting granite and sediment have the textural features of dolerite and porphyritic basalt altered to epidiorite. The former (S 51518) contain sporadic phenocrysts of mid-andesine up to 0·7 mm long and groundmass feldspars of oligoclase-andesine which average 0·3 mm. Potash feldspar is generally absent, but up to 5 per cent of interstitial microcline has been recorded in some sections. Hornblende and deep green biotite make up between 30 and 50 per cent by volume of the rock, and form irregular crystals moulded against feldspar or, occasionally, small clusters. Sphene and apatite are common accessory minerals.

The melamicrodiorite dykes close to the Walls Boundary Fault are sheared and chloritized and individual minerals are broken up or streaked out.

Pegmatite veins

The composition and texture of the pegmatite varies greatly within a single vein (S 51514, 51514A) in which distinct zones can usually be recognized. The inner zone which adjoins an epidote-lined cavity consists of large crystals mainly of microcline and relatively rare plagioclase cut by a network of veinlets of epidote. The epidote is locally coarsely crystalline and may form as much as 50 per cent of the total volume of the rock. Some of the epidote veinlets are parallel to the twin planes of the microcline. The feldspar is also cut by irregular vein-like patches of quartz which are themselves cut by epidote. The outer zone of the pegmatite veins consists largely of crystals of micropegmatite up to 8 mm in diameter. These are composed of graphically intergrown quartz and micro-cline, with a few irregular inclusions of epidote (Plate XXVI, fig. 7)

Ultrabasic rocks

The blocks of ultrabasic rock 180 yd (165 m) WNW of Stump Farm [310 456] (S 31157, 51535, Plate XXVI, fig. 6) contain up to 50 per cent of olivine which forms subhedral to anhedral crystals ranging in size from 4 to 0·5 mm. These are commonly surrounded by a sheath of serpentine, which, together with iron ore, also fills cracks within the crystals. Some olivine crystals are largely altered to serpentine. The olivine is set in a matrix of large subhedral plates of bytownite (An_{70-80}) and large isolated crystals of pyroxene which range up to 3 mm in size. Pyroxene occurs in areas where the concentration of olivines is relatively low. Both enstatite and augite are present. Pyroxene forms nearly 20 per cent by volume of the total rock, bytownite up to 30 per cent. In addition poikilitic crystals of strongly pleochroic reddish brown biotite are a characteristic feature of the rock.

It is not easy to give this rock an appropriate name. It compares fairly closely with the ultrabasic rocks from the east shore of Moora Waters (p. 193-4), which range from harrisite to lherzolite.

Granodiorite

FIELD RELATIONSHIPS

A sheet of coarse-grained granodiorite which is texturally similar to the granite and averages 200 to 300 ft (60–90 m) in thickness, overlies the Culswick diorite, the junction between the two being irregular and steeply inclined to the north. This sheet can be traced from the west shore of Keolki Field, between 650 and 800 yd (595 and 730 m) N of Culswick Broch, eastward as far as Culswick village. At several localities inclusions of baked sediment occur along the diorite-granodiorite boundary, and a number of blocks of sediment, including sandstone with convolute bedding, are entirely in the granodiorite. The junction between diorite and granodiorite is well seen on the west shore of Keolki Field, where the latter contains numerous large xenoliths of diorite. Most diorite xenoliths are straight-sided and angular. Some are horizontally elongated and up to 20 ft (6 m) long. There can thus be little doubt that the intrusion of granodiorite was later than that of diorite.

The granodiorite is cut by a number of felsite dykes, up to 1 ft (30 cm) thick, which trend roughly north–south.

PETROGRAPHY

The granodiorite is coarsely crystalline and composed of broad tabular crystals of calcic oligoclase, usually with a narrow rim of more sodic plagioclase. These range in length from 5 to 0·4 mm and are set in large anhedral poikilitic masses of microcline-microperthite and blotchy microperthite. Plagioclase is only slightly in excess of potash feldspar. Quartz, commonly with small liquid inclusions, which forms about 10 per cent of the total volume of the rock, usually occurs interstitially but in rare cases forms irregular intergrowths with potash feldspar. Dark minerals form 15 per cent of the total volume with thick greenish plates of biotite slightly more abundant than pale greenish hornblende, which is commonly intergrown with the biotite. Sphene occurs in large skeletal crystals, and small euhedral needles of apatite are abundant. The most common ore mineral is pyrite which is present in very small euhedral crystals.

Porphyritic Microadamellite and Associated Rocks

FIELD RELATIONSHIPS

The porphyritic microadamellite is a rock with distinctive field characteristics. In hand specimen it is generally fine-grained with no readily visible quartz, with abundant small plates of black mica and prominent phenocrysts of pink or white feldspar, which may be up to 1 cm in diameter.

Its main outcrop, as shown on Plate XXV, extends from Green Head on the Whites Ness peninsula eastward *via* the Ward of Culswick to the head of the Culswick Valley, where it is displaced southwards by the Culswick Fault. It can be intermittently traced eastward from the east slope of the Culswick Valley to the hillside north-west of Housa Water. Fine-grained granite, texturally similar to the porphyritic microadamellite, forms a sheet-like intercalation in the Culswick diorite, extending from the head of the Stead of Culswick, westward

EXPLANATION OF PLATE XXVI

PHOTOMICROGRAPHS OF THE SANDSTING GRANITE-DIORITE COMPLEX

FIG. 1.　Slice No. S 51550. Magnification × 16. Crossed polarisers. Granophyre sill in Walls Sandstone, composed of graphic intergrowth of quartz and potash feldspar and scattered subrounded crystals of microcline and albite-oligoclase. West shore of Bixter Voe, 240 yd (225 m) SSE of Mosshouse [328 514].

FIG. 2.　Slice No. S 33678. (Analysed specimen No. 1066, Guppy and Sabine 1956, p. 14, No. 653.) Magnification × 16. Crossed polarisers. Diorite, with plates of andesine rimmed with oligoclase, subordinate hornblende, biotite and rare sphene. Interstitial microcline, microperthite and quartz. 1400 yd (1280 m) ESE of Wester Skeld, near Loch of Arg [299 438].

FIG. 3.　Slice No. S 51509. Magnification × 16. Crossed polarisers. Microadamellite. Clusters of near-euhedral plates of zoned plagioclase (andesine rimmed by oligoclase) form 40 per cent of the total feldspar. Large irregular plates of microperthite and interstitial microcline form the remaining 60 per cent. Quartz forms 10 per cent of the total volume. The mafic minerals hornblende and biotite are present in equal proportion. Apatite is an abundant accessory. Coast of Scurdie, 470 yd (425 m) SE of Green Head [253 459].

FIG. 4.　Slice No. S 28878. Magnification × 17·6. Crossed polarisers. Fine-grained hornblende-diorite. Near-euhedral plates of calcic andesine rimmed with calcic oligoclase. Interstitial microcline and quartz. Hornblende and greenish brown mica are present in equal volume. Sphene and small needles of apatite are abundant accessories. Brunt Hamar, 830 yd (760 m) NE of Silwick [299 426].

FIG. 5.　Slice No. S 51523. Magnification × 42. Plane polarized light. Melanocratic microdiorite. Irregular decussate laths of sodic andesine set in interstitial base of ragged plates of deep green biotite forming 30 per cent of total volume. Clusters of small crystals of hornblende. Abundant accessories are sphene and small specks of iron ore. South-east side of Stead of Culswick, 550 yd (500 m) S32°E of south end of Sand Water [272 442].

FIG. 6.　Slice No. S 51535. Magnification × 16. Plane polarized light. Ultrabasic rock resembling harrisite. Olivine is sheathed in serpentine. Pyroxene (augite and enstatite) forms large in part poikilitic plates. Also subhedral plates of labradorite-bytownite and plates of reddish brown mica (?lepidomelane) enclosing skeletal iron ore. 190 yd (175 m) W20°N of Stump Farm [307 457].

FIG. 7.　Slice No. S 51514. Magnification × 16. Crossed polarisers. Pegmatite vein in diorite. Large crystal of micropegmatite (quartz-microcline intergrowth) set in matrix of irregular plates of microcline and quartz with grains and veinlets of epidote. South shore of Taing of Koelkifield, 530 yd (480 m) N18°W of Culswick Broch [252 452].

FIG. 8.　Slice No. S 51528. Magnification × 16. Crossed polarisers. Part of ovoid of pyroxene-monzonite in microdiorite. Characterized by large euhedral sphenes, and smaller euhedral crystals of colourless pyroxene set in base of near-euhedral crystals of clear orthoclase, plagioclase and microcline. West slope of Hestinsetter Hill, just east of road, 340 yd (310 m) S18°E of Giant's Grave [293 455].

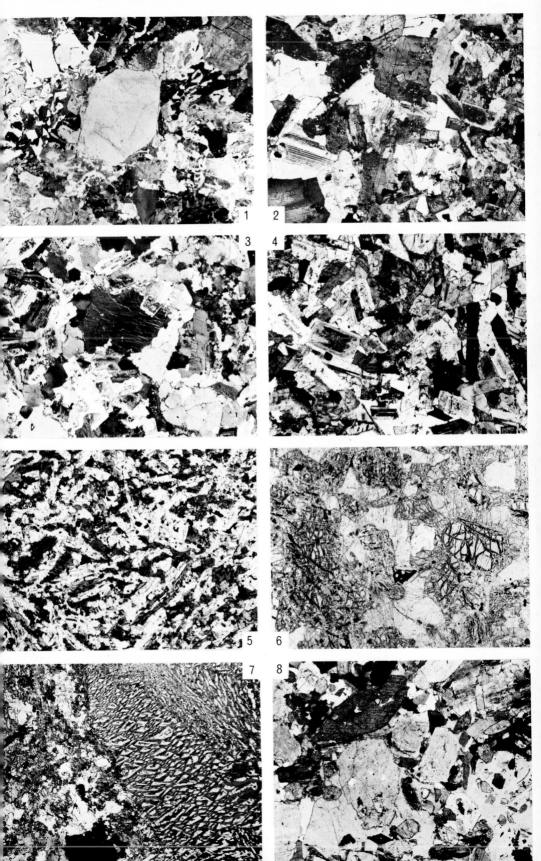

for nearly 1 mile (1600 m) to The Nev. Microadamellite also forms a number of small outcrops on the headland north-west of Culswick Broch. Microgranite, locally porphyritic, forms a high proportion of the veins and major cross-cutting intrusions within the diorite, not only in areas where microgranite adjoins diorite, but also within the diorites of the Hestinsetter and Scarvister–Loch of Arg–Skelda Voe areas (p. 212).

Age Relationships. Microgranite and microadamellite everywhere cut the diorite and, with a few exceptions (p. 215, Fig. 24(1, 2)), the diorite xenoliths in these rocks are angular. Their relationships with the coarse-grained granodiorite and coarse-grained granite are, however, less clear, as junctions are either poorly exposed or faulted.

The junction between granodiorite and microadamellite appears to be sharp, but as there are no good exposures of the contact the age relationship is uncertain. Contacts between microadamellite and coarse-grained granite are well seen in shore sections 1300 yd (1190 m) WNW of the Ward of Culswick and 150 yd (135 m) W of the Broch of Culswick. In the former exposure the junction is vertical and, where undisturbed, it is fairly sharp. There is here a zone, about 4 in (10 cm) wide, in which large feldspars, similar to those in the coarse-grained granite, form abundant porphyroblasts within the microadamellite. The latter is sheared for a distance of 3 to 5 ft (90–150 cm) from the contact, but the granite is unaffected. This shearing may not be connected with the emplacement of magma. The junction is also partially obscured by a 3-ft (90-cm) dyke of felsite. A number of small irregular masses of microadamellite, in one case associated with baked sediment, are present in the granite some distance north of the junction. The junction exposed just west of Culswick Broch is sharply defined, near-vertical, but in detail wavy and irregular, suggesting that the adjoining magmas co-existed in an incompletely consolidated state. Evidence as to age relationships from the contacts is thus inconclusive, but the fact that in most areas the coarse-grained granite is in contact with the overlying sediment (Plate XXV) and that there is nowhere a screen of baked sediment between the inner fine-grained and outer coarse-grained 'granite' would suggest that along the northern (upper) margin of the complex, at least, the coarse granite was emplaced first, but was not fully consolidated when the microadamellite magma was intruded.

Veins. In areas where both microadamellite and granite vein the diorite very few junctions between the two granitic types have been recorded. At the south end of Shalders Taing, 700 yd (640 m) S of the head of the Stead of Culswick, the two types appear to grade into each other, with a complete transition within a distance of 1 ft (30 cm) though the disappearance of the large quartz crystals of the granite is abrupt. There is no evidence of one kind of granite vein truncating the other. Veins of porphyritic microadamellite have been recorded in all diorite outcrops except that of Garderhouse. In some cases the veins have a dark marginal zone up to 1 ft (30 cm) wide in which euhedral feldspar phenocrysts are enclosed in a mafic matrix.

Variation of texture. Though the microadamellite is characteristically porphyritic with subhedral to euhedral potash feldspar phenocrysts, the abundance of phenocrysts varies very considerably. The grain size of the matrix is also very variable, ranging from aphanitic in certain minor intrusions cutting the diorite (as seen in large blocks on the east shore of Keolki Field, 420 yd (384 m) NNW of Culswick Broch), to medium-grained in the main outcrop.

PETROGRAPHY

The salient feature of the microadamellite is the presence of euhedral, generally poikilitic phenocrysts of potash feldspar, which reach a length of 10 mm, but are commonly about 3·5 mm long. They are composed of orthoclase full of small irregularly shaped inclusions of plagioclase, which are in optical continuity throughout the crystal. Phenocrysts of various types of perthite, microperthite, microcline-microperthite and microcline are also

---

EXPLANATION OF PLATE XXVII

PHOTOMICROGRAPHS OF THE THERMAL AUREOLE AND HYDROTHERMAL MINERALIZATION IN THE SANDSTING GRANITE-DIORITE COMPLEX

FIG. 1. Slice No. S 51515. Magnification × 42. Plane polarized light. Hornfelsed sandstone in sedimentary enclave within diorite. Quartz and feldspar grains are welded together. The interstitial matrix is recrystallized into granular epidote, brown biotite and subordinate hornblende. The larger dark patches consist partly of ophitic hornblende. Koelkifield, 500 yd (450 m) N8°E of Culswick Broch [254 452].

FIG. 2. Slice No. S 51782. Magnification × 42. Plane polarized light. Dark grey hornfelsed silty mudstone close to junction with Sandsting Granite. Serrate quartz grains set in matrix of stumpy brown biotite. The darker areas are spongy, highly ophitic crystal aggregates of green hornblende which are up to 0·5 mm in diameter. North-east shore of Bight of Selistack, 650 yd (590 m) E37°N of south point of Green Head [256 466].

FIG. 3. Slice No. S 52534. Magnification × 42. Crossed polarisers. Indurated mudstone overlying limestone, within thermal aureole of Sandsting Complex. Fine-grained hornfelsed calcite-mudstone with oval poikiloblasts (white) of cordierite. The mudstone contains small patches of clinozoisite. Island of Vaila, west shore of Easter Sound, 650 yd (590 m) WNW of Ram's Head Lighthouse [242 463].

FIG. 4. Slice No. S 52536. Magnification × 22. Plane polarized light. Hornfelsed impure limestone within thermal aureole of Sandsting Complex. Composed largely of intensely sieved grossularite enclosing minute grains of diopside. The opaque bands consist almost entirely of finely granular diopside. The white lens in the bottom right of the picture consists of feldspar with diopside grains. Elongate crystals of clinozoisite near bottom left. Island of Vaila, west shore of Easter Sound, 650 yd (590 m) WNW of Ram's Head Lighthouse [242 463].

FIG. 5. Slice No. S 51502. Magnification × 16·8. Crossed polarisers. Junction of scapolite vein with Walls Sandstone. The scapolite forms irregular laths which intersect the vein-margin at an angle of 45°. South shore of Gruting Voe, 820 yd (750 m) E19°N of Hogan [272 473].

FIG. 6. Slice No. S 31121. Magnification × 16. Crossed polarisers. Sheared and partly mylonitized granite with small discontinuous, partly sheared veinlets of scapolite laths. Skelda Ness peninsula. Pundswell, 200 yd (180 m) NNW of summit of Longa Berg [304 422].

FIG. 7. Slice No. S 28732. Magnification × 31. Crossed polarisers. Sheared-out and partly mylonitized scapolite adjoining granite. Hillside, 200 yd (180 m) E of Wester Wick [287 424].

FIG. 8. Slice No. S 31126. Magnification × 31. Crossed polarisers. Intensely scapolitised sheared granite cut by faulted veinlet of potash feldspar. Skelda Ness peninsula, 1000 yd (910 m) SSW of summit of Longa Berg [302 411].

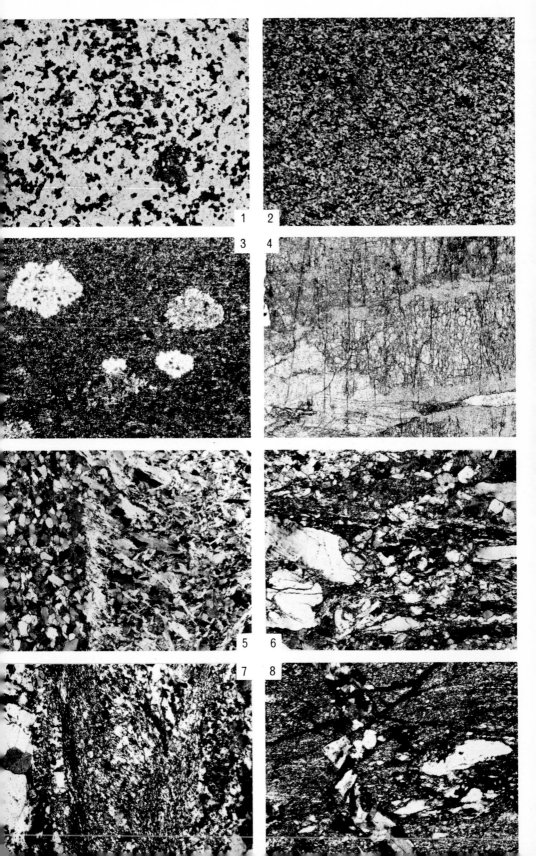

present in various sections (Plate XXVI, fig. 3). The groundmass contains subhedral laths of oligoclase-andesine, generally with a thin rim of sodic oligoclase. Within the main outcrop these range in length from 1·5 to 0·15 mm and are stumpy in some sections but markedly elongated in others. Interstitial microcline and quartz occur in large irregular highly poikilitic areas up to 2 mm in diameter. Quartz forms between 15 and 25 per cent by volume of the rock. The ratio of plagioclase to potash feldspar varies in different specimens from 40:60 to 80:20, the rock type thus ranging from adamellite to granodiorite. In most sections, the ratio approximates to 50:50. Hornblende and biotite form up to 10 per cent of the total. Hornblende exceeds biotite and commonly forms euhedral crystals up to 0·5 mm long. Biotite is partially altered to chlorite. Epidote occurs in rare irregular grains, either associated with biotite-feldspar aggregates or in the centres of altered feldspars. Allanite is relatively common, either associated with epidote (S 51505) or as patches enclosed in potash feldspar phenocrysts. Small needles of apatite are abundant, but sphene is rare.

It will be noted that the grain sizes of the feldspars forming the matrix are in some cases larger than the accepted maximum (0·5 × 0·5 mm) for micro-granites. As, however, the average grain size of the whole group is smaller than this, and as it has not been found advisable to divide the group according to an arbitrary grain-size boundary, all members have been shown on the map as porphyritic microadamellite.

The rock which cuts the diorite cropping out on the north-west shore of the Stead of Culswick (S 28736, 29882, 51519), is a fine-grained sparsely porphyritic, granophyric microgranite, composed of up to 30 per cent quartz and up to 70 per cent feldspar, mainly in the form of blotchy, rather cloudy, perthite or microcline-microperthite which occurs both interstitially and as phenocrysts. The latter are generally poikilitic with small irregular inclusions of plagioclase. The plagioclase is albite-oligoclase with a narrow rim of albite. It never makes up more than 20 per cent of the total feldspar. Biotite and hornblende are present in equal proportions but together do not exceed 5 per cent of the total. The biotite occurs as thick plates and the hornblende as isolated anhedral crystals up to 0·6 mm long.

GRANITE

FIELD RELATIONSHIPS

Coarse-grained quartz-rich biotite-granite forms over half of the outcrop of the Sandsting Complex (Plate XXV). Its grain size is fairly consistent throughout the greater part of the outcrop, but there is a slight eastward decrease in the amount of mafic minerals present. This is coupled with an eastward increase in the proportion of quartz and a decrease in plagioclase. North-north-west of the latitude of Garderhouse the granite forms a series of near vertical sheets in the steeply inclined sediments of the Walls Formation. These thin out northwards and on the shores of Bixter Voe and Effirth Voe they form a number of sill swarms, consisting of a large number of individual sills, which are steeply inclined to the east-north-east and range in thickness from 2 ft (60 cm) to possibly 400 ft (120 m). These sills show a gradual, though uneven northward reduction in grain size, becoming increasingly more granophyric in texture towards Bixter Voe. The thinner dykes are in places felsitic in texture, and in Laxa Burn, a tributary of Bixter Voe, two fine-grained basic dykes respectively 6 ft (1·8 m) and > 2 ft (> 60 cm) thick cut the granite.

The granite and graphic microgranite sills of Bixter Voe are in some respects similar to the sheets of granite and felsite which cut the Walls Formation on the shores of Seli Voe, Gruting Voe and Vaila Sound. On the shores of the Holm of Gruting and Hoga Ness a sill swarm about 800 yd (730 m) wide contains at least 15 sills, of which some consist of very coarse-grained microperthite-granite and others of fine-grained almost aphanitic felsite.

Junctions with sediment. Except at Green Head in the Whites Ness peninsula and for a short distance in the area north-west of the Ward of Culswick the Walls Sandstone is directly underlain by the coarse-grained granite. Good sections of the contact are exposed on the east shore of Gruting Voe, 1850 yd (1618 m) N of Culswick Broch and at Coukie Geo and Vine Geo on the south shore of the Island of Vaila.

On the east shore of Gruting Voe indurated sandstone overlying the granite dips at 28° to 35° to the north-north-east. The granite-sediment junction is sharp, with no marked chilling within the granite and with a 5 mm thick layer of crystalline quartz along the contact. Near sea level the junction is stepped alternately parallel and normal to the bedding, thus producing a steep overall dip to north-north-east. There are a small number of irregular inclusions of sediment within the granite and a number of tongues of granite extend up to 6 ft (1·8 m) into the sediment. The granite in these tongues is in places very fine-grained, almost felsitic in texture. The upper half of the exposed contact is concordant with the dip of the sediment, though minor transgressions of horizon are common.

The junction between sediment and granite at Coukie Geo, Vaila (Plate XXV) is inclined at 65° to NNW, being locally roughly concordant with the dip of the overlying sediment. A number of irregular tongues and veins of granite extend into the sediment. The southern and eastern margin of the Vaila Granite is exposed at several localities in the vicinity of Vine Geo. It is near-vertical, but extremely irregular with many granite veins passing into the adjoining sediment. The indurated and somewhat shattered sandstone adjoining this granite on either side of Vine Geo has a strike which is almost normal to the regional strike of the sediment and contains an intricate network of granite veins. In one locality it contains diorite veins which are in turn veined by granite. It is thought that the intensely veined sediment of this area forms part of a fault-bounded screen of veined and shattered sediment which extends from the south-east corner of Vaila, south-eastward *via* Muckle Flaes and the Taing of Keolkifield to Burri Geo (p. 228).

PETROGRAPHY

Granite of Main Intrusion

The granite forming the main intrusive mass is a typically coarse-grained quartz-rich biotite-granite with 30 to 40 per cent of quartz and with anhedral plates of potash feldspar usually up to 3·5 mm in diameter but in some cases reaching a maximum of 10 mm. Plagioclase forms up to 15 per cent of the total feldspar and normally occurs as euhedral to subhedral plates of oligoclase up to 1·7 mm long, though usually considerably smaller and frequently grouped in clusters. The predominant ferromagnesian mineral is partially chloritized biotite which forms small irregular plates. Hornblende is generally present only along the granite-diorite margins, but is found in certain areas (S 51536) within the main granite, where it forms euhedral crystals up to 3·5 mm long. Sphene and allanite occur as small isolated crystals; apatite is fairly abundant.

Potash feldspar occurs in the form of large plates of microperthite or micro-cline-microperthite and small interstitial patches of microcline. Several forms of microperthite and perthite have been recorded. The most common variety is rod-microperthite (Deer, Howie and Zussman 1963, p. 68), with parallel rods of exsolved sodic plagioclase ranging in width from 0·01 to 0·03 mm, orientated normal to the length of the crystal and cutting the twin planes of microcline and the two main cleavages of the feldspar at 45°. Rods, though most commonly parallel to each other, are branching and generally form a network with the two main sets of rods crossing each other at a very acute angle. String-microperthite with straight parallel fairly widely spaced strings approximately 0·05 mm thick is less common. The strings and rods do not always extend along the whole length of the crystal, in some cases (S 51550) the rods are widest at the edges of the crystal and taper inward, leaving the centre of the crystal relatively free of exsolved sodic plagioclase. Many rods extend for a third to half the length of the crystal.

Of less common occurrence is microperthite with branching vermicular or curved rods up to 0·02 mm thick, which in some cases are roughly parallel to each other and are connected by thin fibrous branches. The plagioclase in micro-perthite of this type commonly follows the cleavages and at intersections it widens out to form beads up to 0·75 mm wide. The sodic plagioclase forming these beads is commonly cloudy.

Another fairly common form of perthite is the type termed replacement perthite by Alling (1938, fig. 2). The plagioclase consists of a number of roughly equidimensional patches of irregular outline which are in places connected with each other and are all in optical continuity throughout the crystal (S 28875, 51521). The patches are usually concentrated near the centre of the crystal, range in diameter from 0·05 mm to 0·1 mm and may form as much as 60 per cent of its total volume.

Antiperthite has not been recorded in the granite of the main Sandsting area, but occurs in the graphic granite forming the sills of Laxa Burn and Bixter Voe. Texturally this is very similar to the replacement perthite described above. All inclusions of potash feldspar within the sodic plagioclase are in optical continuity and the average diameter of these inclusions is 0·075 mm.

Quartz occurs in large anhedral crystals which are usually highly poikilitic. Its margins with adjacent potash feldspar are usually serrate and in a few cases areas of graphic intergrowth are developed. In some cases the marginal area of intergrowth is so irregular that it resembles myrmekite. Graphic intergrowth is less common in the main granite than in the granite of the Bixter Voe sills. The quartz of many specimens from the main granite is characterized by the extreme abundance of liquid inclusions. These are of irregular shape, average 0·005 to 0·01 mm in diameter with a maximum of 0·02 mm and have a slight brownish tinge and a much lower refractive index than the quartz (S 29526, 29518, 28736). In most cases these inclusions are evenly distributed throughout the quartz crystal, but in some thin sections (S 31164, 28875) inclusions are confined to thin narrow roughly parallel lines traversing the crystal. Minerals enclosed in quartz

Q

include apatite and rarely rutile. In many cases the quartz has undulose extinction, due to slight strain.

Sphene is present as small crystals in the more calcic phases of the granite and close to contacts with diorite.

Alteration products. The plagioclase is in places patchily saussuritized into aggregates of chlorite and epidote in the centres of crystals (S 28875). Small grains of epidote are also associated with altered biotite. Potash feldspars are in places patchily altered to sericite.

Graphic Granite of the Northern Sills

The granite sills extending from the eastern end of the granite outcrop near Garderhouse towards Bixter Voe contain coarse-grained granite with microperthite and antiperthite as described above, but only small amounts of ferromagnesian minerals. This granite grades into graphic granite which becomes progressively finer-grained and has more extensive zones of graphic intergrowth as it is traced northwards. The coarse-grained granite of the sills (S 31132, 31133, 51548) differs from the granite of the main mass in that it contains a higher proportion of quartz (up to 40%). This has serrate margins with rudimentary graphic intergrowth. Oligoclase is fairly abundant, forming clusters of euhedral crystals (S 31133). Hornblende is absent and biotite is present as small interstitial scraps.

Graphic intergrowth of quartz and microperthite or microcline ranges from coarse, with the width of adjacent quartz and feldspar phases averaging 1 mm and with an irregular intergrowth pattern (S 31133), to very fine with a regular pattern in which adjacent units are 0·08 mm thick (S 51550, Plate XXVI, fig. 1). The quartz-feldspar ratio in the latter is about 50:50.

GRANITE VEINING IN SHATTERED SEDIMENT WITHIN SANDSTING COMPLEX

A thin north-west trending fault-trough of shattered sediment intensely veined by granite separates the coarse-grained granite forming the cliffs south-east of Culswick Broch from the outcrop of diorite south-east of Keolki Field. The fault-trough extends from Burri Geo, where it is approximately 30 yd (27 m) wide, north-westward *via* the Loch of the Brough to the coast south of the Taing of Keolkifield, where it has widened to over 150 yd (137 m). The north-westward continuation of the fault-trough is uncertain. The islands of Muckle Flaes, 500 to 700 yd (460–640 m) NW of Culswick Broch, and the peninsulas on either side of Vine Geo at the south-east corner of Vaila, are also formed of shattered sediment intricately veined with granite. These areas may be within the north-westward continuation of the above-mentioned fault-trough, whose outcrop has further widened.

At the Taing of Keolkifield and in part of the Green Head peninsula of Vaila, the shattered sediment is veined by diorite, which is, in turn, intensely veined by fine-grained granite. The vein granite throughout this area strongly resembles the porphyritic microgranite exposed on the north-east shore of the Stead of Culswick. The fault crossing the Taing of Keolkifield separates the veined sediment and diorite from diorite without veins and can clearly be seen to truncate the granite veins. The fault clay within the zone of shear is, however, affected by thermal alteration, suggesting that this part of the complex was still hot when the faulting took place.

At Burri Geo the screen of sheared and veined sediment is very steeply inclined to the north-east but on the shore north of Culswick Broch its bounding faults are approximately vertical. It is difficult to obtain a satisfactory picture of the structure of this part of the complex as nothing is known of the seaward extension of the various rock types. It is, however, likely that the coarse granite south-east of Culswick Broch and the coarse granite of Vaila form part of the granite sheet overlying the diorite and that the diorite of Keolki Field has been pushed upward relative to the granite south of the fault belt, possibly during the period of emplacement of the Stead of Culswick microgranite. The screen of sediment between the two masses of igneous rock was shattered and veined by fine-grained granite during an early phase of the movement, but movement along the fault continued after vein emplacement had ceased.

LATE FINE-GRAINED ACID AND BASIC INTRUSIONS

Intrusions cutting sedimentary rocks in the Gruting Voe–Walls district

Field Relationships. Fine-grained acid intrusions which appear to be derived from the Sandsting igneous centre have been recorded in sediments up to 2 miles (3 km) from the granite margin, the most distant being the felsite dykes cutting intensely folded sediments forming the Mara Ness peninsula [269 496] on the west shore of Gruting Voe. Most of these dykes have a trend ranging from W20°N to N25°W. Two felsite dykes have been recorded even further from the Sandsting Granite, on the north shore of Loch of Vadill [227 488], 1 mile (1610 m) WSW of Walls Pier and 530 yd (480 m) E of Bruntskerry [227 502], $1\frac{1}{8}$ miles (1810 m) NW of Walls Pier. Both dykes are composed of spherulitic felsite with corroded phenocrysts of quartz, orthoclase and plagioclase.

On the south shore of Gruting Voe, between the northern margin of the Sandsting Complex and the Hoga Ness sill swarm (p. 226), dykes ranging in thickness up to 36 ft (11 m) cut the highly indurated sediments of the Walls Formation. Their trend in this area ranges from N25°W to N50°W, cutting obliquely across the strike of the sedimentary rocks. All these dykes are felsitic in texture, but at least one has a central porphyritic zone. Some contain fragments of sediment. One dyke exposed on the south shore of Gruting Voe is composite with a central zone of felsite and narrow outer zones of fine-grained basic rock. This is the only recorded composite dyke in the complex.

Petrography. The felsite of these dykes (S 51492, 51505) ranges from aphyric to strongly porphyritic with euhedral phenocrysts of quartz, microperthite and calcic oligoclase set in a groundmass, just over half of which is made up of spherulites composed of radiating fibres of quartz-feldspar aggregate. The remainder of the groundmass consists of small interlocking grains of quartz and feldspar locally showing micrographic intergrowth around corroded quartz crystals. Phenocrysts range in diameter from 2 to 0·3 mm and may form up to 50 per cent of the rock. The ratio of feldspar to quartz phenocrysts is approximately 60:40. Adjacent quartz phenocrysts are commonly in contact and some are in optical continuity. The potash feldspar is usually poorly developed microperthite which may have slightly wavy margins. Plagioclase phenocrysts are relatively rare and composed of calcic oligoclase. The basic outer portions of the composite dyke exposed on the south shore of Gruting Voe (p. 229) are microdiorite composed of hornblende and cloudy plagioclase laths in roughly equal proportion. There are rare plagioclase phenocrysts which are partly enclosed in hornblende. The microdiorite is cut by a number of veinlets, up to 0·4 mm thick, composed almost entirely of fine-grained hornblende. These veinlets also penetrate into the adjacent brecciated sediment.

Intrusions cutting sedimentary rocks at Skelda Voe, Rea Wick and Roe Ness

Dykes and veins of felsite, many highly irregular and very thin, are present in the trough-faulted wedge of Walls Sandstone, which reaches the sea at the head of Skelda Voe, and in the folded and shattered sediment exposed on the east coast between Rea Wick and Roe Ness. In the latter area the felsite intrusions are branching and irregular and locally shattered by later earth movements associated with the Walls Boundary Fault.

Intrusions cutting the Sandsting Complex

A few thin felsite dykes have been recorded in the granite and diorite of the Sandsting Complex, both on the east shore of Gruting Voe between the granite-sediment junction and the Taing of Keolkifield and on the shore between the head of Seli Voe and Rea Wick. The dykes cut both the diorite and the coarse-grained granite, but have not been recorded within the porphyritic micro-adamellite. They are similar to the felsite intrusions cutting the sediment overlying the Sandsting Complex, and it is thought that all these felsites belong to the same period of intrusion.

The Metamorphic Aureole

'The thermal aureole within the sediment adjoining the Sandsting Complex is up to 1 mile (1·6 km) wide in the Gruting Voe area, but is considerably narrower and less well defined along the eastern margin of the granite outcrop, where narrow wedge-shaped granite sheets penetrate the sediment (Plate XII). The zone in which the sediment is completely hornfelsed and the original texture is entirely obliterated is very much narrower, probably nowhere greatly exceeding 550 yd (500 m). All sedimentary enclaves within the Complex (pp. 114–5) are hornfelsed.

Sandstone, siltstones and shales

The alteration of the sedimentary rocks is most intense within the enclaves and xenoliths in both granite and diorite (S 28884, 31156, 51515, 52552) and at the granite contact (S 51507, 53694). In these the sandstone (Plate XXVII, fig. 1) is reconstituted into a mosaic of quartz and feldspar grains with polygonal outlines and with adjoining quartz grains fused to produce irregular masses up to 1 mm long with re-entrant angles. The quartz is in most cases clear and devoid of shadowy extinction and inclusions. In two specimens (S 28882, 28883) however, the quartz grains contain abundant brownish liquid inclusions, like those in the quartz of the Sandsting Granite (p. 227), and in one specimen (S 31156) the quartz contains numerous small cracks intersecting each other at right angles. The feldspars are completely clear in the most intensely hornfelsed sections (S 53694) but more commonly they are slightly turbid, and the decomposition products are in some cases concentrated in the centres of grains. The original chlorite-carbonate matrix (pp. 117–121) is altered to a crystalline aggregate composed of varying proportions of the following minerals:

1. *Green hornblende*, which forms crystals ranging in outline from ragged to almost euhedral, or poikiloblasts up to 2 mm in diameter, which may be sub-hedral or euhedral.

2. *Biotite*, which forms thick flakes in part poikiloblastic, is commonly arranged in a decussate pattern. Close to the granite contact and within the sedimentary enclaves the biotite is commonly strongly pleochroic from straw-coloured to reddish brown. There is, however, considerable variation in the colour, and even in the same thin section it varies from reddish brown to brownish black. In a number of thin sections cut across the granite-sediment contact reddish brown micas are also present within the granite. Biotite and hornblende can occur together throughout the rock or singly along separate bedding planes.

3. *Epidote and clinozoisite*, usually forming anhedral grains partly mantling other minerals, occur in highly variable amounts, probably disposed along original calcareous laminae.

4. *Clinopyroxene*, probably entirely *diopside*, forms small anhedral grains in some xenoliths and along the granite contact, where it is developed to the complete (S 53694) or partial (S 51515) exclusion of amphibole.

Mudstone and siltstone close to the contact (S 51782, Plate XXVII, fig. 2) is altered to a black splintery hornfels composed of angular quartz grains up to 0·03 mm in diameter, set in a decussate network of stumpy strongly-pleochroic reddish brown biotite flakes commonly about 0·03 mm long. The hornfels contains numerous poikiloblasts of green hornblende, which are irregular in shape, in some cases slightly elongated parallel to the bedding and average 0·5 mm in length. Hornblende forms approximately 60 per cent of the volume within these poikiloblasts. In certain bands adjacent poikiloblasts have coalesced into large composite aggregates. Pyroxene has not been recognized in the hornfelsed mudstones.

Away from the granite the mosaic texture of the quartz and feldspar grains disappears fairly rapidly, and the polygonal outlines of the grains give place to more irregular serrate outlines. In thin sections from more than 550 yd (500 m) away from the contact the newly formed coloured minerals have not obliterated the original texture of the sediment but are confined to the interstices, originally occupied by the chlorite-carbonate matrix. Quartz grains are, however, partially fused together throughout the aureole. Biotite, amphibole, epidote and clinozoisite are present throughout the aureole and there is no evidence that amphibole gradually disappears and that its place is taken by biotite, as was suggested by Finlay (1930, pp. 675–6). The changes in the mafic minerals away from the granite contact are as follows:

1. The *biotite* changes colour from reddish brown to brown approximately 200 yd (180 m) from the contact, and to greenish brown or khaki near the outer margin of the aureole. The change from brown to greenish biotite appears to take place closer to the granite in the fine sediment (S 53693) than in the sand-stone. Biotite near the contact forms aggregates or isolated flakes with decussate orientation and, in some cases, irregular poikiloblastic masses, but as the outer margin of the aureole is approached it forms a fine-grained interstitial aggregate grading into the metamorphic biotite of the tectonite belt (p. 122).

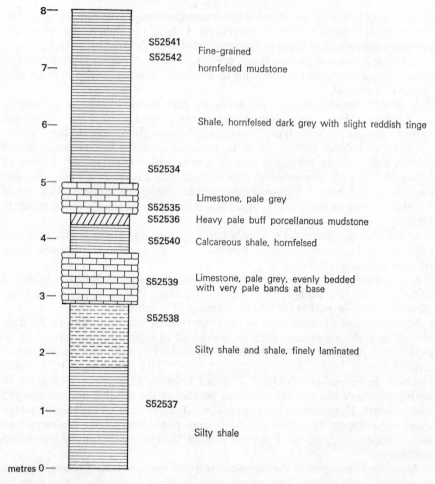

FIG. 25. *Hornfelsed limestones and associated calcareous shales, east coast of Vaila*
[*242 464*]

Mineral composition of specimens: S 52541 finely granular diopside, zoisite-clinozoisite, poikiloblasts of calcic scapolite, acicular amphibole associated with zoisite; S 52542 finely granular diopside, veins and patches of zoisite-clinozoisite; S 52534 cordierite poikiloblasts, calcic scapolite, veins of clinozoisite, calcite and actinolite; S 52535 calcite, calcic scapolite diopside; S 52536 poikilitic grossularite and idocrase, finely granular diopside, veins and patches of zoisite; S 52540 calcite, calcic scapolite, idocrase, amphibole, veins of calcite and zoisite; S 52539 calcite, poikiloblasts of grossularite with diopside grains, calcic scapolite, veins of calcite; S 52538 shale; amphibole laths and poikiloblasts biotite, clinozoisite; siltstone: diopside amphibole, zoisite, calcic scapolite; S 52537 poikiloblasts of calcic scapolite with finely granular diopside, veinlets of clinozoisite.

2. *Amphibole* is present both in the sandstones and the fine-grained sediments throughout the aureole, but individual crystals become progressively smaller and more acicular near its limit. The amphibole from the outermost specimens is pleochroic from straw-coloured to yellowish green.

3. *Epidote* and *clinozoisite* are common in the bands which were originally carbonate-rich and are in many cases associated with fine-grained amphiboles.

No calc-silicates of secondary origin have been recorded in the outer half-mile (800 m) wide belt of the aureole.

Limestone and calcareous mudstone

The thick bed of mudstone and calcareous mudstone with two limestone bands respectively 1 ft 8 in (50 cm) and 3 ft (91 cm) thick (Fig. 25) exposed on the east coast of Vaila, 230 yd (210 m) N of the granite-sediment contact (Plate XXV), is the only sediment containing appreciable quantities of calcite within the aureole. The mineral content of the hornfelsed limestone and calcareous mudstone, in addition to calcite, quartz, mica and clay minerals, is as follows:

Limestone. Grossular, calcic scapolite, idocrase, diopside, epidote and clinozoisite. Grossular occurs as large highly poikiloblastic crystals forming 50 per cent of the total volume of one section (S 52536, Plate XXVII, fig. 4). Scapolite and idocrase form both subhedral to anhedral crystals and poikiloblasts. The diopside is generally finely granular and enclosed by the other minerals.

Calcareous mudstone and shale. Cordierite, usually poikilitic, amphibole, diopside, calcic scapolite and rare idocrase. Cordierite (Plate XXVII, fig. 3) is only present in the mudstone immediately adjoining the limestones, and has not been recorded in hornfelsed mudstones elsewhere in the aureole.

The hornfelsed sediments are cut by veins of calcite containing clinozoisite and tremolite.

Metamorphic Grade

The paragenesis of the sediments within the inner part of the contact aureole and the sedimentary enclaves within the complex suggests that pressure-temperature conditions during the granite and diorite emplacement were those of the hornblende-hornfels facies (Turner 1968, pp. 193–225), grading outwards into the albite-epidote-hornfels facies near the limit of the aureole. There is no evidence of mineral assemblages suggesting pressure-temperature conditions characteristic of the pyroxene-hornfels facies, even in xenoliths enclosed in diorite. Though these contain diopside, they also have ophitic plates of reddish brown biotite, and in most cases, subordinate hornblende, which in some sections mantles the pyroxene. A diagnostic feature of the pyroxene-hornfels facies is the absence of hornblende.

SCAPOLITIZED CRUSH ZONES WITHIN SANDSTING COMPLEX

LINEAR NORTH-NORTH-WEST TRENDING CRUSH BELTS

Field Relationships

Zones of crushed and partially mylonitized rock, which were recorded by Finlay (1930, p. 686), but first recognized as crush zones by Wilson (*in* Summ. Prog. 1933, p. 76), are present in the southern part of the complex. In the Skelda Ness, Silwick and Wester Wick areas they take the form of near-vertical lenticular crush belts, commonly trending west-north-west to north, and ranging in thickness from less than 1 ft (30 cm) to 50 yd (46 m). They consist of sheared granite and locally diorite, and show a gradual transition from relatively uncrushed granite on the outside to a central pale or dark grey zone of variable thickness, which resembles flinty crush and is composed in part of mylonite (p. 235). The crushed rock of the central zone, and, to a lesser extent, the outer zone, is in places metasomatically replaced by sodic scapolite which is itself sheared. The period of scapolite emplacement more or less coincided with the duration of shearing (Mykura and Young 1969).

The following are descriptions of the better-developed crush belts:

1. *Hillside between Wester Wick and Loch of Wester Wick*. The crush belt in this area trends N30°W, and has a maximum width of 35 yd (32 m) of which only the central 5 yd (4·6 m) is intensely sheared. In the wide outer zones individual crystals of feldspar, still more or less intact, are set in a dark greyish groundmass. Locally patches of almost unaltered pink granite are preserved within this zone.

2. *Cliff top, north-west corner of Wester Wick*. The belt of mylonitized rock, which is well seen in the cliffs exposed at the north-west corner of Wester Wick, trends N20°W, is steeply inclined to the west-south-west and attains a thickness of 15 ft (4·6 m). It consists entirely of silicified flinty crush rock and has virtually no marginal zone of less crushed granite.

3. *East shore of Wester Wick 700 yd (640 m) SSE of Wester Wick village*. The granite exposed on the cliffs forming the eastern shore of Wester Wick, 400 yd (366 m) SSE of the head of the bay, is not cut by a single well-defined crush belt, as in the other areas described. It is cut by two intersecting sets of joints along which the granite is locally slightly crushed. In the northern part of the cliff section the granite contains irregular enclaves of shattered diorite and microdiorite. The granite is traversed by a 40 yd (36 m) wide zone containing belts of scapolitization and numerous scapolite and calcite veins. Many of the scapolite veins are emplaced along the two intersecting sets of joints.

4. *Pundswell, near Scarvister, Skelda Ness peninsula*. This crush belt, which trends N15°W and dips steeply to the east, is well exposed 40 yd (36 m) north of the ruin of Pundswell [305 421], where it is nearly 50 yd (46 m) wide. The central strongly sheared zone is 6 yd (5·5 m) wide and is composed of a banded flinty-looking rock, with a slightly wavy schistosity parallel to the trend and inclination of the crush belt. It contains several lenses of white scapolite up to 6 in (15 cm) wide and 2 ft (60 cm) long, as well as an intricate network of minute scapolite veinlets. On either side of the central belt the granite is streaked out, and is generally of cherty aspect, with individual shattered feldspar crystals still visible on weathered surfaces. The effects of shearing are strongly marked within 12 yd (11 m) wide zones bounding the central belt but decrease rapidly away from these zones.

5. *Crush belts in Skelda Ness peninsula*. The southern part of the Skelda Ness peninsula contains a number of parallel, near-vertical crush belts, trending N10°W. The most easterly of these, exposed in Blo Geo, 900 yd (823 m) NE of the southern end of the peninsula is up to 40 yd (36 m) wide and is composed of roughly alternating bands of sheared-out granite and almost normal granite. In this crush belt there is no evidence of extensive scapolitization, but in the one 200 yd (183 m) farther west, which can be traced for 400 yd (366 m) along its strike, and is up to 5 yd (4·6 m) in width, the mylonitized rock is intensely scapolitized and veins of virtually pure scapolite are present. One of these veins reaches a thickness of 8 ft (2·4 m) at the coast.

Petrography

Sheared granite of the outer zones. The first indication of shearing in the outer zones of the shear belts is the distortion of the fabric of the quartz crystals producing undulose extinction and the breaking up of coarse-grained biotite (S 51524). Slightly further in (S 29873, 31121) quartz grains become streaked out with an elongation ratio of up to 10:1 and a highly undulose extinction. In some cases the quartz is finely granulated with grains separated by sutured margins. More commonly the quartz is converted into a very fine mosaic. The plagioclase crystals are usually bent with open cleavages and other cracks. A further effect of strain is the development of twin-planes normal to the original albite twinning of the crystal, producing an intimate intergrowth of two sets of twin lamellae. The potash feldspar shows less evidence of strain, but undulose extinction is seen in some crystals. The cracks within crystals and between adjacent crystals are filled with a very fine-grained aggregate of green mica, which commonly forms up to 30 per cent of the total volume of rock. Fine-grained scapolite forms thin ($< 1 \cdot 8$ mm) discontinuous veinlets within the micaceous aggregate; and some of the veinlets are cut and displaced by small shear planes.

Inner mylonitized zones. In the less intensely mylonitized crush belts and in the zones bounding the central belt of others, sheared granite of the type described above is further broken up into clasts elongated parallel to the schistosity (S 29872). These fragments are enclosed in a matrix of finely comminuted biotite with some larger plates of biotite and rare sphene. There are also thin parallel belts of fine-grained mylonitized quartz which in some cases flows round porphyroclasts of sheared quartz.

The most intensely sheared rock in the crush belts is a dark grey cherty-looking rock with pale patches (S 31126). It consists of roughly oval porphyroclasts of quartz elongated parallel to the schistosity which range in length from about $1 \cdot 5$ to less than $0 \cdot 1$ mm. The quartz clasts have undulose extinction and a number of irregular cracks and are enclosed in an extremely fine-grained, laminated quartzose matrix. The planes of movement within this mylonite are roughly parallel to the trend and inclination of the shear belt but show evidence of minor folding, with axial plane cleavage locally developed. In places the mylonite is broken up into small irregular blocks by shear planes, with individual blocks rotated relative to each other (S 31126). Individual fragments do not exceed 1 mm in diameter and the rock therefore does not strictly correspond to kakyrite (Christie 1960, p. 83) in which individual fragments are megascopically recognizable. Veins of scapolite cutting this ultramylonite are also slightly folded and stepped by later faults.

Associated mineralization. Mylonite from the shear belt exposed on the north-west side of Wester Wick (S 51783) is cut by a network of small irregular patches and linked cross-cutting veinlets of a very fine-grained aggregate of a green unidentified mineral with high refractive index which locally forms up to 40 per cent of the total volume of the rock. This mineral is in turn cut by thin branching veinlets of potash feldspar and carbonate and a few veinlets containing carbonate and sphene, both of which are affected by late small-scale faulting (Plate XXVII, fig. 8).

The mode of occurrence of scapolite (Plate XXVII, figs. 5, 6) and calcite within the shear belts, and that of the scapolite veins in melamicrodiorite dykes within the complex and in the sediment adjoining it, have been described by Mykura and Young (1969), who have suggested that the sodic scapolite was formed both by the metasomatic replacement of feldspars and by the direct deposition in veins from hydrothermal solutions. Scapolitization was more or less contemporaneous with the movement along the crush belts but vein emplacement may have continued for some time after the movement had ceased.

Origin of Crush Belts

The texture and composition of the crush rock within the crush belts indicates that shearing took place at a fairly low crustal level in an environment of fairly high temperature and hydrostatic pressure, possibly before the granite had cooled completely. In the most highly mylonitized and scapolitized areas there is evidence of several phases of movement. There is, however, little evidence of either great vertical or horizontal displacement along these crush belts, as they do not displace any of the junctions between the different rock types within the complex. The lenticular shape and limited length of their outcrop suggests that they may have been formed within 'islands' of consolidated granite which were surrounded by areas of still viscous magma which was able to absorb the movements by flowing rather than by shearing.

IRREGULAR AREAS OF CRUSHED ROCK

In addition to the near-vertical linear crush belts there are three outcrops of crushed granitic and dioritic rock of irregular shape. Two of these are situated just west of Culswick village and a smaller mass crops out on the Knowes of Westerskeld between Wester Skeld and Housa Water. The rock forming these masses is crushed throughout but the grey aphanitic mylonite of the vertical belts is never developed. As these areas are located either close to a junction between diorite and granite or diorite and sediment, it is possible that they were formed during differential movement of two adjacent almost consolidated masses. The irregular shape of the outcrop of these crush zones suggests that they may be lenticular masses whose inclination is almost horizontal.

SHEARING IN AREA ADJOINING WALLS BOUNDARY FAULT

The granite and other members of the Sandsting Complex are intensely shattered within a ¾-mile (1200-m) wide zone adjoining the Walls Boundary Fault (Summ. Prog. 1933, p. 80). The Old Red Sandstone sediments within this zone are both strongly folded and intensely shattered (p. 134). The shearing within the granite is best seen along the south shore between Skelda Voe and the Ayre of Deepdale, where the granite is cut by a large number of closely spaced sub-parallel faults and joints, whose trend varies from N25°E to N10°W and whose inclination is consistently 60 to 80 degrees to the east. Rather less well developed cross-joints, with trends ranging from east–west to E30°S, are locally present. At Aaskberry Taing, 1 mile (1600 m) WNW of Ayre of Deepdale, where the north-north-east trending faults are closely spaced, the granite fabric is distorted within a zone extending from 2 to 4 ft (61–122 cm) from the fault planes, and within a 6-in (15-cm) belt along some fault planes it becomes black and flinty-looking with local pink patches. The sheared rock is, however, shattered and friable, and chlorite is developed along shear planes, suggesting that shearing here took place at a much higher crustal level than in the Skelda Ness–Wester Wick crush belts.

The granite exposed on the west shore of Seli Voe north of Rea Wick is intensely shattered throughout, but there is little consistency in the trend of the fault planes. It contains a number of north-north-west trending near-vertical belts of crush rock, which is very fine-grained, black, cherty in texture and in some places partly replaced and intensely veined by scapolite and calcite (Mykura and Young 1969, p. 5). These crush belts are in some respects comparable with the crush belts of the Skelda Ness–Wester Wick area and as they are themselves affected by the later shearing, they may, like the Skelda Ness crush belts, have originated at a fairly low crustal level and may not be directly connected with the Walls Boundary Fault.

A small wedge-shaped mass of intensely sheared granite, separated from outcrops of Walls Sandstone by two sub-parallel near-vertical, south-south-west trending faults, crops out on the south shore of Rea Wick. The granite is cut by many fault planes and contains a number of faulted slices, up to 15 yd (14 m) long, of intensely shattered sandstone and sandy siltstone. Many of the shear planes in this zone trend roughly parallel to the bounding fault and have an inclination ranging from 25 to 45 degrees to the east.

STRUCTURE AND MODE OF EMPLACEMENT OF THE COMPLEX

As only part of the probable outcrop of the Sandsting Complex may be exposed above sea level (McQuillin and Brooks 1967, p. 14) any speculation as to its structure and mode of emplacement must be tentative. Though the field evidence as a whole is inconclusive, the structure of the north-eastern part of the complex suggests that, rather than being a stock with steeply inclined margins, the complex is a compound sheet which is intruded into the sedimentary rocks of the Walls Formation.

The sheet appears to dip northward at an angle of 40 to 70 degrees in the west, is possibly almost horizontal in the central area between Culswick and Scarvister and dips steeply to the east in the vicinity of Easter Skeld. Its upper junction is sub-parallel to the bedding of the sediment, but steps down slightly to the north-west. The portion of the complex east of Skelda Voe forms a sheet dipping steeply to the east and fingering out northward. In the north the fingers of the sheet are concordant with the bedding of the enclosing sediment.

The first major intrusion of the complex was the diorite which formed a sheet that was roughly concordant with the bedding of the enclosing sediment. This sheet appears to have decreased in thickness from west to east. It contained several large and many small enclaves of sediment. Before the diorite had consolidated completely, two complex sheets of granodiorite, granite and micro-adamellite were intruded along the lower and upper contacts of the diorite. The lower sheet may have been considerably thicker than the upper, and there appear to have been connecting branches which cut across the diorite in the area between Culswick and Housa Water and in the ground north of the Ward of Reawick. In some instances (e.g. $\frac{1}{4}$ mile (400 m) N of Viville Loch) small masses of diorite intervene between the upper margin of the granite and the overlying sediment, while elsewhere (e.g. east of Housa Water) enclaves of sediment occur between the diorite and granite, suggesting that the upper sheet was not everywhere intruded precisely along the diorite-sediment junction.

Nothing is known of the total thickness of this hypothetical compound sheet, as its lower junction is nowhere exposed. The faulted-in portions of net-veined sediment near the Broch of Culswick and at Vine Geo, Vaila (pp. 228–9) could be parts of the floor of the sheet, but they could equally well be parts of sedimentary enclaves or even stoped portions of the roof of the complex.

REFERENCES

ALLING, H. L. 1938. Plutonic perthites. *Jnl Geol.*, **46**, 142–65.

BAILEY, E. B. 1958. Some chemical aspects of south-west Highland Devonian igneous rocks. *Bull. geol. Surv. Gt Br.*, No. 15, 1–20.

—— 1959. Mobilisation of granophyre in Eire and sinking of olivine in Greenland. *Lpool Manchr geol. Jnl*, **2**, 143–54.

—— and McCALLIEN, W. J. 1956. Composite minor intrusions, and the Slieve Gullion Complex, Ireland. *Lpool Manchr geol. Jnl*, **1**, 466–501.

BLAKE, D. H., ELWELL, R. W. D., GIBSON, I. L., SKELHORN, R. R. and WALKER, G. P. L. 1965. Some relationships resulting from the intimate association of acid and basic magmas. *Q. Jnl geol. Soc. Lond.*, **121**, 31–49.

CHRISTIE, J. M. 1960. Mylonitic Rocks of the Moine Thrust-Zone in the Assynt Region, North-West Scotland. *Trans. Edinb. geol. Soc.*, **18**, 79–93.

DEER, W. A., HOWIE, R. A., and ZUSSMAN, J. 1963. *Rock-forming Minerals, vol. 4, Framework Silicates.* London: Longman's.

ELWELL, R. W. D. 1958. Granophyre and hybrid pipes in a dolerite layer of Slieve Gullion. *Jnl Geol.*, **66**, 57–71.

FINLAY, T. M. 1930. The Old Red Sandstone of Shetland. Part II: North-western Area. *Trans. R. Soc. Edinb.*, **56**, 671–94.

GUPPY, E. M. and SABINE, P. A. 1956. Chemical Analyses of Igneous Rocks, Metamorphic Rocks and Minerals: 1931–1954. *Mem. geol. Surv. Gt Br.*

McQUILLIN, R. and BROOKS, M. 1967. Geophysical surveys in the Shetland Islands. *Geophys. Pap. No. 2, Inst. geol. Sci.*, 1–22.

MILLER, J. A. and FLINN, D. 1966. A Survey of Age Relations of Shetland Rocks. *Geol. Jnl*, **5**, 95–116.

MYKURA, W. and YOUNG, B. 1969. Sodic scapolite (dipyre) in the Shetland Islands. *Rep. No. 69/4, Inst. geol. Sci.*, 1–8.

SUMM. PROG. 1933. *Mem. geol. Surv. Gt Br. Summ. Prog. for* 1932.

TURNER, F. J. 1968. *Metamorphic Petrology.* New York.

Chapter 16

MINOR INTRUSIONS

INTRODUCTION

SWARMS of sub-parallel north-north-west to north-north-east trending acid and basic dykes cut the diorite, granite and granophyre of Northmaven and Muckle Roe. Dykes with rather more variable trends are also abundant on Vementry Island and somewhat less abundant within a 2 to 3 mile (3–5 km) wide belt bounding the north coast of the Walls Peninsula. Sills, sheets and intrusions of irregular shape are relatively rare.

The minor intrusions fall into two main groups respectively of acid and basic composition, with a minor group of intermediate composition. The acid dykes comprise felsites, quartz-feldspar-porphyries and feldspar-phyric porphyrites. The basic dykes are principally basalts and dolerites, but many of them contain hornblende as the principal dark mineral instead of pyroxene. Some of the basic dykes have accessory quartz. The rocks shown on Plate XXVIII as pyroxene-porphyrite usually contain some cryptopegmatite. Fresh olivine and orthopyroxene have not been found, but pseudomorphs after these minerals have been observed in a small number of the basic dykes. The intermediate group is represented by only a few individual dykes and includes keratophyres, spessartites and microdiorites. Keratophyric rocks occur also as thin marginal facies to some of the acid dykes.

Distribution. In Muckle Roe and Northmaven both basic and acid dykes form parallel swarms whose trend ranges from north-north-west to north-north-east, and several dykes have curving courses within that sector. The distribution of the basic dykes differs from that of the acid ones in that the former appear in roughly comparable numbers throughout the area, whereas the latter are largely restricted to the outcrop of the Muckle Roe Granophyre. The difference in distribution is readily appreciated in the east–west section along the south coast of Muckle Roe, where between Scarfataing [340 637] and Picts Ness [297 638] at least 60 dykes have been observed. In the central sector of this section, between Ness of Gillarona [327 627] and Burki Taing [317 626], all 22 observed dykes are acid. On either side of this sector of the coast basic and acid dykes crop out in alternating groups, but in the extreme east and west of the shore section basic dykes predominate. Alternate swarms of basic and acid dykes can also be traced across the central part of the island, but this alternation is less evident along the north coast where basic dykes are largely confined to the area immediately east and west of Erne Stack. The concentration of the acid and basic dykes into parallel zones suggests that tensional stress operated in two stages to provide channels along which at one stage acid magma, at the other stage basic magma was irrupted. In the central part of the island the stress effected fissuring, actual or potential, along different but parallel belts. In the west the later stage of stress operated roughly along the same belt as the earlier. This hypothesis of intrusion of acid and basic magmas respectively at different stages of active stress is consistent with the dearth of composite dykes since if both types of

239

magma were susceptible to irruption at the same stage, the production of composite or hybrid dykes would be probable. If, however, the belt of fissures, actual or potential, caused in the first stage were healed by one type of magma, the fissuring strain caused by the second stage of stress would be produced more easily along a less well cemented zone.

The restriction of the swarm of acid dykes to the outcrop of the granophyre appears to indicate that acid magma existed in a state suitable for intrusion at only restricted places in the magmatic hearth during this late stage, whereas the basic magma remained in an intrusible condition below a much wider area.

Along the north coast of the Walls Peninsula and in the islands adjoining this coast basic and intermediate dykes have the same general trends as the dykes of similar composition in Muckle Roe and Northmaven. They are largely confined to a 2- to 3-mile (3–5-km) wide coastal strip which extends from Brough Skerries [216 583] in the west to Papa Little in the east and they are most abundant in the eastern part of this section, particularly in Braga Ness [315 605] where swarms consisting of up to five parallel dykes have been recorded. The acid minor intrusions of this area also have a more restricted distribution and are largely confined to a 3-mile (5-km) wide, roughly circular, zone around the Vementry Granite, though within the granite itself dykes are rare. The trend of these dykes has a roughly radial pattern centred on the Vementry Granite, suggesting that a magma chamber underneath this granite body was the focus of irruption of the dykes.

Age of the intrusions. It is clear that both the acid and the basic dykes are younger than the main plutonic bodies which they cut. There is also abundant evidence that the basic dykes cut, and are therefore later than, the plexus of aplitic, granitic and granodioritic veins which invade the Northmaven diorite. The intrusive relations are particularly clear at Wilson's Noup [300 720] and Lang Head [303 706] and on Egilsay [315 693]. The age of the acid dykes relative to the granitic component of the Northmaven complex is not so readily determined because of the restricted distribution of the acid dykes. An acid porphyrite has been observed to cut the granite-veined diorite on the west shore of Otter Ayre [324 666]. This dyke, like the rocks into which it is intrusive, is crushed (S 44321). In the same area, on the coast west of Lee Skerries [321 670] a dyke of spessartite cuts the granite-veined diorite.

There is only little evidence from the Muckle Roe–Northmaven area of the relative age of the acid and the basic dykes but what there is indicates that the acid are the earlier. A specimen collected from a dyke at the Ness of Gillarona [327 628], for instance, shows a contact of acid and basic pyroxene-porphyrite types. No field observation of the occurrence is available but the microscopic evidence (S 28885) of chilling of the basic rock and partial enclosure of a feldspar phenocryst of the acid rock by the basic indicates the acid as the earlier. Cutting of an acid dyke by an 8-in (20-cm) thick basaltic dyke has also been recorded on the shore [304 733] 200 yd (180 m) SSW of Nibon. In the Walls Peninsula and the adjacent islands there are several exposures of basic dykes cutting acid ones. Good examples are seen on the east coast of Braga Ness [318 602] and on the north-east shore of Linga [285 592]. The age relationship of the intermediate dykes to the acid or basic ones is not known.

Both acid and basic dykes cut the Old Red Sandstone Sandness Formation in the Walls Peninsula and in Papa Little.

J.P., W.M.

BASIC AND INTERMEDIATE MINOR INTRUSIONS

FIELD RELATIONSHIPS

Northmaven and Muckle Roe

Basic Dykes. Few measurements have been made of the widths of individual basic dykes but in general they appear to be less broad than the acid ones (see p. 249). On the south-west shore of Muckle Roe they range in thickness from less than 1 ft (30 cm) to 40 ft (12 m). The basalts and dolerites tend to occur as groups of parallel, roughly north–south trending dykes. The pyroxene-porphyrites, which are the thickest of the basic dykes, on the other hand, have a much more localized distribution and a more variable shape. They are most common in the Busta and Islesburgh peninsulas. The pyroxene-porphyrite exposed on Kat Field [330 687], ½ mile (0·8 km) WNW of Mavis Grind, has a persistent north-westerly trend, which is unusual in the basic dykes, and over its course of 600 yd (550 m) it exhibits an unusual thickening and thinning, which may be due, in part, to minor dislocations along north-north-west trending crush belts. The continuation of the Kat Field dyke can be traced intermittently south-eastward towards Busta, the most southerly exposure being a round boss-like mass on the Ward of Runafirth [343 673], ¼ mile (400 m) NNW of Busta. The presence of a number of small pyroxene-porphyrite dykes which appear to radiate from this boss suggests that the Ward of Runafirth intrusion may be at or near the local focus of a radiating irruption, and that the pyroxene-porphyrites represent a stage of intrusion that differed from the stage of the north–south basaltic intrusions in the character of the magma, the pattern of fissures accessible to the magma, and, consequently, in time. It is suggested that the pyroxene-porphyrite stage is the later and that it is related to the period of development of the north-west trending crush lines which fan out from the Walls Boundary Fault zone (Fig. 27, p. 262).

Only two basic dykes have been recorded in the metamorphic rocks east of the Walls Boundary Fault zone (Plate XXVIII). One of the basalt dykes trends north–south, the other has an east-north-easterly trend. Neither basalt is foliated.

Dykes of intermediate composition. The most abundant intermediate dykes are keratophyres, three of which crop out on Mid Field [313 638], Quilt Ness [297 644], and Strem Ness [292 658] on Muckle Roe. Other keratophyre dykes occur on the south-west corner of the Isle of Nibon [296 727] and on the Mainland west-south-west of Nibon House. In the field the latter looks like a felsite and displays thin colour lamination with spherulitic and non-spherulitic banding. A number of acid dykes, composed principally of felsite and/or quartz-feldspar-porphyry, have dark keratophyric marginal facies. Good examples of these are seen in the multiple acid dyke exposed at the west end of Raavi Geo [310 675] (see p. 250 and Fig. 26), and in several other dykes exposed along the north coast of Muckle Roe between North Ham [303 665] and Lothan Ness [316 676]. These marginal facies are darkish grey, green or purplish in colour and have a superficial resemblance to basic dykes.

Only two lamprophyre dykes are known in the area. Both are spessartites; one trends north-north-west on the north-east coast of Muckle Roe [322 670], midway between Lothan Ness and Otter Ayre and the other is exposed on Steinawall [330 651], 700 yd (640 m) WSW of the south end of Kilka Water.

The only microdiorite dyke known in the Muckle Roe–Northmaven area crops out at Picts Ness [297 637].

J.P.

Walls Peninsula and adjacent islands

Within the southern area the basic and intermediate minor intrusions are largely confined to Vementry Island, Papa Little and a 2 to 3-mile (3–5 km) wide belt along the northern coast of the Walls Peninsula. They are here less abundant than in the area to the north. Most of the intrusions are dykes with a north-north-westerly to north-north-easterly trend and a mean trend of N5°E. Comparatively few trend parallel to the strike of the bedding or foliation of the country rock.

Basalts and dolerites and unclassed basic and sub-basic rocks. Though the one-inch geological map of Western Shetland distinguishes between basalts and dolerites on the one hand and unclassed basic and sub-basic rocks on the other, it is likely that all these intrusions belong to a single dyke-suite. The distinction between the two rock-types is largely artificial, being due to the fact that many of the unclassed basic dykes have not been examined in thin section and that the remainder are too highly altered for accurate determinations to be made. The basalts and dolerites are dark grey fine-grained and, in some instances, very sparsely microporphyritic rocks, with the feldspar phenocrysts roughly equi-dimensional and not exceeding 3 mm in size. The north–south trending dykes range in width from a few inches to 20 ft (6 m) and a high proportion are between 5 and 8 ft (1·5 and 2·4 m) thick. The density of dykes increases markedly towards the eastern part of the area, and in the Braga Ness peninsula they form swarms of up to five closely-spaced dykes with average thickness of 5 ft (1·5 m). The basic dykes cut the acid and sub-acid dykes but field evidence relating to the age relationship of the basalts and dolerites to the porphyrites, spessartites and quartz-dolerites is lacking. All dykes are displaced by the east-north-east trending faults.

There are a number of dykes and sheets with trend roughly parallel to the strike of the country rock. These are more variable in thickness and generally finer grained than the north–south trending dykes. One sheet of highly shattered aphanitic basalt which crops out along the junction between the metamorphic rocks and Old Red Sandstone sediments, 400 yd (360 m) ESE of Vementry House [313 595], is up to 65 ft (20 m) wide. Concordant sheets exposed on the shores of West Burra Firth [251 570] and the Voe of Snarraness [237 567] (Plate XXIV) are respectively 10 and 4 ft (3 and 1·2 m) thick.

Quartz-dolerite. Only one dyke of quartz-dolerite has been recorded in the Walls Peninsula (Plate XXVIII). This is up to 60 ft (18 m) thick and has a sinuous north-east trending outcrop which passes the north-west shore of Mousavord Loch [223 554] and can be traced for 1300 yd (1200 m). It consists of pale grey medium-grained, non-porphyritic dolerite, with pale greenish grey subhedral feldspars up to 1 mm in size. The sediments adjoining this dyke are markedly indurated.

Pyroxene-porphyrite. In the Walls Peninsula pyroxene-porphyrite forms a very small number of wide north to north-north-west trending dykes. The most westerly of these dykes can be followed along a slightly sinuous course from the shore [219 583] between Brough Skerries and the Skerry of Stools southwards for 1000 yd (910 m) to the west slope of the Hill of Bousta [220 572]. The trend of this dyke is continued south-south-eastward by three dykes which are arranged *en echelon* to each other and can be traced for nearly 2 miles (3 km) almost to the latitude of Burga Water. These three dykes are 20 to 30 ft (6–9 m) thick throughout the greater part of their outcrop. The northern two consist of pale grey highly porphyritic rock, with abundant fresh, almost glassy euhedral to subhedral plagioclase phenocrysts, up to 10 mm in size, set in a somewhat darker grey aphanitic groundmass. In the south-eastern parts of the dyke-complex the volume-percentage of glassy feldspar phenocrysts gets gradually less and the matrix becomes somewhat finer-grained and contains small patches of pyrite and veinlets of calcite. Irregular pale greenish epidotic patches and small areas in which the feldspar phenocrysts are pink or pale red-stained are present throughout the dyke-complex.

The only other porphyrite dyke recorded in the Walls Peninsula crops out on the south shore of West Firth Burra [250 569]. Because of indifferent exposures it has not been possible to determine the exact thickness and trend of this dyke, which cannot be traced inland. Two porphyrite dykes, both less than 10 ft (3 m) thick, crop out on the north-west shore of the island of Papa Little. Of these, one trends north–south and the other east–west. Both contain very platy phenocrysts of fresh, almost glassy feldspar, up to 10 mm long, set in a very fine-grained dark grey matrix.

Microdiorites and spessartites. Three spessartite dykes crop out along the north and east shores of the Neeans peninsula. The most westerly of these, exposed on the east shore of the Geo of Djubabery [263 589] is 12 ft (3·6 m) wide and has a N15°E trend. It is pale reddish grey, holocrystalline and is composed of randomly orientated needles of pink feldspar up to 4 mm in size, set in a greenish grey very fine-grained matrix. The two dykes exposed on the east shore of the Neeans peninsula are less than 5 ft (1·5 m) thick and are considerably finer-grained than the Geo of Djubabery dyke.

A dyke of quartz-microdiorite crosses Swarbacks Skerry [290 622] off the north-west coast of Vementry. This trends W20°N and is closely associated with a parallel dyke of spherulitic felsite.

W.M.

PETROGRAPHY

Basic Minor Intrusions

The basic dykes are described under two headings: (1) basalts and dolerites, and (2) pyroxene-porphyrites. The former are further divided into quartz-bearing types and quartz-free types including a few which carry pseudomorphs after olivine.

Quartz-bearing dolerites and basalts. A small number of the basalt and dolerite dykes contain primary quartz. There are two types. In one the plagioclase and augite are both idiomorphic forming, respectively, tabular and stout prismatic crystals which range up to 4 mm in length and are seriate to groundmass dimensions, 0·2 to 0·4 mm. The plagioclase, broadly zoned from central labradorite with about 65 per cent An, is extensively sericitized and is the dominant constituent. The interstitial material is mainly chlorite in which brown flakes of biotite and anhedral grains of quartz are present in accessory proportions (S 30877), and in which potassium feldspar can be locally identified as small stout prisms in quartz (S 30881). Magnetite is abundant as a minor constituent, apatite is accessory as thin prisms, and calcite occurs interstitially. A rock, which is

R

referred to this type because of the similarity of texture, contains quartz mainly as a secondary mineral in association with chlorite and iron ore in an aggregate which might have arisen by destruction of olivine (S 29976). The other type consists essentially of plagioclase and ophitic or subophitic augite or hornblendic replacement of augite. The feldspar (labradorite) occurs as diversely orientated lathy prisms, rarely exceeding 1 mm in length, but augite, usually of comparable grain-size, may build shapeless grains, up to 4 mm across, enclosing many plagioclase laths (S 30747). Some examples contain scattered phenocrysts of plagioclase which rarely exceed 2 mm in length (S 43776, 43780). All the feldspar is zoned and though in some rocks it is completely albitized to oligoclase or indeterminable because of turbidity, its composition in fresher rocks is centrally labradorite in the range An 55 to 65 per cent. Except in the coarsely ophitic rock referred to above, the pyroxene is partially (S 43780) or wholly (S 43779) transformed into semi-opaque brown hornblende and into later clearer green and scarcer bluish green uralitic hornblende. The brown hornblende retains the subophitic shape and dimensions of the original pyroxene, which locally remains as small ill-defined relics (S 43776). The green uralitic amphibole merges into the interstitial material. The latter consists of chlorite and fibrous brown amphibole, granular quartz and feldspar, probably oligoclase, and biotite. In the rocks with fresh augite, chlorite is the important mineral of the mesostasis (S 30877, 47793) and is accompanied by flaky ragged biotite. In the uralitized rocks chlorite is scanty and fibrous hornblende with which is associated biotite in aggregates of minute clear brown or brownish green scales is the dominant component. These differences suggest that in the uralitization process a reconstitution of interstitial chloritic material to biotite which is akin to thermal metamorphism takes place (S 43776, 43778) (see also p. 204). Analogously in the fresher rocks the abundant iron ore remains black and opaque, whereas in the uralitized rocks it is thoroughly leucoxenized. Quartz is an abundant accessory or a minor constituent in these rocks and usually has a primary aspect but in the altered rocks (S 30920, 30996) much of the quartz may be secondary. Because of the reconstitution of the plagioclase to oligoclase and the pyroxene to amphibole some of the dykes mentioned could be classed as microdiorites, but they all seem to have been originally essentially basic mesocratic rocks. Albitization and epidotization have proceeded so far in one example that it is practically an oligoclase-epidote rock with some quartz and uralite (S 43758). In this rock epidote has largely replaced the pyroxene.

Quartz-free basalts and dolerites. Basalts composed essentially of plagioclase, augite or derived hornblende, and green mafic mesostasis form the largest group of the basic dykes. They include porphyritic, microporphyritic, and non-porphyritic varieties with groundmass in which the plagioclase rarely exceeds 0·5 mm in length. Only one dyke is coarse enough to be classed as dolerite. It is composed of shapeless crystals of augite, up to 3 mm across, partly or completely transformed to pale green, monocrystalline hornblende, which enclose zoned plagioclase prisms 0·2 to 1·5 mm long (S 43774). The latter consists of labradorite-bytownite (An_{70}) in the centre and oligoclase on the margin. Pale green cryptocrystalline aggregate, with tufts of acicular, very pale green amphibole and locally grains and prisms of epidote is interstitial to the plagioclase of the groundmass, and large (0·5 mm) crystals of leucoxenized ore which enclose plagioclase ophitically are accessory. The usual basaltic dykes when little altered show zoned plagioclase, centrally as calcic as An_{65}, enclosed ophitically by colourless or very pale brown augite which is mantled by turbid brown hornblende fringed in its turn by acicular green hornblende which merges into the mesostasis (S 53570). Some dykes are porphyritic with plagioclase more calcic than the groundmass laths (S 43754, 43772). Feldspar-phenocrysts so greatly altered in comparison with the rest of the rock that they are probably xenolithic are also present (S 43773). Iron ore in idiomorphic and shapeless grains is abundant. The mesostasis is entirely mafic being composed of chlorite (S 30742) or, more usually, chlorite and green fibrous hornblende and is commonly sufficiently abundant to isolate the plagioclase in single crystals or groups of only a few prisms. Rocks in which augite is still an important mineral though subordinate to hornblende (S 43754) are transitional to those in which hornblende is the essential ferromagnesian mineral. In these rocks the subophitic texture is maintained by the amphibole enclosing lengths of the feldspar prisms; usually some relics of augite are to be found in the hornblende (S 43756, 43771, 44326). The plagioclase maintains its zoned, centrally calcic character. In none of the basaltic dykes has fresh olivine been observed but some contain small round or oval pseudomorphs on which the adjacent feldspar is moulded. The pseudomorphs are composed of different minerals—clear green hornblende aggregate with (S 55176) or without (S 30021) ore granules, carbonate aggregate (S 44319), pale green serpentine (S 45028)—but all may represent olivine. Most are small, the largest being 0·5 mm long, and they do not form a major constituent of the rocks. More common and larger than this type of pseudomorph are aggregates of cryptocrystalline material and of microcrystalline to moderately coarse (0·2 mm) equant anhedral masses of clear green hornblende which have no definite shape or margin against the other constituents. They may in some cases be of a vesicular nature (S 43754, 43756, 43771), but it is probable that in others (S 43767, 43769–70) they represent olivine or othopyroxene crystals or crystal groups from an early precipitation or from xenoliths. Many of the dykes are greatly altered by post-magmatic processes, such as sericitization (S 49335), epidotization (S 44626, 50130), chloritization and albitization. Recrystallization of sericite has led to almost complete muscovitization of some porphyritic plagioclase crystals (S 31025) and in this rock the pyroxene has been calcitized. Chloritization and albitization have converted some dykes into chlorite-oligoclase rocks which could be classed as basic keratophyres (S 30706, 30708). Two albitized basalts from South Voe, Vementry Island [306 605] and West Ness, Muckle Roe [298 658] (S 49323, 53578) in which the augite has remained quite fresh contain a mesostasis of chlorite with which abundant granular epidote and garnet are associated. The garnet forms round or oval grains 0·02 to 0·05 mm but exceptionally 0·2 mm long, entirely enclosed in chlorite or associated with epidote in aggregates in which chlorite may be only minor. In one of the rocks the garnet always has a semi-opaque peripheral zone which seems anisotropic (S 53578). Some grains of the garnet show alteration to cryptocrystalline matter of high birefringence which is thought to be sphene since sphene occurs enclosed in chlorite as grains of the size and shape of the garnet. X-ray examination by R. J. Merriman has confirmed the mineral as garnet, the d-spacings of which indicate a composition in the range almandine-spessartine. Scarce thin zeolitic veins, 0·5 mm wide, seen in a hornblende-basalt exposed just north of Kilka Water [333 659], Muckle Roe, cut the rock irregularly and enclose small sinuous fragments of hornblende as if they occupy slight shear fractures. The zeolites are analcime and thomsonite, identified optically and confirmed by X-ray examination by Merriman who also identified leonhardite in a second vein (S 53570, 53570A).

Two of the quartz-free basalts are exceptional in containing brown biotite in the groundmass along with chlorite and microgranular oligoclase. Both contain fresh microporphyritic calcic labradorite and stout prismatic augite, interfering with the plagioclase but locally subophitic. In one of these rocks the larger crystals as well as the groundmass grains of pyroxene are almost completely carbonated and the brown biotite, which occurs as idiomorphic flakes in an obscure interstitial base of microcrystalline calcite, idiomorphic and skeletal ore, and indistinctly interleaved chlorite and green biotite, has the appearance of neoformation (S 30998). In the other rock the augite is unaltered and the brown biotite, partly moulded on plagioclase and ore and grading into the interstitial chlorite, appears to be a late primary mineral (S 47782).

Two dykes of basalt or fine-grained dolerite call for special mention because they are the only basic dykes found in the Lunnister–Haggrister area of metamorphic rocks (Plate XXVIII). Both are composed of zoned labradorite prisms and ophitic, colourless or faintly wine-coloured augite, fringed by green or brown hornblende with abundant accessory magnetite or leucoxene and green interstitial cement which is composed largely of microcrystalline green amphibole and minor chlorite and biotite. Minor differences exist. One of the dykes contains larger crystals of bytownite, rarely over 1 mm long (S 54280), the other is non-porphyritic (S34942); in the former biotite is crystallized in small idiomorphic flakes, while in the other rock it is intermixed with fibrous amphibole. Like the dyke of quartz-porphyry from the Bight of Haggrister (p. 256) these dykes petrographically belong to the dyke-suite of the intrusive complex west of the Busta–Haggrister Fault but unlike the acid dykes they are not deformed.

EXPLANATION OF PLATE XXIX

PHOTOMICROGRAPHS OF THE BASALT-GRANITE BRECCIA AND MINOR INTRUSIONS

FIG. 1. Slice No. S 55676. Magnification × 14. Plane polarized light. Breccia-form basalt cemented by granodiorite. The basaltic rock has a microgranoblastic base of andesine, hornblende and biotite in which lie small phenocrysts of zoned calcic plagioclase, marginally recrystallized, and recrystallized groups of amphibole and biotite prisms pseudomorphous after ferromagnesian phenocrysts. The rock resembles the thermally altered dyke of Plate XXII, fig. 1. Lang Head, 49 yd (45 m) inland from Geo of Drengi [303 704].

FIG. 2. Slice No. S 43772. Magnification × 14. Plane polarized light. Hornblende-basalt. Tablets and stout prisms of zoned plagioclase (centrally An 70+) are subophitically related to xenomorphic green hornblende and cemented by pale green fibrous amphibole; minor augite (NW of centre) is ophitic to plagioclase. South-east shore of Soolmisvird Water [320 726].

FIG. 3. Slice No. S 30598. Magnification × 11. Plane polarized light. Basic pyroxene-porphyrite. Phenocrysts of labradorite (An_{65-70}) and of yellow augite lie in a base of strongly zoned plagioclase laths, subophitic to purplish augite, and minor magnetite which are cemented by a turbid mixture of chlorite, biotite, alkali-feldspar and some quartz. Near Skerry of Stools, 710 yd (650 m) NW of Bousta [219 582].

FIG. 4. Slice No. S 49323. Magnification × 65. Plane polarized light. Garnet in basalt. Small garnets, about 0·02 mm across, are enclosed in clear chlorite (centre) and turbid plagioclase (SE of centre). Grains in chlorite (SW corner) include one garnet and two epidotes. North shore of South Voe, Vementry Island [305 605].

FIG. 5. Slice No. S 28885. Magnification × 16. Plane polarized light. Feldspar-porphyry and basalt. The basalt is chilled; its margin partly enwraps an alkali-feldspar phenocryst of the porphyry. Ness of Gillarona, Muckle Roe [327 627].

FIG. 6. Slice No. S 30732. Magnification × 14. Plane polarized light. Nodular felsite. The nodules consist of feldspar-phyric glass, variably devitrified. The base is microcrystalline to cryptocrystalline quartz-feldspar aggregate, variably sericitized, with frayed slivers of glass. Heill Head, Vementry Island [282 605].

FIG. 7. Slice No. S 30711. Magnification × 13. Plane polarized light. Quartz-feldspar-porphyry with microporphyritic hornblende, biotite and ore. The feldspar phenocrysts include orthoclase (usually mottled) and more abundant albite (usually turbid). Hornblende is replaced by chlorite-calcite aggregate; biotite is pseudomorphed by chlorite. Egga Field, 83 yd (75 m) N of Maa Loch, Vementry Island [301 604].

FIG. 8. Slice No. S 28908. Magnification × 22. Plane polarized light. Allanite. One end of the zoned crystal is embedded in an idiomorphic crystal of hornblende (SE of centre; almost completely destroyed in grinding the section); the other end is held in a rounded crystal of quartz. Such a cumulophyric group, though cognate, is xenocrystic, possibly derived from an early drusy stage of crystallization. The rock is a hornblende-quartz-porphyry. North of Murbie Stack, south-west shore of Muckle Roe [303 630].

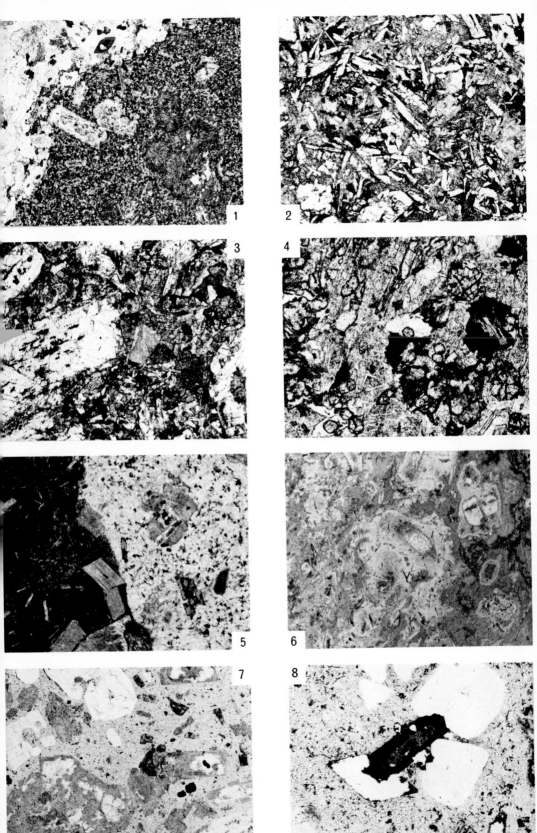

Pyroxene-porphyrite. The pyroxene-porphyrites carry conspicuous tabular phenocrysts of plagioclase up to 1 cm long in a dark grey, fine-grained matrix which in the fresher examples is composed of lathy plagioclase prisms, subophitic augite, idiomorphic iron ore leucoxenized in parts, and interstitial chlorite, quartz, alkali-feldspar, biotite and local epidote. The grain-size of the main minerals in the base is usually about 0·4 mm but augite may form ophitic grains up to 1 mm across. Seriate grading of plagioclase from phenocryst to groundmass prism occurs but is rare (S 56503). The porphyritic plagioclase commonly ranges from basic andesine to labradorite (An_{48} to An_{55}), but some crystals appear to be more calcic (up to An_{68} in S 30598). Compositional zoning in the crystals is limited to a narrow peripheral zone. The plagioclase of the groundmass is usually turbid and altered centrally and shows zoning from labradorite (An_{55}) outwards to oligoclase. Augite usually forms equant hypidiomorphic prisms or shapeless subophitic grains but occurs also in stellate groups (S 28911); it varies in colour from rock to rock, being rose (S 28911), brownish yellow (S 29417, 30598), or colourless (S 56503). An early ferromagnesian mineral is represented by green pseudomorphs which may be too shapeless to provide an indication of their original species (S 44625) but may show definite olivine (S 28911) or orthopyroxene (S 56504) sections. The mineral of the pseudomorphs is usually chlorite but in some cases, where the original is indicated as orthopyroxene by its habit, the replacement is by clear pale green, cryptocrystalline or microcrystalline acicular actinolitic aggregate. Pseudomorphs in granular quartz cemented by oxidized ore and of section suggesting original olivine occur in rocks which also carry pseudomorphs after orthopyroxene (S 30598, S 67503). The feature of the dykes of this group which distinguishes them from the basalts is the nature of the mesostasis. This is usually abundant and in the coarse-grained rocks is seen to consist of chlorite, quartz, alkali-feldspar, biotite, and locally epidote. In the finer-grained rock the constituents are not readily distinguishable but together have the appearance of the hypocrystalline cement of andesites (S 28911, 29417). The alkali-feldspar is marginal to the plagioclase of the groundmass where the laths abut against quartz (S 30598) and is present also in a feathery cryptocrystalline micropegmatite (S 44625, 56504). Chlorite, usually interstitial in shapeless aggregate, also forms microcrystalline intergrowths with quartz and, though in rare cases derived from biotite, is predominantly a late primary product. The character of the mesostasis thus clearly allies these basic porphyrites to quartz-dolerite. Several dyke-rocks which are much altered are included in this group. In some the alteration is peritectic resulting in the conversion of augite to turbid green hornblende (S 56503–4) as in the augite-bearing diorites. Deuteric alteration is widespread in partial to almost complete replacement of plagioclase by muscovite aggregate (S 30599, 30883), in extensive but irregular albitization to oligoclase (S 52077), and in replacement of augite by cryptocrystalline material along a riddle of cracks (S 47780) or by epidote and chlorite (S 30599, 52077). In rare cases the rocks show evidence of slight deformation in twisted chlorite along shear lines (S 47780). The existence of shear stress during intrusion is indicated by curved twinning and microscopic faulting of twin lamellae in plagioclase (S 56503–4).

The Ward of Runafirth porphyrite (p. 241), represented by a single specimen from the centre of the outcrop (S 44329), is in most respects similar to the pyroxene-porphyrites described above, but in its mesostasis contains abundant microcrystalline biotite in aggregates which peripherally disperse into both the sodic mantle of the plagioclase prisms and the interstitial quartz and alkali-feldspar. Locally the biotite aggregate is densely packed round grains of epidote and occasionally forms a monocrystalline pseudomorph. These relations suggest that the rock has undergone some degree of thermal recrystallization and it is possible that this boss-like outcrop of porphyrite is relic from a pre-diorite intrusion. It is, however, so similar to the basic porphyrite dykes that it is accepted here as a late-stage intrusion; the peculiar crystallization of its mesostasis remains unexplained.

Dykes of Intermediate Composition

This group includes the keratophyres, microdiorites and spessartites.

Keratophyre. The keratophyres are in general brownish or greenish grey, structureless, almost aphanitic rocks which are composed essentially of diversely arranged laths of turbid sodaclase feldspar, of composition about 8 to 15 per cent An, in an abundant cement of chlorite which may be only interstitial (S 44332, 53575) or may form a continuous infilling (S 53571), or may be disseminated uniformly or patchily through a cryptocrystalline base (S 53573, 56716). The feldspar laths are usually 0·2 mm or less in length: rarely they show a stellate or radiate arrangement (S 44332). Microphenocrysts of the same composition and idiomorphic stout prismatic habit, about 0·4 to 1 mm long, are usually present in small numbers and may be replaced by calcite, chlorite and quartz. The chlorite of the cement is usually in shapeless flakes and aggregates but occurs also as thin fibres or plates (S 53571); it does not possess a pseudomorphous habit. Grains of ore or leucoxene, limonitic aggregates, and grains of epidote and calcite are common accessory minerals, and apatite is sparse. Quartz is present in some members of the group as a primary mineral (S 53573, 56716); in others it is secondary, at least in part (S 53571). Interstitial potassium feldspar has not been identified with certainty but is thought to be present in some specimens (S 44332, 53575). Varieties with a chloritic base but also containing chlorite as pseudomorphs after a ferromagnesian mineral (S 43759, 56739) are transitional to the feldspar-porphyries or acid porphyrites and the first-cited contains quartz micrographic with feldspar which, extinguishing with the adjacent laths, is probably sodaclase. It is noteworthy that several of the specimens referred to in these descriptions (S 44332, 56716, 56739) occur as the dark marginal facies of quartz-feldspar-porphyries (p. 250). Keratophyres of this kind thus are petrogenetically very different from those of a spilite-keratophyre association.

An anomalous type, difficult to classify, crops out as a dyke on the coast 150 to 200 yd (135–180 m) south by west of Nibon House [304 730]. The rock resembles a felsite and is colour-banded in pale and purplish grey, some bands containing tiny spherulites. Microscopically it is composed essentially of sodic plagioclase and quartz which form a cryptocrystalline to microcrystalline base through which the mafic constituents, green biotite, green hornblende, epidote, and minor chlorite, magnetite, leucoxene, and decomposed trichites tend to be severally concentrated in bands 1 mm to 1 cm broad (S 30007, 33747–8). The plagioclase is albite (An 5–8); no potassium feldspar was positively identified. The texture of the rock is controlled by a mesh or a banded variation in the degree of crystallinity. The coarsest bands contain subprismatic albite of about 0·5 mm length. The banding though probably original is accentuated by recrystallization accompanying redistribution or accession of silica so that some bands are composed mainly of dusty and granular quartz and grains and prisms, locally radiating, of epidote. In these bands relics of plagioclase persist and some contain biotite recrystallized in tiny, clear, equant grains. The original composition of the rock is uncertain but it is probable that the bands had differing mineral compositions during intrusion. The rock can be classified as quartz-albitite or as a quartz-keratophyre.

Spessartite. The few lamprophyric rocks of this group are composed essentially of idiomorphic plagioclase, hornblende and biotite with minor ore and interstitial chlorite, feldspar and quartz (S 56643). Augite is abundant in some specimens as stout prisms, up to 0·3 mm long, and as clusters of granules (S 47744), but rare and relict as thin prisms enclosed in plagioclase in others (S 44323). In general the plagioclase is zoned and greatly sericitized or chloritized but has been determined as of intermediate labradorite-andesine composition in the core (S 15107) while the margins range to alkalic feldspar. The crystals are idiomorphic prisms less than 0·4 mm long and microporphyritic stout prisms up to 1 mm long are common. Hornblende and biotite also are idiomorphic in thin brown prisms which are usually chloritized. Hornblende occurs also as microporphyritic crystals up to 0·5 mm across. Secondary minerals include epidote, which is usually abundant, and calcite. Xenocrystic quartz is rare

(S 56643). A micro-ocellar structure formed by a crude tangential orientation or concentration of small hornblende prisms round spaces filled by more coarsely crystallized minerals is apparent in one member of the group (S 44323). The coarser mineral may be plagioclase which forms hypidiomorphic prisms with a tendency to fan arrangements, or xenomorphic chlorite which encloses pale green fibrous amphibole. In both types granular epidote is abundant. The structure is perhaps due to reconstitution of feldspar or pyroxene xenocrysts, or crystals precipitated from the magma at a deeper level, during the ascent of the dyke.

Microdiorite. The few rocks representing this group are composed essentially of plagioclase and hornblende which, however, do not show the sharply idiomorphic habit characteristic of the spessartites. Varieties transitional to spessartite have turbid idiomorphic plagioclase diversely arranged in lathy prisms, averaging 0·5 mm long, against which prismatic green hornblende is moulded and in places subophitic (S 47757). Obscure small aggregates which are probably altered pyroxene are present and quartz is interstitial.

The more usual microdiorites resemble the plutonic diorites in texture but are of finer grain and the plagioclase constitutes by far the greater volume of the rock. It is consistently dusky with a brownish impregnation and from poorly determinable optical properties appears to be an acid oligoclase. Its hypidiomorphic prisms interfere with one another and with equant grains of colourless augite (S 30880) or prismatic to subophitic green hornblende (S 28610, 50136). Titaniferous magnetite, partly leucoxenized, is a minor essential mineral forming crystals which are generally idiomorphic but form also shapeless growths of black ore which enclose augite and prismatic plagioclase. Interstitial material, abundant in patches, consists of chlorite, epidote, colourless feldspathic or zeolitic mineral, and granules of sphene and ore. Apatite in long slender prisms which locally form bundles, is an abundant accessory. The colourless mineral is greatly obscured by small plates of chlorite which in places are disposed in a pattern so regular as to suggest micrographic intergrowth (S 50136A). It has low birefringence, is optically negative with moderate or low optic axial angle, and has refractive indices distinctly lower than the balsam cement; no cleavage was observed in its allotriomorphic grains. The mineral is provisionally assigned as scolecite. If this is correct the occurrence is similar to that of thomsonite in the altered gabbros (p. 186).

J.P.

ACID MINOR INTRUSIONS

FIELD RELATIONSHIPS

Northmaven and Muckle Roe

The acid dykes within the northern part of the present area form parallel swarms which have a general north-north-west to north-north-easterly trend. Unlike the dykes of Vementry (p. 251) they show no definite radial pattern in their distribution which would imply a central focus of irruption. Except for a small number of dykes around Nibon [305 731] and at the mouth of Gunnister Voe [308 738], the acid dykes are restricted to the outcrop of Muckle Roe Granophyre and within that outcrop they form swarms of roughly north–south trending dykes which alternate, in an irregular fashion, with swarms of basic dykes of a similar trend (see p. 241). Around Lothan Ness [316 676] on the north-eastern margin of the granophyre outcrop there is a change in the trend of the dykes to the north-east, which may indicate a greater ease of fissuring in a radial direction at the margin of the granophyre.

The acid dykes are generally from 1 ft (30 cm) to about 60 ft (18 m) wide, widths between 4 ft (1·2 m) and 20 ft (6 m) being most common along the Muckle

Roe coast. Inland they weather more quickly than the granophyre and here many of the fine-grained acid rocks appear to have short lenticular outcrops up to 60 yd (55 m) wide. Though these outcrops could be interpreted as small bosses it seems more probable that they represent local thickenings of dykes. This view is supported by the observed changes in width along the course of a dyke traced from Raavi Geo [310 676], where it is 15 ft (4·5 m) wide, southwards for 250 yd (230 m) to a small loch where it has increased to 90 ft (27 m) in width; while a still broader dyke, lying 200 yd (180 m) to the west, splits at the mouth of Raavi Geo into two members of less total width. The length of the dykes is probably comparatively great but where so many occur in proximity, connection of isolated outcrops is uncertain. A quartz-feldspar-porphyry exposed at Grusterwick Geo [296 649] is traceable for 800 yd (780 m), and several in the north-west part of the island for 600 yd (550 m).

The acid group of dykes includes porphyritic and non-porphyritic, spherulitic and non-spherulitic felsites, porphyries containing conspicuous quartz and feldspar phenocrysts, and acid porphyritic rocks without visible quartz. In colour they range from grey through flesh-coloured to brick red. There is no apparent spatial distribution of types and several types occur in one dyke. A multiple acid intrusion exposed at the west end of Raavi Geo, has six members and a section across this intrusion is shown in Fig. 26. The members, from east to west, have the following thickness and petrographical characteristics:

A   5 ft (1·5 m)   Purplish   biotite-feldspar-porphyry;   coarse-grained   felsitic
                     groundmass with some decomposed mafic minerals (S 46409).
B   12 ft (3·6 m)   Red and yellow felsite banded by the presence and absence of
                     microporphyritic feldspar and mafic crystallites and by varying
                     degrees of devitrification (S 56410).
C   15 ft (4·5 m)   Pale red unbanded microcrystalline felsite with microphyric
                     feldspars and sparse decomposed mafic minerals set in a
                     cryptopegmatitic base (S 56411).
D   10 ft (3 m)   Aphanitic, dull red felsite; non-spherulitic, cryptopegmatitic
                     groundmass banded by varying proportions of decomposed
                     mafic material (S 56412); one band is feldspar-phyric.
E   6 ft (1·8 m)   Brick-red biotite-feldspar-porphyry; spherulitic, quartz-rich,
                     with a patchily chloritic groundmass (S 56413).
F   2 ft (0·6 m)   Purplish grey, feldspar-microphyric acid porphyrite or quartz-
                     keratophyre (S 44332) (see p. 251).

The banding illustrated by members B and D is a common feature of the acid dykes. It is conspicuous in many dykes because of the colour change induced by the decomposition of the mafic minerals which exist in varying proportion in the bands. In others it is made evident by variation in the proportion and size of spherulites or of quartz and feldspar crystals. In general it is parallel to the course of the dykes; around Lothan Ness, however, banding oblique to some dyke margins has been recorded. The nature of this structure, which is recorded as a vertical foliation, is not certain; it may result from the NW–SE crushing which affects the rocks of this locality. The general small-scale banding of the dykes probably represents fluxional structure due to differential movement of unequally crystallized portions of the upwelling magma. In some cases in which it is of only millimetre to centimetre scale and related to small variations in mafic content it must indicate a degree of heterogeneity in the intrusive material. It is probable, however, as in the case of the multiple dyke described above that

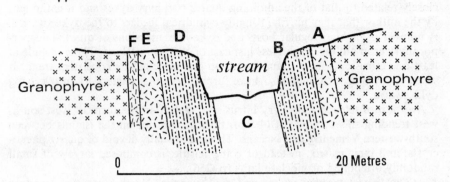

FIG. 26.   *Multiple dyke, west end of Raavi Geo, Muckle Roe*

some kinds of banded structure are caused by pulsatory intrusion of magma of slightly differing composition from pulse to pulse.

Columnar structure transverse to the course of a dyke has been recorded in only one case, a thick dyke of feldspar-porphyry exposed in the scars [303 670] between Erne Stack and Limpet Geo. This has columnar joints at right angles to the walls and meeting along a median surface.

The narrow dark member (F) of the dyke figured in Fig. 26 exemplifies a common characteristic of the acid dykes which can be readily observed in the exposures along the north coast of Muckle Roe between North Ham and Lothan Ness. These marginal facies have the composition of keratophyre and are described in p. 248.

                                                                    J.P.

North shore of Walls Peninsula and Adjacent Islands

The acid dykes cutting the metamorphic rocks adjoining the Vementry Granite can be divided into two major groups: (1) quartz-feldspar-porphyries and (2) spherulitic or fluxion-banded felsites. The former give rise to extensive north-north-east trending dyke swarms, small rounded bosses and irregular sills; the latter form west-south-west and east-west trending dykes and dyke swarms. The quartz-feldspar-porphyry and felsite dykes together have a roughly radial pattern with a focal point within the Vementry Granite (Plate XXIV, p. 207).

Quartz-feldspar-porphyries and feldspar-porphyries

Irregular bosses, dykes and sills. Quartz-feldspar-porphyry with a medium- to fine-grained holocrystalline matrix forms three small circular outcrops, 30 to 50 yd (27–45 m) in diameter, close to the north and south shores of Maa Loch, where they are closely associated with a swarm of thick quartz-feldspar-porphyry dykes. T. Robertson (manuscript notes) has tentatively suggested that these are bosses projecting from the roof of the Vementry Granite and it seems possible that they are on the line of a southward underground projection of the granite, like that exposed at the head of Suthra Voe and just west of Cow Head (p. 207).

The composition and texture of the rock forming these bosses is, however, more closely related to that of the adjoining quartz-porphyry dykes and it is thought by the author that they may be roughly cylindrical feeders to the dyke swarms. A number of highly irregular boss- and dyke-like intrusions of quartz-feldspar-porphyry crop out on Braga Ness just east of Vementry Island. These include at least one intrusion with a circular outcrop and none have a consistent trend. It seems likely that these may also have formed irregular feeders to an overlying dyke swarm.

A group of feldspar-porphyry intrusions, mainly in the form of west-south-west trending sills up to 15 ft (4·5 m) thick, crops out on the islands between south-western Vementry and Neeans. These are usually devoid of quartz pheno-crysts, and their matrix, instead of being felsitic, is composed largely of small randomly orientated plagioclase laths (p. 256).

North-north-east trending dykes and dyke swarms. North-north-east trending dyke-swarms of quartz-feldspar-porphyry are abundant in a 1-mile (1·64-km) wide zone south-west and south of the Vementry Granite. Isolated dykes thought to belong to this group have been recorded as far west as the skerries north of the Neeans peninsula, 1½ miles (2·4 km) SW of the granite margin.

A dyke-swarm, ranging in width from ¼ to ⅓ mile (0·4–0·5 km) and containing a large number of closely spaced dykes, extends from the south-western corner of the granite outcrop at the head of Suthra Voe south-south-westwards across the Brindister Hill peninsula. The dykes within the swarm have a trend which varies from N25°E to N36°E, but individual dykes cannot usually be followed for long distances. Many are irregular in outline, but others have straight margins for distances of up to 200 yd (180 m). In certain areas within the swarm, as, for instance, at the head of Suthra Voe, porphyry dykes make up over 50 per cent of the total exposed rock and along the coast on either side of Trea Wick, where the dykes dip steeply to west-north-west, adjacent intrusions are separated only by thin screens of metamorphic rock. The thickness of individual dykes within the swarm varies from a few inches to 30 yd (27 m). Members of the swarm cut the nose of the Vementry Granite exposed on the headland projecting into Suthra Voe, but very few appear to extend far into the main body of the granite farther north.

Porphyry dykes up to 15 ft (4·5 m) thick are common in the southern part of the western peninsula of Vementry, between the western granite margin and Heill Head. Their trend changes gradually from N30°E close to the granite to E30°N on the coast north of Heill Head. In the latter locality dykes of both quartz-feldspar-porphyry and spherulitic felsite (pp. 254–5) are present, the two together forming an intricate network.

Concentrations of quartz-feldspar-porphyry dykes also occur on Vementry Island, between the granite and Maa Loch, and along the south shore of Uyea Sound just north and north-east of Vementry House. In the former area there are several very thick NNW to NNE trending dykes of a very coarse-grained variety of porphyry, which locally grades into porphyritic microgranite. In the latter area dykes of variable trend, many of them trending roughly east–west, locally form nearly 50 per cent of the total rock outcrop.

Within the Vementry Granite quartz-feldspar-porphyry dykes are much less common than in the surrounding country rock. On the shore of South Voe (Plate XXIV) there is a vertical composite dyke which, from west to east, has the following members:

	ft	m
Fine-grained basic rock (olivine-basalt)	4–6	1·2–1·8
Quartz-feldspar-porphyry	15	4·5
Basic rock, locally sheared	2	0·6

All junctions are vertical and there is marked chilling of the basalt against the porphyry. There is also a close association of acid and basic dykes on the cliffs forming the south-west shore of Lamba Wick [303 607]. No other composite acid-basic dykes have been recorded.

Felsite

Dykes of sparsely porphyritic or non-porphyritic felsite, some of which are spherulitic and/or banded, have been recorded on Vementry Island, both within the granite and in the adjoining metamorphic rocks. They are particularly abundant in the north-western peninsula between Heill Head and Swarbacks Head, where they have a trend which varies from south-west (exceptionally south-south-west) in the south to roughly east–west in the north. There is a swarm of east–west trending dykes in the vicinity of Corbie Geo, in which individual dykes vary greatly in width, averaging 6 ft (1·8 m) and locally reaching 25 ft (7·5 m). Another swarm of spherulitic felsite dykes, steeply inclined to the south-east, cuts Heill Head, where some quartz-feldspar-porphyries are also present. Felsite dykes also cut The Heag [279 602], the island just west of Gruna. Within the Vementry Granite only two dykes of spherulitic felsite have been recorded.

Felsite intrusions, many trending north-east, parallel to the strike of the country rock, cut the basal sediments of the Sandness Formation, which forms the high ground south of Vementry House between the Stead of Aithness and North Voe of Clousta, as well as the Ness of Nounsbrough. They are also present in the basal sediments of the Sandness Formation farther west and in the strongly folded strata of the Walls Formation between the Voe of Clousta and the head of Bixter Voe. The geographical distribution, trend and petrographical character of the intrusions within the Sandness Formation suggest that they may be unrelated to the felsite dykes described above, and that they are probably associated with the phase of extrusive volcanism which gave rise to the Clousta Volcanic Rocks (p. 83). The north-north-west trending intrusions in the Walls Formation between Clousta and Bixter Voe cut strata which are thought to be younger than the Clousta Volcanic Rocks and must thus have been intruded at a later date. These dykes may have their source in a granite which underlies the sedimentary rocks and links the Vementry and Sandsting granites at depth. Finlay (1930, p. 685) has suggested that the Vementry and Walls granites are part of one intrusive complex whose top is depressed by the Walls Syncline, and that this belt of felsite dykes marks the zone in which the granite is nearest to the surface.

W.M.

PETROGRAPHY

This large group of minor intrusions is subdivided into felsites, quartz-porphyries and feldspar-porphyries including acid porphyrites.

Felsites. The term 'felsite' can be usefully applied by both the field geologist and the petrographer to a range of rocks which are macroscopically non-porphyritic, aphanitic, poor in mafic minerals, and which are composed essentially of quartz and alkali-feldspar. Other terms, such as microgranite or aplite, are inappropriate to these rocks. They may be macroscopically structureless or show banded, nodular, or spherulitic structure. Microscopically the rocks which show macroscopic structure range from hypocrystalline, through microcrystalline to cryptocrystalline, the degree of crystallinity varying in banded (S 28593), cellular (S 29503), honeycomb (S 29409, 53588), or nodular (S 28609, 28613) fashion. The banding is commonly combined with one of the other structures, for example as banded-nodular (S 30709). The nodules are formed of single or clustered, isotropic, cryptocrystalline, or spherulitic bodies, the latter being composed mainly of alkali-feldspar. Though usually these bodies represent early plastic, less or more deformed autolithic globules of glass moulded on feldspar crystals, they appear in some cases to be xenolithic since they contain or are coated by chloritic and opaque material which is not present in the matrix (S 28907). One specimen is specially noteworthy because of the presence of very thin lines of dark inclusions and silicified cracks which run generally parallel through the nodules but are not seen in the matrix. These lines are deviated locally through about 30 degrees in nodules elongated along the macroscopic flow structure, indicating a period between peneconsolidation and renewal of intrusive stress (S 28609). Devitrified glass may be present also as orientated fragments which show both angular and plastic outlines (S 30731). All the rocks show extensive sericitization, and in some cases calcitization, which affects the more poorly and the better crystallized parts unequally but not preferentially.

About half of the felsites from the area under description consist of a uniformly cryptocrystalline, microgranular, or microlitic aggregate of quartz and feldspar identical with the more crystalline parts of the matrix of the banded and nodular rocks. In this structureless ground microporphyritic turbid feldspar is sporadic and quartz rare (S 30583, 30709, 30732). The feldspar phenocrysts are not so numerous as in the nodular varieties mentioned above and, as in them, they are mainly orthoclase, soda-orthoclase and microperthite but include also acid plagioclase, with An probably about 10 per cent. They may be single idiomorphic, glomeroporphyritic, or cumulophyric (S 28612, 30732, 30743, 31000, 53580). Generally autolithic the feldspar phenocrysts include also xenolithic crystals (S 53588), and in one specimen fragmented crystals with attached devitrified glass indicate flow-brecciation (S 30887). Crystals forming the nucleus of nodules tend to extend skeletal processes into the devitrifying envelope. The microcrystalline portions of the quartz-feldspar groundmass show a variety of textures from rock to rock and in the one rock. The two minerals may be mixed in allotriomorphic 'felsitic' aggregate; feldspar microlites may be enclosed singly or micropoikilitically in quartz; they may be intergrown as micrographic grains, as small as 0·05 mm, or as spherulites, radiating cones, or feathery growths of micropegmatite in which quartz and feldspar are not individually observable. In some rocks the proportion of determinable quartz is low (S 28612, 30723), in others, perhaps because of late magmatic or post-consolidation growth or redistribution, it is high (S 29503, 30710, 31028). The relative proportions of potassium feldspar and plagioclase are not normally determinable microscopically owing to fineness of grain and close dissemination of alteration products. Either may appear, on criteria of shape or aggregate refractive index, to be dominant (in S 3023 potassium-feldspar, in S 30724 plagioclase) or both equally present (S 30585). Mafic minerals, present only in accessory proportions in most felsites, are represented by shreds and ragged or mossy patches of chlorite or limonite, or by trichites and microlites replaced by chlorite, epidote and ore dust which in some cases (S 28907, 30710) resembles the riebeckite needles and acmite grains of the granophyric blue felsites of the North Roe–Ronas Hill area (Phemister and others 1950). In a few examples in which the dark minerals are present in minor essential proportion, biotite and green hornblende can be recognized among the microlites (S 33742) but in others replacement by chlorite, epidote and oxidized ore is complete (S 43755, 53580, 53588). Chlorite and ore may be so abundant and universally distributed in the groundmass that the rock is transitional to the quartz-keratophyres (S 30585). Allanite occurs as a scarce accessory mineral (S 30709).

Quartz-porphyries. In this group, quartz, potassium feldspar and plagioclase are always present as phenocrysts though in relatively varying amount. Biotite and hornblende occur in most examples as idiomorphic microphenocrysts occasionally exceeding 2 mm in length but are always pseudomorphed, biotite by chlorite, ore and sphene, hornblende by chlorite, sphene, ore, and calcite or epidote (S 30711). In a few, more leucocratic, specimens the mafic material is represented only by trichites and ore granules in a groundmass which is patchily cryptocrystalline, microspherulitic, minutely micropegmatitic, and micropoikilitic (S 30860, 54282); they are in effect quartz-feldspar-phyric felsites. Transitional to the more numerous biotite-hornblende-bearing porphyries are rocks which contain small microphenocrysts of chloritized biotite and chlorite longulites (S 53576) or chlorite dispersed through the base (S 29420). All the leucocratic members show patchy quartzification which is perhaps no more than recrystallization of the silica of the base. In the porphyries containing mafic minerals as minor essential constituents (S 28611, 30735-6, 53577) biotite and hornblende occur as idiomorphic pseudomorphs seriate from 2 to about 0·05 mm in length. The accessory minerals include iron ore, and chlorite, the former in small equant crystals, up to 0·3 mm, normally black magnetite but also leucoxenized (S 30763), and in granules and short rods; chlorite occurs mainly in aggregates of flakes replacing biotite and hornblende, and rarely in spherulitic form (S 50140). Apatite is scarce as is sphene except as an alteration product of biotite and hornblende. Allanite is sporadic in small crystals (S 30711, 30763) but is conspicuous as a zoned microphenocryst, 1 mm in length, with one end of the prism embedded in a quartz phenocryst, the other end in a hornblende crystal (Plate XXIX, fig. 8). The quartz-feldspar groundmass of these rocks varies in texture as described under the felsites, that is it may be felsitic, microlitic, micropoikilitic, minutely micrographic, or microspherulitic.

The quartz phenocrysts of the porphyries are idiomorphic and practically always corroded, with deep embayments filled by material identical with the groundmass. Most have turbid microcrystalline to cryptocrystalline coronas in which the quartz is in optical continuity with the quartz of the phenocrysts. The same optical continuity does not appear in the quartzo-feldspathic aggregate enclosed in the phenocryst (S 30763, 47328). In one specimen (S 28910) the coronas are cryptopegmatitic whereas the groundmass is microlitic; the resulting structure giving the corona and phenocryst a xenolithic aspect. In this case the coronas may represent magma locally enriched in silica by the process of corrosion but prevented by consolidation from incorporation into the bulk. Similar cryptopegmatite coronas are observed around feldspar phenocrysts but in that case spherulitic micropegmatite is abundant in the groundmass (S 28885).

The porphyritic feldspar of the porphyries always includes both potassium feldspar and plagioclase. The former is represented by both orthoclase and microperthite as idiomorphic crystals, patchily or completely argillized, or hypidiomorphic groups which with associated plagioclase can reach 1 cm in diameter. The proportion of plagioclase varies from approximately 1:1 downwards and when low the potassium-feldspar is entirely microperthite (S 53577). Orthoclase phenocrysts with no discernible perthite admixture may enclose small perfectly idiomorphic crystals of plagioclase, partly chloritized biotite and rarely (?)pyroxene (S 30763). The porphyritic plagioclase is greatly sericitized and when determinable is consistently of oligoclase composition (about An 10–12 per cent). Unlike the potassium feldspars it shows seriate diminution from several millimetres in prism length down to groundmass dimensions. Both feldspars are in general autolithic but some glomeroporphyritic groups contain flakes of biotite, only partially chloritized (S 28910), which indicate crystallization from a more deep-seated stage. Patches of turbid orthophyric (S 29420) and microgranular quartz-plagioclase-cryptopegmatite aggregates (S 53577) similarly are early crystallizations.

Some phenocrysts however show peculiarities which prove them xenolithic though cognate. For example, plagioclase sieved or cavernous with chloritic aggregate carry, contrary to the usual relations, idiomorphs of orthoclase (S 53574); cryptopegmatite replaces the feldspar, as a crystal or a coarse granular group, marginally or completely (S 28908); corrosion cavities in oligoclase are filled with material of coarser grain than the surrounding matrix (S 30711); fragmental plagioclase, spotted with quartz inclusions and showing a very close twinning which contrasts with the usual habit, is mantled by microperthite (S 53576).

A dyke of porphyry from the Ness of Haggrister, east of the Walls Boundary Fault, is of the same type as those described above. It carries whole and fragmental crystals of quartz and perthite and a xenolith of coarsely granophyric rock in a turbid, crypto-crystalline, partly silicified base which contains limonitized pseudomorphs of an acicular mineral (S 54282). Curved threads of ore granules suggest original perlitic structure and cryptocrystalline shells round the phenocrysts and xenolith may represent chilled envelopes of glass.

Feldspar-porphyries and acid porphyrites. The rocks of this group repeat the charac-teristics of mineral composition and texture shown by the quartz-porphyries but contain no porphyritic quartz. As in the quartz-porphyries two subgroups are distinguished. In one, which is referred to here as feldspar-porphyry, the members are highly leuco-cratic and contain phenocrysts of potassium feldspar and acid plagioclase which are conspicuous in hand specimens but do not usually exceed 2 mm in size. The other subgroup, the acid porphyrites, contains an essential proportion of ferromagnesian minerals, usually seriate from 2 mm downwards, in addition to the porphyritic feldspar which is predominantly acid plagioclase (An about 8 per cent) and forms single crystals up to 3 mm long and glomeroporphyritic groups up to 6 mm across.

In the feldspar-porphyries phenocrysts of microperthite or orthoclase containing small patches of plagioclase are as numerous as those of acid plagioclase. Both feldspars occur as single crystals and in glomeroporphyritic groups and also interfere in cumulo-phyric clusters with which may be associated ore grains and chloritized ferromagnesian minerals (S 30748) of larger size than any in the groundmass. The latter consists essentially of quartz and alkali-feldspar the nature of which is uncertain owing to fineness of grain, cryptocrystalline spherulitic structure, and decomposition. In two specimens, however, skeletal prisms of fresh orthoclase, only 0·05 mm at most across, pierce the turbid spherulitic material and enclose it where the prisms are hollow (S 30745, 53581). It is of interest to note that in one member of the group (S 31161) evidence of stress during crystallization is afforded by bent twinning and the strain extinction pattern of plagioclase.

In the acid porphyrites, which are more abundant than the feldspar-porphyries, potassium feldspar is sparse and may not be present at all (S 53572, 53579) among the phenocrysts; it may be only microporphyritic and mantled by plagioclase or in one and the same rock appear as a microphenocryst within plagioclase or as a phenocryst moulded on plagioclase. Seriate down from about 3 mm to groundmass dimensions, plagioclase is always idiomorphic in single crystals and in the free faces of glomero-porphyritic groups. Large spongy shapeless crystals which are apparently xenolithic also occur but are rare (S 30740). Owing to the universal turbidity and common sericitization the composition of the porphyritic plagioclase is difficult to ascertain. The ferromagnesian minerals are always altered but their outline in section and the style of replacement by the secondary chlorite and ore, epidote and sphene show that both hornblende and biotite have been primary minerals. Iron ore is abundant as octahedra and large grains in the groundmass which contains also much interstitial chlorite and ore granules and grains of epidote scattered through the turbid base of microcrystalline, in some rocks partly cryptopegmatitic, alkali-feldspar and quartz. Apatite in thin prisms is a common accessory mineral and zircon is rare; trichites of chlorite are numerous in members containing cryptopegmatite (S 29504).

Some features of the space relations of the potassic and sodic feldspars call for attention. The non-seriate character of the potassium feldspar contrasts with the seriate crystallization of the sodaclase. Large phenocrysts of potassium feldspar are moulded on plagioclase (S 56413) and in this rock and others (S 44318, 56409) microporphyritic potassium feldspars are enclosed within idiomorphic plagioclase usually with a separating zone of groundmass material. These features considered together seem to imply the existence, in a dominantly sodic feldspathic magma, of potassium-rich gouts which in some cases are trapped by the growing plagioclase.

J.P.

REFERENCES

FINLAY, T. M. 1930. The Old Red Sandstone of Shetland. Part II: North-western Area. *Trans. R. Soc. Edinb.*, **56**, 671–94

PHEMISTER, J., SABINE, P. A. and HARVEY, C. O. 1950. The riebeckite-bearing dikes of Shetland. *Mineralog. Mag.*, **29**, 359–73.

Chapter 17

MAJOR TRANSCURRENT FAULTS

THE FAULTS described in this chapter are the Walls Boundary Fault, which forms the eastern boundary of the area described in this memoir, and the Melby Fault which separates the Melby Formation from the Walls Sandstone. Both faults may have had considerable post-Devonian dextral transcurrent movement along them and these movements may explain the present close juxtaposition of the Melby Formation, the Walls Sandstone and the East Shetland Old Red Sandstone, which not only differ from each other in age, but have such diverse sedimentological, volcanological and structural histories that it must be assumed that each evolved in a separate depositional and structural province.

WALLS BOUNDARY FAULT

The Walls Boundary Fault is a complex, locally branching, dislocation zone, which separates two areas of widely differing lithology and structural trends. It has a complex geological history, having been involved in at least three phases of movement (pp. 263–5). Flinn (1961, p. 589) has suggested that it may be the northward continuation of the Great Glen Fault.

In the Walls Peninsula the fault is accompanied by a crush belt of considerable width, but the actual fault plane is everywhere narrow and clearly defined. Fault-bounded enclaves of rock derived from opposite sides of this fault plane have, however, been recorded within the shatter belt exposed on the west shore of Aith Voe. In the area north of Papa Little the fault splits into a number of branches which enclose a complex dislocation zone. This zone was developed during two phases of dislocation long before the final brittle transcurrent fault movements took place.

OUTCROPS OF FAULT

Southern end of outcrop: Roe Ness, Rea Wick and Sand peninsulas (Plate XII). Though the actual fault plane is not exposed where it crosses the shore at the head of Seli Voe, there are excellent sections of the shattered and faulted rock around the coasts of the Roe Ness and Sand peninsulas. West of the fault there is a 1¼-mile (2-km) wide zone in which the sandstone, granite and diorite have been affected by intense brittle shattering which has produced closely spaced faults and joints and has made the rock very friable. The faults in this zone include a number of major sub-parallel N10°E to N30°E trending dislocations which appear to be branches of the Walls Boundary Fault. Some north-north-west trending belts of mylonite are present on the coast between 600 and 1200 yd (550 and 1100 m) N of Rea Wick [328 448]. These belts are scapolitized and they do not have the same trend as the crush belts associated with the Walls Boundary Fault. It is thought that their formation may have preceded the brittle movements along the fault (see pp. 236–7, Mykura and Young 1969, p. 3).

East of the Walls Boundary Fault, a ¾-mile (1·2-km) wide belt of shattered schist is exposed just east of Kirka Ness [337 464] along the south shores of the

Sand peninsula. The rock is here composed almost entirely of soft micaceous fault-gouge derived mainly from mica-schist with fault-bounded lenses of shattered quartzite which form the headlands.

The soft and friable nature of the crush rock suggests that in this area the movements within the fault-zone took place at a high tectonic level. There is here no evidence of earlier shear or thrust movements along the line of the fault. *Bixter Voe.* On the shores of Bixter Voe [336 506 and 337 513] the fault plane is clearly defined and inclined at 85° to the west. The zone of intense shearing within the Walls Sandstone immediately west of the fault is only 200 yd (180 m) wide, though the sandstone is shattered for a further 70 yd (65 m). East of the fault the zone of shearing is well defined and 150 to 160 yd (140–150 m) wide. The sheared rock is again very friable and cut by many small closely spaced faults which are sub-parallel to the main fault, as well as by many irregular shear planes which are variable both in direction and inclination. As in the area further south, there is no mylonite or compact crush rock which can be attributed to earlier movement within the fault zone. All the dislocation may thus have taken place at a high tectonic level.

W.M.

West shore of Aith Voe (Plate IX). On the western side of Aith Voe the fault-zone is wedge-shaped, reaching several hundred yards in width on Aith Ness but thinning southwards towards Breawick [338 577]. On Aith Ness the Walls Boundary Fault itself has two main branches which at Keen Point [336 593] and the bay to the east are about 40 yd (36 m) apart. The faults are very steep to vertical. Inland, to the south, these two branches cannot readily be traced on the ground but their positions may be seen from the eastern bank of Aith Voe. At the points where the two branches of the fault crop out on the coast the rocks are seen to be much crushed. They include a rock of felsitic appearance which is found in thin section (S 47940, from the coast 100 yd (90 m) S of Keen Point) to be a cataclastically deformed argillaceous feldspathic sandstone. It is a shattery streaky greenish grey sediment having numerous curved faces coated with red to chocolate-coloured iron oxide. The quartz grains are mainly 0·1 mm or less in length although occasionally much coarser. The groundmass contains abundant chloritic granules and shreds and flakes of green mica. There are plentiful heavy mineral grains, including zircon and green tourmaline, and epidote is plentiful in grains and aggregates. There is abundant actinolitic amphibole, abundant ore granules including hematite, and ilmenite rimmed by sphene. The rock is traversed by numerous veinlets along lines of movement of chlorite, calcite, hematite, and alkali feldspar showing complicated twinning or granulation. The veins show a continuation of the rock fabric transversely and sygmoidally across them; this applies to the fine detail of the fabric, although in the veins the fabric is represented by different minerals from those of the host rock. Their texture may be interpreted as being in part, at least, due to replacement following the imposition of foliation.

South of Breawick the two branches of the Walls Boundary Fault cannot readily be separated. The fracture is intermittently marked by a line of seepages and springs, and by occasional exposures of the rocks on each side.

East of the Walls Boundary Fault on Aith Ness the crushed rocks consist of schist and sandstone with a higher proportion of schist occurring nearer to Point of Sletta [339 591]. Limestone bands not more than a few feet (1 m) thick

S

crop out close to the Walls Boundary Fault on the coast and inland. The schists are variable, including semipelitic and psammitic types, garnetiferous mica-schist and hornblende-schist, with pegmatite veins and quartz veins. The strike is variable but predominantly north-easterly, and the dip is generally high. Isolated observations were made of rodding which is to the north-east at 7° and 45°.

Northerly trending faults cut the main bay and the promontory between the Loch of Aithness [334 583] and Aith Voe, two of them dipping eastwards at 45°. All of the ground in this part of Aith Ness is extensively broken and crushed. The main fracture which marks the eastern boundary of this crush zone cuts the coast just north-north-west of Stiva [341 575]. This fracture belt which is some 40 yd (36 m) wide may conveniently be referred to as the Stiva Branch. The rocks in this Stiva fault-belt include green and chocolate-coloured sheared and crushed quartzite and sandstone, apparently of Old Red Sandstone age, as well as schist, pegmatite and some fault-breccia. A shattery greenish rock with hematite-coated joints having some resemblance to mylonite is seen in thin section (S 51266) to be a cataclastically deformed sandstone, traversed by veinlets mainly of calcite. The easterly fault bounding the Stiva Branch brings crush-rock against broken limestone 60 ft (18 m) thick. The fault plane is well exposed and dips west at 80°. Some of the faults within the Stiva fault belt are vertical or steep, but one prominently exposed fault plane is nearly horizontal and undulating.

P.A.S.

The Walls Sandstone adjoining the western branch of the Walls Boundary Fault at Keen Point is hardened and colour-laminated (yellowish green, brownish purple and buff) for a distance of 30 yd (27 m) from the fault plane. Individual colour bands are 2·5 to 6 mm thick. They trend north-north-east to north-east and their inclination ranges from 73° to WNW to vertical, (i.e. sub-parallel to the main fault). Within a 15-yd (14-m) wide zone adjoining the fault the sandstone is cut by many thin 'veins' of pink mylonite, which cut and locally slightly displace the colour lamination.

The induration, colour banding and local mylonitization of the rocks within the fault zone can be attributed to one or more phases of shearing or thrusting which took place at a fairly low tectonic level and created a zone of weakness which was later utilized by the faulting along the Walls Boundary Fault Complex. The latter, with its large number of sub-parallel shear planes and shatter zones, has been formed by dislocation at a much higher crustal level.

Papa Little. (Plate IX). The two branches of the Walls Boundary Fault cut the island of Papa Little. The western branch is exposed on the north and south shores of the island and in the Little Burn [338 604]. In all these exposures it is associated with bands of pink, commonly colour-laminated, highly indurated crush rock and mylonite which in hand specimen resembles felsite. Mylonite is also developed in the schists just east of the fault plane. The eastern (main) branch of the fault is seen on the east coast, 600 yd (560 m) E of North Ward [346 614]. The superimposed brittle crushing within the western fault planes is less intense than in the exposures along the shore of Aith Voe and the effects of shattering in the rocks on either side of the fault are much less extensive.

W.M.

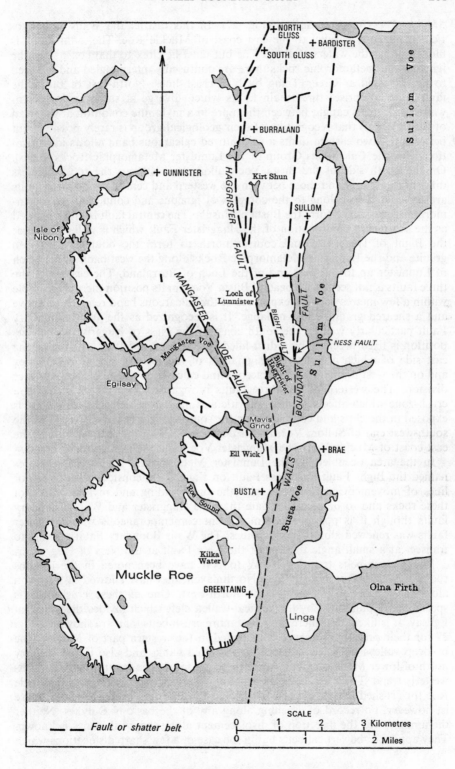

FIG. 27. *Major faults in the area north of Busta Voe*

Muckle Roe–Gunnister–Sullom (Fig. 27). In this district the Walls Boundary Fault-zone runs parallel to the east coast of Muckle Roe. The main break is hidden under the waters of Busta Voe but must lie close to that coast since the igneous and metamorphic rocks there are confusedly intermingled and are seen to be traversed at Green Taing by N–S crush-lines. North-east of Busta the fault zone comprises three main faults which diverge at small angles north-wards. The rocks caught between them are in a mylonitic condition at the head of Busta Voe so that recognition of their geological group is rarely possible, but between the two eastern faults a mylonitized calcareous band allows identifica-tion with the Calcareous Group of the Lunnister Metamorphic Rocks (p. 15). On the north side of the Busta Voe–Sullom Voe isthmus the rocks are less difficult to identify and those between the western and central of the three faults are considered to belong to the complex of igneous and contact altered meta-morphic rocks which form the Busta peninsula. The central fault is thus regarded as the southward continuation of the Haggrister Fault which is well exposed at the Bight of Haggrister and courses north to form the boundary between granite and the Lunnister Metamorphic Rocks along the west bank of the Loch of Lunnister up to the west end of the Loch of Burraland. The eastern of the three faults is not seen at the head of Busta Voe, but its position there is locatable within a few metres between exposures of the calcareous band referred to above and a sheared granite-schist mélange. It is recognized as the Walls Boundary Fault particularly well on the south coast of the Ness of Haggrister where its position is fixed precisely by a drift-filled vertical cleft $16\frac{1}{2}$ ft (5 m) wide on the east side of which are sheared migmatites of the Delting Injection Complex and on the west tightly folded micaceous and calcareous schists of the Lunnister division. The western of the three faults is regarded as a major break in a crush-zone which affects granitic, dioritic and metamorphic rocks. This zone is exposed in the cliffs and coast forming the south-east corner of Ell Wick, at the south-west end of Sullom Voe, and is comparable with that exposed along the east coast of Muckle Voe. The Mangaster Voe Fault may be a similar break.

In the area occupied by the Lunnister Metamorphic Rocks the fractures named the Bight Fault and Ness Fault on Fig. 27 are considered to be major lines of movement synchronous with the tight folding and mylonitization of these rocks and to be of earlier date than the Haggrister and Walls Boundary faults though it is possible that movement contemporaneous with the latter faults was renewed along the older lines. The Walls Boundary Fault appears to transect at a small angle the line of the Ness Fault at the Ness of Haggrister.

Many crush-belts trending NNW to NW have been noted in the igneous rocks of Muckle Roe and southern Northmaven; they have allowed easy erosion along the geos which open to the north-west. One of these crush-belts is spectacularly illustrated by the vertical-walled cleft which divides the island of Egilsay. It is likely that there are many more crush belts than are shown on Fig. 27 for their probable presence is indicated in the western part of Muckle Roe by deep valleys bounded by steep scree-coated flanks and also in the country north of lower Mangaster Voe where there are long steep features with a north-westerly trend. The crush-belts are probably later than the acid dykes of Muckle Roe for crushed and brecciated felsites have been noted by D. Haldane; there is, however, no record of crushing along any of the basic N–S dykes. Neither the amount nor the direction of displacement along the crush-zones is known. They appear to be vertical belts but in the case of a few sharp dislocations which have been mapped by D. Haldane as faults he has shown hade in both east and west directions.

In addition to the faults and crushes in the N to NW quadrant a few faults trending ENE to NE have been recorded by D. Haldane. One of these displaces the northern part of a composite acid–basic dyke eastwards for a distance less than the width of the dyke at a locality about ⅓ mile (0·5 km) W of the north end of Kilka Water.

The fan pattern formed by the faults and crushes in this area (Fig. 27) is noteworthy. The Walls Boundary Fault is directed a little east from its N–S Aith Voe–Busta Voe trend but the N–S line is continued by the Haggrister Fault. West of the latter the faults and crushes have in general a more north-westerly trend the farther they lie from the main zone of translation. It appears that the crustal fracture meeting a massive block of gneiss in which old structural lines were established and which was buttressed by massive crystalline igneous rocks, had its major displacement deflected while a minor component followed old structural lines or was dissipated in the structureless igneous massif.

J.P.

TECTONIC HISTORY

Early Thrusts. It is believed that the early mylonitization within the northern wedge is associated with a series of low-angled thrusts, which have been rotated by subsequent folding into a vertical position (pp. 21–23). The rather narrower mylonitized belts in Papa Little and Aith Ness may lie within the southward extension of this wedge. As no mylonite or phyllonite have been recorded along the fault south of Aith Voe, it is assumed that the early thrusting either did not extend further south, or followed planes which were not utilized by the later movements along the fault.

It is tempting to correlate this early phase of thrusting and associated mylonitization and phyllonitization with the major thrust movements postulated elsewhere in Shetland. Examples of such movements are the thrusting responsible for the first phase of retrograde metamorphism in the Valla Field Block of Unst (Read 1934, p. 650; 1937), the emplacement of the 'Nappe Complex' of Unst and Fetlar (Flinn 1958), and the major phase of dislocation which brought the East Mainland, Quarff and Lunnasting successions of Shetland Mainland into tectonic contact (Miller and Flinn 1966, p. 114; Flinn 1967, pp. 287–8). The age of all these movements has, however, been tentatively set at or before 420 m.y. by Miller and Flinn (1966, p. 114). The mylonitization along the Walls Boundary Fault, on the other hand, affects the Old Red Sandstone of Aith Ness and Papa Little, which indicates that in these areas deformation leading to mylonitization was active at some time after 360 m.y. or, at most, 370 m.y. B.P. There is, however, no evidence which could be used to prove that all the mylonitization along the Walls Boundary Fault is of roughly the same age and it is possible that there were several periods of dislocation during which mylonite was produced.

Folding. Evidence for the folding or, at least, rotation of the belts of mylonite into a vertical position prior to the final faulting has been recorded in the Lunnasting area (p. 22) and parts of Northmaven (Phemister, 1976).

Late brittle dislocation. The latest movements along the fault produced a belt of brittle shearing and shattering which contains clearly defined steep fault planes. The belt appears to be narrowest at Papa Little. It widens out southward to a width of over 2 miles (3·2 km) at the south shore of the Walls Peninsula and northwards to over 1 mile (1·6 km) at the northern margin of the Sheet. A number of branching faults splay out southwards from the main fault south of the latitude of Bixter Voe, and northwards from focal points near Aith [339 564] and in Busta Voe just east of Busta House [348 665]. In the most northerly splay most of the dislocation appears to have been along the most easterly branch of the fault, though some of the movement was taken up by faults branching off westward and extending into the Northmaven Granite-Diorite Complex (pp. 261–2). These faults are clearly younger than the Old Red Sandstone sediment-ary and intrusive rocks of the area, but the true age of the faulting remains a matter for conjecture.

DIRECTION AND AMOUNT OF MOVEMENT

Both Flinn and the author have put forward evidence which supports the concept of a considerable post-Old Red Sandstone dextral transcurrent move-ment along the fault. Flinn (1969, p. 291) has suggested that this movement may have been of the order of 40 miles (65 km). He has based this conclusion on evidence obtained from his interpretation of the IGS geomagnetic map and has no doubt found further support from the fact that the smaller sub-parallel Nesting Fault, which lies some 6 miles (9·5 km) to the east of the Walls Boundary Fault, has a dextral displacement of 10 miles (16 km). The author (Mykura 1972a, p. 51; 1972b, p. 30; Mykura and Young 1969, fig. 1) believes that the amount of post-Old Red Sandstone dextral transcurrent movement along the fault may have been between 35 and 50 miles (60 and 80 km). This conclusion is based on the following evidence:

1. Sodic scapolite is a common vein and replacement mineral, both in and around the Sandsting Complex of the Walls Peninsula and in the somewhat indurated sedi-mentary rocks of south-west Fair Isle. Fair Isle is thought to lie a few miles east of the southward continuation of the Walls Boundary Fault.
2. There is evidence for the presence of a granitic body a short distance to the south-east of Fair Isle (see Mykura 1972a, p. 51). This granite could possibly be part of the displaced eastern portion of the Sandsting Granite-Diorite Complex.
3. The sedimentary rocks of Fair Isle are of Lower and/or Middle Old Red Sandstone age and in the south-west of the island they are strongly deformed and intruded by acid and basic dykes (Mykura 1972b). The only other Old Red Sandstone rocks of similar age within the Orkney–Shetland area which are intensely folded and cut by late-Caledonian dykes are those forming the Walls Sandstone. There is thus a distinct possibility that these two sequences were deposited and deformed in the same depositional and structural basin (see p. 124) and were later separated by dextral transcurrent movement along the Walls Boundary Fault.
4. The Walls Sandstone is completely different, in age, content of volcanic and intrusive rocks and tectonic style from the high Middle to Upper Old Red Sand-stone (Upper Givetian) sediments of south-east Mainland (see Mykura 1976). Both formations rest unconformably on the metamorphic basement and the two formations must have evolved in completely separate sedimentary, magmatic, and tectonic environments. Yet their outcrops are now only 7 miles (11 km) apart. This suggests that they have been brought into virtual juxtaposition by considerable transcurrent movement along the two intervening north–south trending faults: the Walls Boundary Fault and the Nesting Fault. As a dextral

sense of movement can be demonstrated in the case of the Nesting Fault it is reasonable to assume that the movement along the much larger Walls Boundary Fault was also dextral.

5. There is inconclusive evidence that some of the branch-faults of the Walls Boundary Fault within the Walls Peninsula are tear faults with a dextral displacement. The Reawick Fault, which cuts the south-east corner of the peninsula, for instance, separates strongly folded and locally cleaved strata to the east from considerably less intensely folded strata to the west. The nearest exposures of sediments to the west of the fault belt, which are deformed to the same extent as the strata east of the Reawick Fault, are found 6 to 7 miles (9–11 km) further north.

None of the evidence so far quoted is in any way conclusive. Perhaps the strongest evidence is that provided by the presence in both Fair Isle and West Mainland of scapolite associated with a granite complex. The Fair Isle scapolite, however, could just as well be connected with a granite mass which was at the time of emplacement located well to the south of the Sandsting Complex and which may be a member of a north–south trending suite of granite masses of which the granites cropping out between North Roe and Sandsting form only a part. All the other points of evidence above can be similarly questioned. The coincidence of all these factors, however, adds up to a strong impression that the latest movement along the Walls Boundary Fault was a major dextral transcurrent displacement.

MELBY FAULT

The Melby Fault cuts the north-west corner of the Walls Peninsula, where it has a south-westerly trend and separates the Melby Formation in the west from the Walls Sandstone and Walls Peninsula Metamorphic Rocks in the east (Plate XII). It appears to continue north-north-eastwards across the centre of St Magnus Bay and re-appear on the Esha Ness peninsula of Northmaven, where it separates the Eshaness Volcanic Rocks in the west from the North-maven Granite in the east. The south-westward extension of the Melby Fault is uncertain. The data from Sheet 16 of the IGS Aeromagnetic Map (1968) suggest that it passes 5 to 6 miles (8–9·6 km) E of Foula.

The fault plane is well exposed at Hesti Geo [172 556] on the west shore of the Walls Peninsula (Fig. 16, p. 145), where it is inclined at 60° to 70° to the south-east and associated with a 300 yd (270 m) wide zone of intensely sheared and folded strata. The intense folding is largely confined to the Melby Formation west of the fault plane and in these beds disturbance on a diminished scale can be traced for a further 200 yd (180 m) to the west as far as Pund Head. Some of the folds within the disturbed zone appear to have been affected by more than one phase of deformation and some have axes which trend almost at right angles to the Melby Fault. The fault forms a prominent feature along the north-west slope of the Hill of Melby but is not exposed on the north coast of the peninsula, just north-east of Melby Church.

If the Melby Fault were a fault along which only differential vertical movement had taken place, it would be a reversed fault along which the Walls Sandstone, which is here steeply inclined to the south-west, has been thrust over the south-eastward dipping Melby Sandstone. The intensity and extent of deformation in the adjoining Melby Sandstone would suggest that this was a fault with a very considerable throw. There are, however, a number of factors which suggest that the Melby Fault, like the Walls Boundary Fault, is a transcurrent fault with considerable dextral displacement. It has been argued in Chapter 10 (p. 153) that the predominantly fluvial Melby Formation may have been deposited near the north-western margin of the extensive Orkney–Caithness depositional basin and that one of the Melby fish beds might even have been formed in the same extensive lake as the Sandwick Fish Bed of Orkney and the Achanarras Limestone of Caithness. Though such a correlation must remain highly speculative there is no doubt that in its fauna, lithology and structure the Melby Sandstone bears a much closer resemblance to the Givetian sediments of Orkney than to the adjoining Walls Sandstone, which has not only been intensely folded but has been intruded by a granite complex and by suites of basic and acid dykes. If the Melby Fault were a simple reversed fault the Melby Formation would be expected to rest unconformably on the strongly folded Walls Sandstone which would have been extensively eroded before the deposition of the former. As the base of the Melby Formation is nowhere exposed, nothing is known about the rocks underlying it either in Mainland or Papa Stour. In the Island of Foula, however, there is both geophysical evidence (McQuillin and Brooks 1967, p. 15) and a certain amount of field evidence (p. 175) that the Foula sandstone rests directly on the metamorphic basement.

It is the author's opinion that the Melby and Foula formations were deposited along the north-western margin of the open Orcadian basin, in a position that was a considerable distance south-west or south-south-west of the basin in which the Walls Sandstone was laid down and that the two formations were subsequently brought together by dextral transcurrent movement along the Melby Fault. The actual amount of displacement along this fault cannot, as yet, be calculated.

Conclusion

Large post-Old Red Sandstone dextral transcurrent shifts along the Walls Boundary Fault and Melby Fault can most satisfactorily account for the virtual juxtaposition of three diverse formations of Old Red Sandstone rocks in Shetland. Each formation appears to rest directly on the metamorphic basement, but each differs from the others in its age, in its depositional and volcanological development, its tectonic history and in the extent to which it has been affected by igneous intrusions. The recognition of the presence of two major transcurrent faults also permits us to attempt a reconstruction of the Old Red Sandstone palaeogeography of the Shetland–Orkney area. Such a reconstruction assumes that there were three distinct, possibly fault-bounded intermontane basins, each of which had a completely independent evolution. The most southerly basin, which extended over Orkney and part of Caithness, was flat and open and tectonically relatively stable. In middle Givetion (?Eday) times it was affected by a phase of extrusive volcanism which was most extensive in the north (i.e. Eshaness) and became thin and intermittent at the latitude of Orkney. The central basin, in which both the thick sequences of the Walls Sandstone (pp. 123–124) and the Fair Isle Old Red Sandstone (Mykura 1972b) may have been laid down, lay within a tectonically and magmatically active belt. This basin was probably elongated in a west-south-westerly or east–west direction and was affected at the end of Middle Old Red Sandstone times by compressive forces which acted first in a north–south direction and later in an east–west direction. The most northerly basin, of which only the western marginal deposits are now seen in eastern Shetland, was probably bounded in the west by mountainous terrain carved partly in a granitic complex similar to that now exposed in Northmaven and North Roe.

W.M.

REFERENCES

FLINN, D. 1958. The nappe structure of North-East Shetland. *Q. Jnl geol. Soc. Lond.*, **114**, 107–36.

—— 1961. Continuation of the Great Glen Fault beyond the Moray Firth. *Nature, Lond.*, **191**, 589–91.

—— 1967. The metamorphic rocks of the southern part of the Mainland of Shetland. *Geol. Jnl*, **5**, 251–90.

—— 1969. A Geological Interpretation of the Aeromagnetic Maps of the Continental Shelf around Orkney and Shetland. *Geol. Jnl*, **6**, 279–92.

McQUILLIN, R. and BROOKS, M. 1967. Geophysical surveys in the Shetland Islands. *Geophys. Pap. No. 2, Inst. geol. Sci.*, 1–22.

MILLER, J. A. and FLINN, D. 1966. A Survey of Age Relations of Shetland Rocks. *Geol. Jnl*, **5**, 95–116.

MYKURA, W. 1972a. Igneous intrusions and mineralization in Fair Isle, Shetland Islands. *Bull. geol. Surv. Gt Br.*, No. 41, 33–53.

—— 1972b. The Old Red Sandstone sediments of Fair Isle, Shetland Islands. *Bull. geol. Surv. Gt Br.*, No. 41, 1–31.

—— with contributions by D. FLINN and F. MAY. 1976. Orkney and Shetland. *British Regional Geology, geol. Surv.* in press.

—— and YOUNG, B. 1969. Sodic scapolite (dipyre) in the Shetland Islands. *Rep. No. 69/4, Inst. geol. Sci.*

PHEMISTER, J. 1976. The Lunnister Metamorphic Rocks, Northmaven, Shetland. *Bull. geol. Surv. Gt Br.* in press.

READ, H. H. 1934. The Metamorphic Geology of Unst in the Shetland Islands. *Q. Jnl geol. Soc. Lond.*, **90** 637–88.

—— 1937. Metamorphic Correlation in the Polymetamorphic Rocks of the Valla Field Block, Unst, Shetland Islands. *Trans. R. Soc. Edinb.*, **59**, 195–221.

Chapter 18

RECENT AND PLEISTOCENE

MAINLAND, MUCKLE ROE AND PAPA STOUR

INTRODUCTION

THOUGH SHETLAND, like the rest of Scotland, must have been covered by ice during all four glacial maxima of the Pleistocene period, it is probable that the glacial deposits and features now seen are largely attributable to the last (Weichselian) maximum. It has long been accepted (Peach and Horne 1879; Finlay 1926b, p. 180) that the Shetland Islands were at some stage covered by ice, which had reached the area from the east and is thought to have originated in Scandinavia. After the connection with the eastward moving Scandinavian ice had ceased to exist the islands were covered by a local ice-sheet which spread out seawards in all directions from the central range of Mainland (Robertson, Geological Survey records 1935; Flinn 1964, p. 338; Hoppe and others 1965, p. 110). This may have involved a series of advances and retreats. The final episode in the glacial history of Shetland was a period of corrie glaciation during which local glaciers were confined to relatively small corries and poorly defined nivation hollows (Charlesworth 1956, pp. 887–91). Only the corries of Foula form marked topographical features (pp. 4 and 171).

In West Mainland of Shetland the westward and north-westward movements of the ice-sheets have produced a less pronounced smoothing-out of the relief than in other parts of Shetland. In the northern parts of the Walls Peninsula, Muckle Roe and the adjoining islands the craggy topography shows some effects of glacial moulding, such as north-west trending ridges and intervening glacially-gouged depressions. The glacial scouring has produced numerous hollows, most of which now form small inland lochs. These are particularly abundant in Northmaven and the northern half of the Walls Peninsula. In the southern half of the Walls Peninsula, where the only recognized ice movement was to the south-west, glacially moulded features are less prominent.

The glacial deposits of Western Shetland consist largely of grey to brownish, generally sandy, clay with abundant stones. This drift occupies many hollows and fills some possible pre- or inter-glacial valleys (p. 272), but appears nowhere to attain a thickness greater than 30 ft (9 m). Ice-transported boulders indicating a west to north-westward ice movement are common in the northern part of the Walls Peninsula, Papa Stour, Muckle Roe and Northmaven. On the island of Papa Stour the remnants of a possible terminal moraine are preserved (p. 272, Plate XXXB).

The evidence for more than one period of glaciation, separated by an inter-glacial period of milder climate, is at present confined to one section on the west coast of the Walls Peninsula, where boulder clay is locally underlain by a deposit of sand and gravel which fills a pre- or inter-glacial depression and contains a bed of peat (p. 273) up to 18 in (45 cm) thick.

Features and deposits formed by glacial meltwater during the ablation of the ice (pp. 277–8) are sparse.

Raised beaches are not present in the Shetland Islands, and submerged peat deposits have been recorded at depths of up to 30 ft (9 m) below High Water Mark (pp. 280–1). The coastline of Shetland with its long, gentle-sided voes has the characteristic features of a submerged landscape, and a detailed study of the submarine topography around Shetland by Flinn (1964) suggests that in this area submergence may have continued since the last glacial maximum. During the last 6000 years, the sinking of Shetland seems to have been more rapid than in most of the remainder of Britain and it is thought that this may have been, in part, isostatic (pp. 281–2).

GLACIAL DEPOSITS

Boulder Clay

Rather less than half the area of Western Shetland is covered by a thin irregular layer of fairly sandy and generally stony drift (Fig. 28). In the northern and central parts of the Walls Peninsula and in the western parts of Muckle Roe and Northmaven, drift deposits are confined to small irregular patches filling depressions between rocky and, in places, ice-moulded hillocks. The drift cover is more continuous in the less rugged terrain along the eastern margin of the sheet, on the relatively flat 'plateaux' of Sandness and Papa Stour, and in the lower ground along the western and south-western coasts of the Walls Peninsula.

Except on the Island of Papa Stour (p. 272) and on the north and west slopes of Sandness Hill (p. 271) where there are morainic mounds and ridges, the drift deposits form thin sheets with a relatively smooth surface and 'feather-off' against the hillsides. The thickness of drift is in most cases less than 10 ft (3 m) and the maximum recorded thickness is 25 ft (7·5 m) (p. 272). The deposit is fairly homogeneous throughout the area and it has not been possible to differentiate between deposits of different ages as has been the case in Northmaven (Chapelhow 1965, pp. 65–6). The matrix of the boulder clay consists of a greyish brown sandy clay, with up to 40 per cent of sand grains set in a clayey base. The deposit is everywhere very stony, with a high proportion of pebbles less than 3 in (8 cm) in diameter, though larger boulders are locally abundant. The pebbles are never well rounded and are seldom strongly striated. Both the matrix and pebbles vary in texture and composition according to the underlying bedrock.

There is invariably a high content of pebbles derived from rocks exposed a short distance to the south-east or east, and in a number of drift sections the debris of the immediately underlying rock is confined to the basal 3 ft (1 m). Thus in the drift-filled hollow cut in rhyolite at Geubery Head on the east shore of Papa Stour, the upper 10 ft (3 m) of the sandy boulder clay contains a high proportion of pebbles of hornblende-schist, vein-granite and sandstone, some of it of Walls type. Local rhyolite clasts are confined to the basal 3 ft (1 m) of the section.

Sections in the sandy boulder clay are seen in a number of roadside pits and in a limited number of coast sections, most of which are not easily accessible. The following were the best and most easily accessible drift sections at the time of writing:

1. *Large quarry in drift (50 × 30 yd; 45 × 27 m) on west side of road 2000* yd *(1800 m) NE of Bridge of Walls Hotel* [277 524]. The quarry is situated near the western edge of a sheet of blanket-drift which occupies the low ground containing the Loch of Murraster and thins out westwards on the slopes of the Hill of

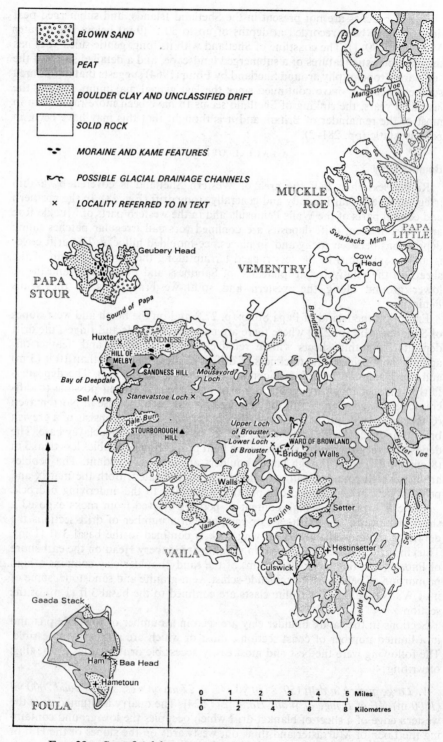

FIG. 28. *Superficial deposits and possible late-glacial drainage features*

Murraster. A section of up to 18 ft (5·5 m) of drift is exposed. The matrix is ochre-brown, composed of 35 per cent sand grains of fine to medium grade set in a clayey base. Of the abundant pebbles, 80 per cent are less than ½ in (15 mm) in diameter and are composed mainly of grey and purplish sandstone and felsite with rarer clasts of vein quartz, basic lava, black shale and siltstone. The larger pebbles range up to 3 ft (90 cm) in size, and are subangular to subrounded with rare smoothed and striated surfaces. Most of these are of Walls Sandstone; the remainder are, in order of abundance: felsite, porphyry and quartz-porphyry, mica- and staurolite-schist, amphibolite, calcareous mudstone. In the topmost 4 ft (1·2 m) of the exposure only sandstone, microgranite and vein quartz have been recorded.

The lower part of the section exhibits rudimentary stratification and contains indistinct layers of large pebbles with long axes inclined at 25° to the east, which suggests that the ice movement was to the west.

2. *Small quarry in drift close to east shore of Seli Voe 200* yd (*180* m) *NE of Setter* [294 483]. The quarry is cut in stony drift which fills irregularities in the rock surface and a section of up to 10 ft (3 m) is exposed. The matrix is of grey silty clay in which there are pebbles up to 1 ft 6 in (45 cm) in diameter of sand-stone, dark grey siltstone, and black microfolded shale. There is no indication of stratification and no obvious pebble alignment.

3. *Drift sections in road cutting 800 yd (720 m) N of Hestinsetter* [290 464]. The cutting exposes 4 to 6 ft (1·2–1·8 m) of stony drift resting on granite. The matrix is ochre-brown silty clay and the clasts range up to 3 ft (90 cm) in diameter. All the larger angular blocks are granite and diorite of the Sandsting Complex. Sandstone pebbles, with some ice-polished faces, form up to 20 per cent of the clasts. Pebbles of metamorphic rock, particularly feldspathized hornblende-schist and amphibolite, are scattered throughout.

4. *Quarry at roadside, 550 yd (500 m) ENE of Huxter* [180 573]. This small quarry is excavated in stony blanket drift up to 5 ft (1·5 m) thick which overlies Melby Sandstone. The matrix contains a fairly high proportion of sand grains in a clayey base. The clasts range up to 2 ft (60 cm) in diameter and consist of equal proportions of Walls Sandstone and reddish purple siltstone and shale of the Melby Formation. Rhyolite and rhyolitic breccia of local derivation are also common and there are a number of pebbles of metamorphic rock, particularly hornblende-schist, feldspathized hornblende-schist and vein quartz. All of these types are common in the metamorphic rocks of the Walls Peninsula.

Morainic Mounds and Ridges

1. *North slope of Sandness Hill.* A series of small morainic mounds and sinuous ridges are developed on the lower slopes of Sandness Hill [196 564]. The mounds, which occupy the higher ground around the 250 ft (82 m) contour, are up to 300 yd (275 m) long and 150 yd (135 m) wide and are mostly elongated in a north-westerly direction. The ridges, which appear to be the downstream continuations of these mounds, are sinuous and, in some cases, branching, 10 to 20 yd (9–18 m) wide and up to 250 yd (230 m) long. Both mounds and ridges project less than 10 ft (3 m) above the surface of the surrounding drift and one branching ridge partially encloses a kettle hole. The material forming both mounds and ridges is pale brown unsorted gravel with only a slight clayey admixture. The included pebbles are all angular, up to 15 in (37 cm) in diameter,

and composed entirely of grey and reddish sandstone. It is significant that pebbles of metamorphic rock, which crops out a few hundred yards to the east, are absent and there is thus little doubt that these deposits are the englacial and sub-glacial deposits of a late local glacier occupying the north slope of Sandness Hill, which had a drainage in a north-westward direction (p. 276).

2. *West Slope of Hill of Melby*. On the west slope of the Hill of Melby [174 557] there are two small converging dry channels which are up to 15 ft (4·5 m) deep, with intakes which commence as a number of small depressions (Plate XXXc). Immediately below the exits of these channels there is a system of very small highly sinuous ridges and mounds up to 7 ft (2·1 m) high, composed mainly of sandstone debris. It is thought that this small system of channels and ridges may have been formed by sub- and englacial streams within a small residual glacier. The streams were erosive in their upper part where they formed sub-glacial chutes, but deposited their load as eskers in crevasses within the lower part of their course in the ice.

3. *Papa Stour*. A north-north-east trending belt of morainic mounds extends across the centre of the island of Papa Stour from Sholma Wick [162 618] in the north *via* the ground between Dutch Loch [162 607] and Gorda Water [167 607] to the eastern shore of Hamna Voe [166 604] (Plate XXXв). The moundy belt is up to 500 yd (460 m) wide and can be traced for over a mile (1·6 km) along its length. Individual mounds have a north–south elongation and vary in size from 30 × 50 yd (27 × 45 m) to 100 × 400 yd (91 × 365 m). Most have a slightly steeper slope facing east than west. The mounds are composed of sandy unbedded drift, which contains some large boulders of sandstone, and the whole morainic belt is covered with scattered blocks of sandstone, conglomerate and rhyolite. It seems possible that this belt is a terminal moraine, marking the position of a halt during the eastward recession of the ice or the westward limit of a local re-advance of the ice sheet centred on Mainland, but the form of the mounds is unusual and in some cases reminiscent of drumlins. Much additional work would, however, be necessary to determine if this deposit can be separated from the 'blanket drift' and if there is any justification for its possible correlation with the drift overlying the peat-bearing gravel of Sel Ayre.

INTERGLACIAL DEPOSIT

An exposure of bedded sand, gravel and peat, overlain by boulder clay, occurs at the top of a cliff, approximately 350 ft (107 m) above sea level, at Sel Ayre [177 541]. The sand and gravel, which appears to dip gently to the north-east and of which up to 25 ft (7·6 m) are exposed, fills an east-north-east trending drift-filled valley. The junction between the gravel and the overlying boulder clay is more or less horizontal. The sequence is as follows:

	ft	in	m
10. *Peat*	1	0	0·3
9. Boulder clay, with clayey matrix and angular to subangular pebbles and boulders of Walls Sandstone and rare basic lavas	9	0	2·7
8. Gravel, well-bedded with predominantly angular pebbles set in a silty to clayey matrix	6	0	1·8
7. Sand, pale brown with patchy brown iron staining. Sparse pebbly bands up to 9 in (23 cm) thick, more common at sides of channel	4	6	1·4

	ft	in	m
6. Sand, black to dark brown, limonite-impregnated and peaty	0	1	0·025
5. Sand, pale ochre-brown, pebbly, locally bleached white; patchily ochre-stained at base	1	4	0·4
4. *Peat* with scattered round grains and some sandy lenses .. .. up to 0		1½	0·038
3. Soft pale clayey sand with thin laminae of clay and peat .. .. 9 in to 0		10	0·23–0·26
2. *Peat* with scattered sand grains in lower part. Passes laterally into sand with thin peat bands and thickens to 1 m	1	6	0·45
1. Sand with scattered pebbles, base not seen ..	1	4+	0·4+

It is not known whether the sand and gravel of this exposure is underlain by a lower boulder clay.

The only other recorded peat bed underlying boulder clay in the Shetland Islands occurs at Fugla Ness [312 913] on the north-west coast of North Roe, where peat, up to 5 ft (1·4 m) thick, occurs between two layers of boulder clay, the upper of which is 17½ ft (5·3 m) thick (Chapelhow 1965, pp. 65–8). The detailed pollen analysis of this peat by Birks and Ransom (1969, pp. 777–96) has shown that it was probably formed during the Hoxnian interglacial stage. The Sel Ayre peat deposit differs from that of Fugla Ness in that it underlies a thick bed of sand and gravel. The overlying boulder clay contains pebbles of basic lavas as well as sandstone and must have been derived from a south-easterly direction, and not from the late local ice centre around Sandness Hill which lies to the north-east. This suggests that the Sel Ayre sand, gravel and peat may also be part of an interglacial deposit. Samples of the peat have been examined by Dr. Birks, who has suggested that the samples are interglacial. They contain large amounts of Ericaceae pollen and several oceanic genera like *Osmunda*, *Ilex* and *Polypodium*. There is also pollen of *Picea*, a tree not native in Britain today. The pollen analytical similarities between the Sel Ayre and Fugla Ness deposits are very strong and it is possible that both are of Hoxnian age. However, as there are no other interglacial deposits within a few hundred miles, any correlation must be very tentative.

Radiocarbon dates from material collected from the Fugla Ness and Sel Ayre peats have been provided by Dr. D. D. Harkness of the Scottish Reactor Centre. These are as follows:

SSR 59 Fugla Ness (wood within peat): $40\ 000\ ^{+\ 2000}_{-\ 1600}$ BP

$$\delta^{13}C = -26\cdot4\%.$$

SSR 60 Sel Ayre (peat): $36\ 800\ ^{+\ 1950}_{-\ 1560}$ BP

$$\delta^{13}C = -25\cdot7\%.$$

The date from the Fugla Ness wood is in rough agreement with the dates obtained from this deposit by the Radiological Dating Laboratory, Trondheim (Page 1972, p. 136). These are T—1092 (wood within peat): $34\ 800\ ^{+\ 900}_{-\ 800}$ BP and T—1093 (peat): $37\ 000\ ^{+\ 1200}_{-\ 1100}$ BP.

EVIDENCE FOR DIRECTIONS OF ICE FLOW

The information regarding the directions of ice flow has been obtained from the orientation of glacial striae, roches moutonnées, drumlins and other glacially moulded features (Plate XXXA), as well as from the distribution of those drift pebbles and erratic blocks whose source can be located with reasonable accuracy. Striae are abundant and well preserved in the northern half of the Walls Peninsula, on Vementry Island, Papa Stour, Muckle Roe and North-maven. They indicate that in these areas the ice moved to the north-west, with local swings to north-north-west and, in Northmaven, to the west (Fig. 29). In the southern and western parts of the Walls Peninsula striae are much less abundant. They show that the last ice moved to the south-west in the area extending from Skelda Ness to Gruting Voe and to the west-south-west to south-south-west sector in the ground between Vaila Sound and Wats Ness.

The direction of ice-movement as indicated by striae appears thus to have been not only westward but radially outward from the land area, with ice crossing the present coastline more or less at right angles. The 'ice-shed' between north-west and south-westward moving ice appears however to have lain somewhat south of the centre of the peninsula, more or less along a line extending from Garderhouse in the east via the centre of Gruting Voe and Walls, to Setter in the west. The south-west trending striae could be due to a later or local period of ice movement. The absence of any trace of north-west trending striae in the southern part of the peninsula, however, makes this unlikely.

The west and north-westward direction of ice movement in the greater part of the area is amply confirmed by the distribution of drift pebbles and erratics. Blocks of schist and Walls Sandstone from the Mainland are common on Papa Stour. Clousta conglomerate is abundant on Vementry Island and Braga Ness. Serpentine from the outcrop at Houlland at the eastern margin of the area is common in the drift immediately to the north-west and staurolite-schist from the Scallafield Ridge, located just east of the present area, occurs on the west side of Aith Voe. Pebbles of the porphyritic epidote-bearing granite, which crops out east of the Walls Boundary Fault from Aith southwards, are commonly found throughout the eastern part of the Walls Peninsula.

The evidence within the area under description does not give any definite indication of more than one direction of ice-movement. No crossing striae, or striae showing marked divergence within a limited area, have been recorded. Nor is there any indication of two widely distributed boulder clays overlying each other. In some areas, however, the direction of the most likely source of stones and erratics within the drift does not match with the direction of ice-movement deduced from striae. Thus the drift from quarries near Huxter contains pebbles and boulders of hornblende-schist and metamorphic rocks, which crop out east and north-east of their present position, whereas the direction of striae in the area indicates ice-movement from the south-east. The divergence between ice-movement directions deduced from striae and drift content is, however, nowhere greater than 30° and can be explained by possible local topographically controlled variations in the ice-movement or by possible local differences in the direction of movement of various levels within the ice-sheet. Local re-advances of the ice-sheet, such as that which may have given rise to the possible moraine belt of Papa Stour (p. 272) may also account for such variations in the direction of the ice flow.

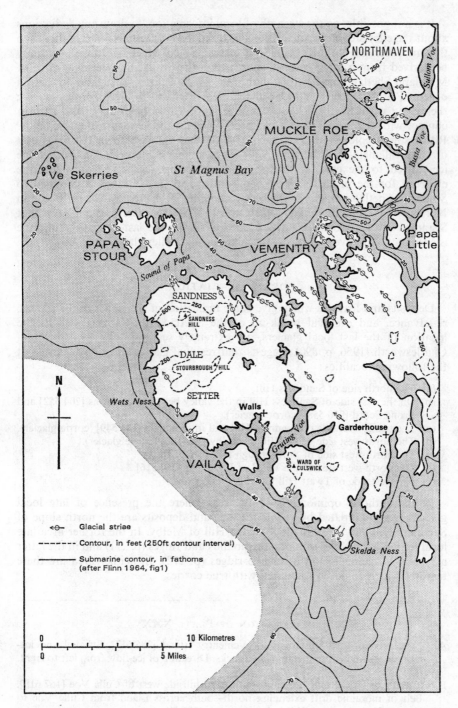

FIG. 29. *Glacial striae in Western Shetland, also topography of land and surrounding sea floor*

T

It is not possible, on the strength of local information, to draw any deductions about the effects of earlier glaciations or directions of ice movement, other than the one observed. No drift pebbles or erratic blocks which could with certainty be derived from a more easterly source than the Scallafield–Weisdale ridge of Mainland Shetland have been found in the Walls Peninsula, suggesting that all observed drift deposits and glacial features in this area were formed during the period when Shetland had its own ice-cap. This ice-cap may have been in contact with the Scandinavian Ice Sheet which lay to the east of the Shetlands. Robertson (Geological Survey records 1935) has shown that in the areas east of the Scallafield–Weisdale ridge all the available evidence indicates that the ice moved to the east, north-east, and north-north-east. The only concrete evidence for the presence of Scandinavian ice in these islands is confined to Southern Shetland where a tönsbergite boulder of a type known to occur in Norway was found (Finlay 1926b, p. 180). It is therefore, unlikely that Scandinavian ice crossed Central Mainland during the last glaciation, and the island may also have had its own ice-cap during earlier glacial maxima.

CORRIE GLACIATION

Deposits and features which give some indication of the stages in the retreat, re-advance, and eventual break-up of the Shetland Ice-Cap, as well as the location of the last local glaciers, are extremely scarce in Western Shetland. Charlesworth (1956, p. 890) suggests that residual glaciers may have existed at the following localities:

(a) The north side of Sandness Hill.
(b) On the east side of Sandness Hill in the valley between the Burns [204 552] and Trona Scord [210 558] (corrie glacier).
(c) The south-west side of Sandness Hill and in Deepdale [184 549] (corrie glacier).
(d) The north-west side of the Hill of Melby [180 560] (corrie glacier).
(e) The north-west side of Stourbrough Hill [213 527].
(f) The north-west side of the Ward of Browland [267 516].
(g) The west side of Twatt Hill [289 519] (uncertain).

In the author's opinion, the only areas where the presence of late local glaciers is substantiated by visible features and deposits are the north slope of Sandness Hill and the west slope of the Hill of Melby. In the former area late local ice has left a series of morainic mounds and ridges (p. 271) and in the latter a system of channels and esker-like ridges (p. 272). In neither area are there features which could be compared with true corries.

EXPLANATION OF PLATE XXX

A. Crooie Geo, north-west coast of Vementry Island [286 617]. Striated and ice-moulded pavement of metamorphic rocks. Direction of ice-flow from left to right. (D 916).
B. Interior of Papa Stour, looking south from hillside west of Culla Voe [167 618]. Belt of morainic drift extending north–south across island from Culla Voe to Hamna Voe. Sandness Hill in background. (D 931).
C. Western slope of Hill of Melby [172 555]. Ramifying system of small meltwater channels. (D 901).

GLACIAL RETREAT PHENOMENA

With the exception of those detailed below Western Shetland is virtually devoid of glacial drainage channels and meltwater deposits which might have formed during the wasting of the last ice-sheet. During the last glacial maximum, the sea level in Shetland may have been 40 to 50 fm (75–90 m) lower than at present (Flinn 1964, p. 337) and as the products and features of ice-wastage are usually concentrated on the lower ground, it can be assumed that in Shetland these are now largely submerged. There are, however, a small number of dry valleys in the area and some of these may have originated as glacial drainage channels.

Examples are (Fig. 28):

1. A 328 yd (300m) long N20°E trending peat-filled channel between Mousavord Loch [225 552] and Djuba Water [226 560] with its present thalweg having a gentle fall to the north.
2. A series of branching channels 1250 yd (1140 m) NNW of the Ward of Browland [268 525].
3. A prominent dry curving valley which passes through the village of Culswick to the Stead of Culswick [274 448]. This is about 1 mile (over 1½ km) long and has a flat alluvium-filled base, locally up to 656 ft (200 m) wide.
4. A roughly horse-shoe shaped dry valley cut in rock on the eastern side of Cow Head on Vementry Island [308 606]. This is over 656 ft (200 m) long and locally up to 115 ft (35 m) wide. It has an undulating floor which contains at least one small lake.

The lack of glaciofluvial deposits, and other late-glacial features in these areas, makes it impossible to use these channels in any attempt to reconstruct a sequence of deglaciation.

PERIGLACIAL AND POST-GLACIAL DEPOSITS

Residual deposits formed by frost action and weathering of the bedrock, such as the block fields (Felsenmeere) on the summit plateaux of Sandness Hill and to a lesser extent on the Ward of Culswick were probably formed during the tundra regime which prevailed immediately after the retreat of the ice.

Deposits which have been formed since the final retreat of the ice include blown sand, lake and stream alluvium, peat, as well as land-slip deposits and scree.

Blown Sand

Wind-blown sand covers an area of about three-quarters of a square mile (2·19 km²) in south-east Papa Stour and a somewhat smaller coastal strip between the Ness of Melby and the Neap of Norby at Sandness. In both areas the sand is relatively thin [probably less than 10 ft (3 m)] and in Papa Stour a number of rocky ridges and hillocks project through the otherwise smooth surface.

Lake and river-alluvium

The core sampling of post-glacial deposits of Shetland lochs by Hoppe and others (1965; pp. 5–6) has shown that the thickness of unconsolidated silts and organic mud in the three sampled Western Shetland lochs is as follows (Fig. 28):

Stanevatstoe Loch [216 545] 3·0 m
Upper Loch of Brouster [262 520] 4·5 m
Lower Loch of Brouster [260 514] 3·5 m

The sediments consist of muds and silts with a fairly high but variable organic content. Radiocarbon measurements of the basal post-glacial lake sediments have been carried out by Engstrand (1967, pp. 413–5) and the following dates were obtained:

Stanevatstoe Loch (Altitude 80 m OD). Core sample from 382 to 389 cm below upper surface of sediment. 9725 ± 265 BP.
Upper Loch of Brouster (Altitude approx. 3 m OD). (a) Core sample from 400 to 404 cm below upper surface of sediment. 8760 ± 250 BP. (b) Core sample from 440 to 450 cm below surface of sediment 9670 ± 540 BP.
Lower Loch of Brouster. Core sample from 345 to 350 cm below upper surface of sediment. Originally dated as 15 080 ± 850 BP, but revised by Ollson and others (1967, p. 460) to 7700 ± 600 BP and 7600 ± 600 BP (two samples).

These dates are as yet unsupported by pollen analyses or other evidence. Hoppe and others (1965, p. 111) came to the provisional conclusion that the date of the earliest lake deposits in Shetland is approximately 10 000 BP or 'a thousand or thousands of years older' and suggested that this may indicate that Shetland became free of ice during the Alleröd interstadial or possibly slightly earlier. The revised age of the Lower Loch of Brouster sample and the re-dating of the C14 ages of other Shetland loch samples may, however, necessitate a re-assessment of this estimate.

Owing to the absence of any sizeable rivers in Western Shetland, fresh-water fluvial alluvium is confined to small areas flanking streams such as the Burn of Mirdesgil [186 577] north of Melby, the Burn of Trona-Scord [204 577], the Burn of Turdale [198 500], the Burn of Setter [205 497] and the Burn of Selivoe [302 489]. The valley at Culswick (p. 277), now virtually without a stream, is also floored by alluvium. No detailed examination of these sediments has been carried out.

Peat

Peat, almost entirely of the blanket bog type, covers nearly 40 per cent of the area of the Walls Peninsula, Vaila and Papa Little (Fig. 28). In the part of Northmaven within the area under description and on Vementry Island it is restricted to a few areas of limited extent. On Papa Stour the thin peaty turf which once covered parts of the island has long since been completely removed for fuel.

Blanket bog covers extensive areas in the west and south-east of the Walls Peninsula. Its thickness does not normally exceed 5 ft (1·5 m), but in hollows and valleys up to 19 ft (5·8 m) of peat have been recorded. The stratigraphy and vegetational history of this peat has been studied by Lewis (1907, pp. 50–4; 1911, pp. 796–803) who established a correlation of the vegetational zones within the peat of Shetland with those in other parts of Scotland, and was able to draw a number of conclusions about the post-glacial climatic and vegetational history of the area. No recent research has been carried out on the pollen chronology of the Shetland peats and it is not yet possible to correlate Lewis's sequence with the Scottish pollen sequence.

Lewis noted that in Western Shetland the peat can be divided into the following two major groups:

1. A lower stratified peat which occurs in certain valleys and hollows and on the plateaux of the higher hills, and
2. an upper unstratified peat, which has a very much greater extent and which in many areas rests directly on weathered boulder clay.

Lewis established the following sequence of vegetational zones within the lower peat:

5. *Second Arctic Bed* with *Salix reticulata* Linnaeus, *S. arbuscula* Linnaeus, *Empetrum nigrum* Linnaeus and *Betula nana* Linnaeus. Locally overlain by a bed of diatomaceous earth.
4. *Lower Peat Bog*, up to 7 ft (2·1 m) thick, with *Sphagnum sp.*, *Scirpus caespitosus* Linnaeus, *Eriophorum vaginatum* Linnaeus, *E. angustifolium* Honckney and *Molinia caerulea* (Linnaeus) Moench.
3. *Forest Bed*, with remains of large trees of *Betula alba* Linnaeus and *Corylus avellana* Linnaeus. Correlated by Lewis with the lower Forest Bed of the Scottish Mainland.
2. *First Arctic Bed*, 1 ft to 1 ft 6 in (30–45 cm) thick, with *Salix reticulata*, *S. herbacea* Linnaeus and *Betula nana*.
1. *Basal peat* with aquatic plants including *Potamogeton pectinatus* Linnaeus, *Menyanthes trifoliata* Linnaeus, *Viola palustris* Linnaeus, *Ranunculus repens* Linnaeus and *Equisetum sp.* Thought to have formed in small marshy pools scattered over the surface of the tundra, and believed to be contemporaneous with the First Arctic Bed.

Lewis recorded remains of the Forest Bed as far west as Simli Field [188 513] and Blouk Field [179 536], close to the western cliffs of the peninsula, and concluded that the climatic regime at that time must have been very different from the present climate with its abundant strong westerly winds.

The upper, unstratified peat consists of the remains of *Sphagnum sp.*, *Calluna sp.*, *Scirpus sp.*, and *Eriophorum sp.*, and Lewis believed that this has formed without interruption to the present day.

At present, peat is rapidly being eroded by wind and water. In areas with an almost complete peat cover, the peat is cut by large numbers of sub-parallel channels up to 10 ft (3 m) deep and 15 ft (4·5 m) wide, which in many cases form branching networks. Lewis has found that the upper peat is less affected by denudation than the lower peat.

The only area of peat in Shetland Mainland which has been surveyed by the Moss Survey of the Scottish Peat Committee with a view to possible commercial exploitation is the 'Kame Bog' in Pettadale, situated between Sand Water and Petta Water, 1¼ miles (2 km) ENE of Aith and outside the area described. The peat of Pettadale occupies a depression and is rather thicker than most of the blanket bog of Western Shetland, but its profile, plant associations, moisture content, degree of humification and other properties are no doubt similar.

The following description of the Kame peat is based on the account on pp. 85–100 of Scottish Peat Surveys, vol. 4 (Department of Agriculture and Fisheries for Scotland, 1968).

'The peat of Kame bog ranges in thickness from 1 m to 6 m. Like all Shetland peats it has an upper section, up to about 15 cm thick, of light brown peat which has a low degree of humification. The greater part of its profile is uniform amorphous black peat with no recognizable structure and with a high degree of humification. Wood has been recorded in the basal ½ to 1½ metres of the peat in a number of sections. The unstratified peat is composed largely of the remains of bog mosses (*Sphagnum spp.*) together with (in order of abundance): cotton grass (*Eriophorum spp.*), heather (*Calluna sp.*), sedges (*Carex spp.*) and, locally, woolly hair moss (*Rhacomitrium lanuginosum* (Hedwig) Bridel). The remains of the latter are generally present in the upper 1·6 m of the peat only. They are relatively unhumified and are not a significant peat former elsewhere in Scotland. The species of wood at the base of the Pettadale Moss were not determined. The degree of humification in the bog was found to vary from H_2 to close on H_{10}, the average for the whole bog being H_7. The lower values occurred near the surface and, occasionally, at the bottom of the peat. The average moisture content was estimated to be 853 g of water per 100 g solids, which is drier than might be expected in the prevalent climatic conditions and may be due to the good natural drainage of the bog. The average ash content of the bog was found to be 3·78 per cent, increasing slightly towards the base. In one borehole the peat was found to be underlain by 2 m of diatomite, some of which was also present in the basal 1 m of the peat.'

Diatomite has been found elsewhere in small pockets below peat but none has so far been recorded below the peats of Western Shetland.

POST-GLACIAL CHANGES IN SEA LEVEL

Finlay (1930, pp. 672–3) referred to the continuous slope of the sea-floor to the fifty fathom contour, the lack of a submerged shore platform and the presence of peat bogs flooring the voes and shallower straits, all of which are evidence for a continuous submergence of the land in post-glacial times. He also quoted local traditions which indicate that submergence has continued during historic times. Thus the sound between Muckle Roe and Mainland, now spanned by a bridge 100 yd (90 m) long, was said to be at one time fordable at low water, and a crofter's steading on its shores is now covered by shingle.

Flinn (1964, pp. 321–39) has suggested that the Shetland Islands are the summits and upper slopes of a range of hills rising from the gently undulating plain of the North Sea floor, some 67 fm (122 m) below sea level. The present coastline of Shetland has many features characteristic of recent submergence, the most obvious being the ria-like voes which are drowned drainage basins. The abundance of spits and bars, known locally as 'ayres', in many voes (where some have cut off the head of the voe and formed a fresh-water lake) as well as between islands, is a feature of both emerging and submerging coastlines.

A characteristic feature of the sheltered shores of the voes is the presence of submerged peat, in many instances covered by a thin layer of shingle or sand. The best examples in Western Shetland occur in the bays on either side of the headland of Mara Ness, on the west shore of Gruting Voe [268 495]. This peat is exposed between the high and low water marks and is partially covered by shingle.

Submerged peat has been recorded during excavations in other parts of Shetland at Lerwick Harbour (Finlay 1930, p. 673), Scalloway, Bressay, Graven (Sullom Voe) and Symbister in Whalsay (Hoppe 1965, pp. 195–7). At Lerwick Harbour peat with tree stumps occurs at a depth of 21 ft (6·4 m) below high water mark and in Symbister Harbour peat was found between 28 and 29 ft (8·6 and 8·9 m) below high water mark. Five radiocarbon dates obtained from the peat samples from Symbister Harbour (Hoppe 1965, p. 201) range from 5455 BP to 6970 BP. Hoppe considers that the Symbister peat was formed *in situ* and that some 5500 years ago sea level in Whalsay was *at least* 29 ft (9 m) lower than at present.

In an evaluation of published data relating to the world-wide post-glacial eustatic rise in sea level, Hoppe concludes that at 6000 BP the average sea level throughout the world was probably about 16½ to 23 ft (5–7 m) lower than at present. The considerably lower sea level 5500 years ago in Shetland thus indicates that there has been either,

(a) a eustatic rise in sea level as great as the highest recorded elsewhere, or (b) more likely, an isostatic contribution to the submergence.

The concept of the continued isostatic sinking of Shetland after 6000 BP is also favoured by Flinn (1964, p. 338). He states that the isostatic depression of Scotland and Scandinavia during the glacial maxima was probably accompanied by peripheral uplift which affected the Shetland area, as this lay between the two ice-sheets, and that part of the submergence observed in Shetland is probably due to recovery from this peripheral uplift.

Flinn (1964, fig. 1) has produced a contour map of the sea-floor around Shetland, which is based on Admiralty charts and Admiralty 'fair charts' which cover areas of limited extent and show about 100 soundings per square mile, supplemented by a number of echo traces. He has recognized the following major topographic features in this sea area (Fig. 29):

1. A 45 fm (82 m) shelf which forms an extensive platform south-west of the Walls Peninsula and includes the sea-floor around Foula. It also forms a ledge extending from Papa Stour and the Ve Skerries northwards to Esha Ness, which cuts off the deep St Magnus Bay basin.
2. A 25 fm (45 m) shelf, which forms several shoals on the 45 fm (82 m) shelf and includes the bank extending from the Ve Skerries to Foula and the arcuate shelf extending for 8 miles (13 km) eastward from the latter island.
3. A 13 fm (24 m) shelf, which is generally narrow and not always recognizable on the 'fair charts'.

Though there is also a possibility that near some semi-sheltered coasts a shelf exists at 5 fm (9 m), it is probable that on most coasts the cliffs plunge at a reduced angle, but without an intervening shelf, to about 13 fm (24 m).

Flinn believes that these shelves are not drowned wave-cut platforms, as they do not have smooth upper surfaces. He concludes (1964, p. 332) that there is probably only a single surface of marine erosion formed during a lengthy period of continuously rising sea level and that the shelves may mark the existence of wave-cut erosion surfaces that were already present before the last submergence of the land took place. These shelves would have been modified by sub-aerial erosion and glaciation. Flinn tentatively suggests that platforms of this type occur also at 200 to 250 ft (60–76 m) and 500 to 550 ft (152–167 m) above the present sea level.

On the west side of the Shetland sea area, the 45 fm (82 m) shelf is bounded by a gentle slope whose margin extends along a straight line from a point 3 miles (5 km) W of Esha Ness to a position 7 miles (11 km) NW of Foula. This slope descends gently to the 80 fm (150 m) contour and then more steeply to 120 fm (220 m). Between Papa Stour and Esha Ness the 45 fm shelf is bounded on the east by a steep drop into the St Magnus Bay basin which has a maximum depth of 87 fm (159 m). This basin has steep sides between 50 and 70 fm but a more gently sloping floor below that depth. Flinn originally thought that, like the 'deep' between Whalsay and Yell, the St Magnus Bay 'deep' is the result of glacial overdeepening, but he has more recently (Flinn 1970, p. 131) considered it to be a possible meteor impact crater. Flinn also recorded the presence of a number of possibly glacially overdeepened troughs just east of this basin. The Swarbacks Minn trough between Vementry and Muckle Roe is up to 50 fm (90 m) deep. Another depression in the sea-floor commences 5 miles (8 km) S of Skelda Ness (the southern tip of the Walls Peninsula). This is a N10°E trending trough, which is up to 70 fm (128 m) deep and marks the probable southward continuation of the Walls Boundary Fault along the sea-floor.

FOULA

The island of Foula has considerably greater relief than West Mainland and was probably never completely overridden by the ice from the east (Finlay 1926a, p. 564). It contains two high 'hanging' corries on the northern face of the ridge extending from the Kame *via* the Sneug to Hamnafield, and a lower corrie forming the Netherfandal just east of Wilsie. The west-north-west trending depression between Hamnafield and the Noup, known as the Daal, may be another glacially deepened valley whose corrie-head, possibly originally sited just south-west of the present shore, has been removed by the rapid retreat of the sea cliffs due to marine erosion. These corries were the sites of local corrie-glaciers which flowed first in a general eastward direction and were then diverted to north-north-west or south-south-east, probably by ice approaching Foula from the east as indicated by erratics (p. 283).

DRIFT DEPOSITS AND GLACIAL ERRATICS

Sandy boulder clay forms a cover of very variable thickness along the northern and eastern coastal platforms of the island. The drift blanket extends inland at Ham and along the low ground around Hametun into the floor of the Daal.

Excellent sections in the drift are seen along the north shore of the island between Ness and Logat Head [959 415], on the east shore just south of Baa Head [976 385] and on the north shore of Ham Voe [975 388]. In the first two localities the drift rests on a rock surface with well developed striae trending W5°N to W10°N.

The section on the north shore is as follows:

	ft	m
Fine silt, thinly bedded, dark brown, with small pebbles, probably water-deposited	4	1·6
Sandy boulder clay, reddish brown, stony	25	7·6
On striated pavement of sandstone.		

In this section, the larger boulders are almost entirely of grey and brownish sandstone. Among the cobbles up to 1 ft (30 cm) in diameter, metamorphic rocks are common. These include mica-schist, hornblende-schist, granulitized gneiss, pegmatite and microgranite. There are also small well-rounded quartz pebbles (some red and amber-coloured) derived from a pebbly sandstone. Rare pebbles of epidote-granite are present. The pebbles of metamorphic rock occur only in the lower 15 ft (4·6 m) of the section. Many deep scratches in the striated pavement taper out eastwards.

On the south-east shore of Baa Head 10 ft (3 m) of tough reddish brown sandy boulder clay are exposed. This boulder clay rests on metamorphic rocks, but in contrast to the drift seen on the north shore, its most abundant pebbles are grey and purplish sandstone. There are also numerous small rounded pebbles of orange and amber-yellow quartzite, black chert and pink microgranite derived from the local pebbly sandstone (p. 173). Pebbles of metamorphic and igneous rock, also present throughout the section, include hornblende-schist, pegmatitic granite, microgranite and greenish porphyritic epidotic granite.

Erratic blocks are abundant along the eastern and south-eastern coastal stretch of Foula. The most conspicuous are of epidotic granite, which bears a close resemblance to the Spiggie Granite which crops out just east of the Walls Boundary Fault on Shetland Mainland and forms a series of islands extending from Sandsound Voe southwards along the west coast of Shetland to Spiggie. The quartz and chert pebbles derived from the pebbly bands in the Foula sandstone are abundant on the surface of the drift, particularly in the Hametun area.

All drift deposits of Foula contain both metamorphic rocks and sandstone. As the distribution of rock outcrops on the sea floor around Foula is not known in any detail, no deduction as to direction of ice-movement can justifiably be made from the pebble content. Finlay (1926a, p. 564) suggested that the pebbles and erratics of porphyritic epidotic granite of the type found at Bixter Voe (i.e. Spiggie Granite) indicate that ice from Shetland Mainland reached Foula. The abundance of epidotic granite within the drift, however, makes it more likely that this rock was derived from an outcrop, now submerged, which is very much closer to Foula.

CORRIE GLACIATION

The three corries in north-west Foula have more characteristic corrie shapes than any others in the Shetland Islands and were without doubt formed by local ice. The Overfandal Corrie has an arcuate terminal moraine which dams the Overfandal Loch on its hanging eastern side. A shallow north-north-west trending depression extends from Logat Head on the north coast to the shore at Hametun in the south-east and is bounded by truncated spurs and hanging corries on the west and by a series of low north-north-west trending ridges on the east. This, together with the distribution of erratics, suggests that the ice from the corries was deflected both north-westwards and southwards probably by the ice which occupied the sea area to the east of Foula.

ALLUVIAL DEPOSITS

No alluvial sand and gravel is present on Foula, but on the low ground the boulder clay is locally overlain by a thin deposit of bedded silt, clay and sand (p. 284) which appears to be of fresh-water origin. A patch of fresh-water alluvium of considerable size is present in Hametun.

PEAT

Peat covers most of the lower part of the island and a very thin peat cover is present at altitudes of up to 1200 ft (366 m). Lewis (1911, pp. 805–6) examined sections in the neighbourhood of the Flick Lochs [950 402], at Bark Hill (Overfandal) [957 398] and on the lower ground in the eastern part of the island. He found that the remains of birch are present in nearly all sections where the peat is more than 8 ft (2·4 m) deep. The general sequence in the peat is the same as that found in Shetland Mainland (pp. 278–9), but in a section near the Flick Lochs Lewis recorded a distinct layer containing *Juniperus commumis* in the peat between the Second Arctic Bed and the more recent peat above. This Juniperbearing horizon is said by Lewis to occupy the same relative position as the Upper Forest (*Pinus sylvestris*) bed in the south of Scotland, and Lewis suggests that the presence of the remains of the Upper Forest bed only in Foula within the Shetland Group may possibly be attributed to the greater height of the hills in the island.

Between Ham and the Mill Loch the peat is locally underlain by a thin deposit of diatomaceous clay.

REFERENCES

Birks, H. J. B. and Ransom, Maree E. 1969. Interglacial Peat at Fugla Ness, Shetland. *New Phytol.*, **68**, 777–96.

Chapelhow, R. 1965. On glaciation in North Roe, Shetland. *Geogrl Jnl*, **131**, 60–70.

Charlesworth, J. K. 1956. The Late-glacial history of the Highlands and Islands of Scotland. *Trans. R. Soc. Edinb.*, **62**, 769–928.

Department of Agriculture and Fisheries for Scotland. 1968. *Scottish Peat Surveys, Vol. 4, Caithness, Shetland and Orkney.*

Engstrand, L. G. 1967. Stockholm Natural Radiocarbon Measurements VII. *Radiocarbon*, **9**, 387–438.

Finlay, T. M. 1926a. The Old Red Sandstone of Shetland. Part I: South-eastern Area. *Trans. R. Soc. Edinb.*, **54**, 553–72.

—— 1926b. A Töngsbergite Boulder from the Boulder-clay of Shetland. *Trans. Edinb. geol. Soc.*, **12**, 180.

—— 1930. The Old Red Sandstone of Shetland. Part II: North-western Area. *Trans. R. Soc. Edinb.*, **56**, 671–694.

Flinn, D. 1964. Coastal and Submarine Features Around the Shetland Islands. *Proc. Geol. Ass.*, **75**, 321–39.

—— 1970. Two possible submarine meteorite craters in Shetland. *Proc. geol. Soc. Lond.*, **1663**, 131–5.

Hoppe, G. 1965. Submarine peat in the Shetland Islands. *Geogr. Annlr.*, **47A**, 195–203.

—— Schytt, W. and Strömberg, B. 1965. Från Fält och Forskning Naturgeografi vid Stockholms Universitet. *Särtryck ur Ymer*, H. 3–4, 109–125.

Lewis, F. J. 1907. The Plant Remains in the Scottish Peat Mosses. III. The Scottish Highlands and the Shetland Islands. *Trans. R. Soc. Edinb.*, **46**, 33–70.

—— 1911. The Plant Remains in the Scottish Peat Mosses. IV. The Scottish Highlands and Shetland, with an Appendix on Icelandic Peat Deposits. *Trans. R. Soc. Edinb.*, **47**, 793–833.

Olsson, Ingrid U., Stenberg, A. and Göksu, Y. 1967. Uppsala Natural Radiocarbon Measurements VI. *Radiocarbon*, **9**, 454–70.

Page, N. R. 1972. On the age of the Hoxnian interglacial. *Geol. Jnl*, **8**, 129–42.

Peach, B. N. and Horne, J. 1879. The Glaciation of the Shetland Isles. *Q. Jnl geol. Soc. Lond.*, **35**, 778–811.

Chapter 19

ECONOMIC GEOLOGY

IN THE AREA covered by the Western Shetland One-inch Geological Sheet the only mineral which has been exploited commercially for export is the magnetite from Clothister Hill, near Sullom. Bulk materials quarried and removed in the area during the last 20 years include schists of the Green Beds Assemblage, granite, diorite, sandstone and boulder clay, which have been used locally for road construction, and shore gravel and sand used for the manufacture of concrete blocks and other building purposes. Peat is cut by the crofters as a source of fuel, but is not exploited commercially on a large scale. Though outcrops of limestone are present in the area they are too small and scattered to be quarried as a source of agricultural lime or cement.

CLOTHISTER HILL MAGNETITE

The orebody at Clothister Hill, Sullom [342 729] (Fig. 30) consists of massive magnetite of a high degree of purity and with an exceptionally low phosphorus content. It is situated in the north-east corner of the area, 850 yd (770 m) due north of Mavis Grind. It was discovered in 1933 by Mr. D. Haldane of the Geological Survey of Great Britain (Summ. Prog. 1934, p. 71). Between 1941 and 1943 it was investigated in detail and developed with a view to exploitation by the Scottish Home Department (Groves 1952, pp. 263–95). This development work included the driving of a shaft, a longitudinal mine, an adit mine, as well as several cross-cut mines. At the same time a magnetometer survey covering a considerable area around the mine, and a drilling programme, involving the sinking of 13 bores, were carried out on behalf of the Home Ore Department of the Ministry of Supply.

The exploration programme commenced with the extensive trenching of the known exposures of the orebody in order to determine its exact outcrop and to obtain channel and bulk samples for analysis. As the ore was found to contain 60 to 67 per cent iron and less than 0·006 per cent phosphorus it was decided to sink a shaft at the widest part of the orebody to prove its extension at depth. Cross cuts were driven at depths of 32 and 72 ft (9·75 and 22 m). A horizontal shaft with cross cuts leading from it was then driven along the length of the ore-body at the 72 ft (22 m) level. The exploratory work which was supplemented by a magnetometer survey and by three inclined boreholes was completed in July 1942, and, as an estimated 12 000 to 20 000 tons of ore were thought to be available, an adit was driven with a view to commencing commercial exploita-tion. The project was abandoned at that stage because of the lack of skilled labour in Shetland, and because the available coastal shipping to transport the ore could not be guaranteed.

The exploratory development has provided precise data regarding to the size, shape, quantity, quality and geological environment of the orebody (Groves 1952, pp. 285–82). These are summarized below:

Form of Orebody. The orebody is elongated in a north–south direction and is lenticular in horizontal section as well as in E–W vertical section. Its outcrop

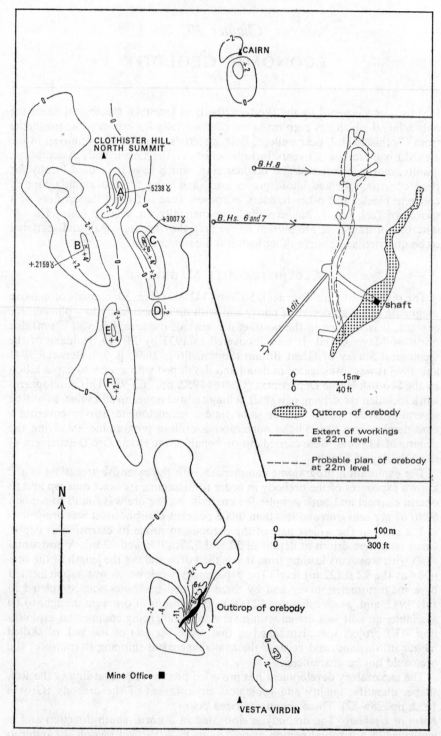

FIG. 30. *Magnetic anomaly map of the area around and north of Clothister Hill magnetite mine. Contour interval 2 gammas except in areas of high magnetic gradient. Inset: Plan of outcrop of orebody at surface and position of orebody at 22m level. Also positions of three inclined bores sunk to prove orebody at depth.*

trends north-north-east, has a length of 174 ft (53 m) and is slightly sinuous, several of the bends being apparently due to small faults. Its width varies from a few feet to a maximum of 22 ft (6·7 m). The orebody has a westerly inclination, which averages 51° at its northern end and 85° at its southern end. In consequence of this 'twist' it trends exactly north–south at the 72 ft (22 m) level, where it is 169 ft (51 m) long. Its maximum width at the 72 ft (22 m) level is 21 ft (6·4 m) and its average width is 11 ft (3·35 m). Groves believes that the orebody plunges northwards in a series of steps which may in part be due to the presence of a number of small faults.

Three inclined boreholes (Fig. 30) were put down to prove the extension of the known orebody in depth. One borehole (No. 6) intersected the projected position of the orebody at about 67 ft (20·4 m) inclined distance below the 72 ft (22 m) level but no ore was encountered. A second borehole (No. 7) was directed to intersect the orebody at an inclined depth of 13 ft (3·9 m) below the 72 ft (22 m) level and encountered only patches of ore which did not exceed 5 ft 6 in (1·67 m) in total width. These boreholes suggest that below the 72 ft (22 m) level the orebody is dying out rapidly downwards. The third borehole (No. 8) intersected the northern part of the orebody some 15 ft (4·5 m) below the 72 ft (22 m) level and proved two bands of compact magnetite respectively only 2 ft 6 in and 1 ft 3 in (0·76 and 0·38 m) thick.

Estimated tonnage of Orebody. It was estimated that the orebody has an average width of 10 ft (3 m) at the surface and 11 ft (3·35 m) at the 72 ft (22 m) level. Groves calculated the average height of the orebody down to the 72 ft (22 m) level to be 85 ft (26 m), its average length 171 ft (51 m) and its average width 10 ft (3 m). This gives an estimated volume of 144 500 cub. ft (4000 m³) of ore. By assuming that 8 cub. ft of the Clothister Hill ore produce one ton of magnetite, and allowing 32·8 per cent by weight for dilution by skarn, Groves estimated that the proved ore reserve above the 72 ft (22 m) level was 18 000 tons. There are also a probable 2000 tons below this level, making a total of proved and probable ore of 20 000 tons.

Quality of ore. Analyses of samples of the Clothister Hill ore indicate that its iron content varies from 60 to 67 per cent. The sulphur content, which has been lowered by weathering at the outcrop, rises on the average to about 0·5 per cent at depth. It occurs as pyrites and is very unequally distributed. Except at the extreme northern and southern ends of the orebody, where it rises slightly but is still within the limits for hematite iron, the phosphorus content is uniformly 0·006 per cent.

Post-war Development and Exploitation. The magnetite was mined between 1954 and 1957 by Deering Shetland Mining Limited, a subsidiary of Deering Products Limited. This firm supplied the ore to the National Coal Board for use in the manufacture of heavy mud used in coal flotation. The ore was extracted both by mining through the existing adit and by opencast methods. It is believed that between 6000 and 10 000 tons of ore were obtained and that the rate of extraction during 1955 amounted to 300 to 400 tons of crushed rock per month. An appraisal made in 1956 suggested that the total quantity of ore remaining and extracted is higher than the amount suggested by the wartime investigations.

Possible additional orebodies on Clothister Hill. During 1941 a magnetometer survey was carried out over the Clothister Hill magnetite body, and this survey was subsequently extended further northwards where it proved a number of magnetic anomalies similar to that produced by the known orebody. In all, four anomalies were found at distances 800 to 11 000 ft (0·24–3·4 km) N of the known ore mass. Two of these are positive and two negative (Fig. 30). All the anomalies are on the ridge of Clothister Hill and several of them, including the strongest, appear to be in alignment with the elongation of the outcrop of the proved orebody. In 1942 several of the anomalies were tested by diamond drilling. Altogether nine vertical and inclined boreholes were put down in positions calculated to intersect the supposed magnetite orebodies responsible for the anomalies C, A and B. Although zones of calcite and intermittent epidote veining were encountered, no magnetite ore was found.

The systematic magnetometer survey was continued northwards to the Loch of Kirt Shun [343 738] and southwards to the north end of the Loch of Lunnister [345 718]. Though two small magnetic anomalies were found, the ground had no major anomalies. Similarly, scattered traverses and isolated observations made between the south end of the Loch of Lunnister [343 711] and the Bight of Haggrister [348 701] gave no suggestion of the presence of possible large anomalies.

Road Metal

Though most of the rock used at present for surfacing on the roads of Western Shetland is obtained from Scord Quarry [412 400] situated on the west slope of Wind Hamars, ¾ mile (1·2 km) NE of Scalloway, bulk material used in road reconstruction is derived locally from roadside quarries scattered throughout the area. The quarry on the east slope of Clothister Hill [343 734] within the Green Beds Assemblage has supplied road metal for local needs. The granite and the diorite of the Northmaven and Sandsting complexes as well as the Walls Sandstone have been extensively used. Boulder clay obtained from the quarry [272 157] near Murraster, ¾ mile (1·2 km) NE of Bridge of Walls, and from many smaller quarries is used as fill material.

Building Materials

Owing to the lack of raw material for brick making in the Shetland Islands the most commonly used building material at the present day is light concrete blocks. These are made locally from either crushed schist or from beach gravel. In recent years the gravel from the beach of the Ness of Little-Ayre [318 629] on the south shore of Muckle Roe has been used for the manufacture of such blocks.

Sandy beaches are rare on the mainland of Western Shetland, and most do not contain sufficient sand to permit its removal for building purposes. Small quantities of sand have been obtained from the beach at The Crook [197 577], north of Norby.

Limestone

The bands of crystalline limestone within the metamorphic rocks of Western Shetland (pp. 44–45, Plate II) have never been worked either as a source of agricultural lime or as a raw material for cement. This is probably due to the fact that limestone is more accessible and available in much greater quantities in Eastern and Central Shetland. The bulk of the agricultural lime at present used in the Shetland Islands is obtained from a quarry in the Girlsta Limestone at Girlsta [429 505] 6¼ miles (10 km) NNW of Lerwick. Several of the limestones in the area between Norby and Bousta reach a considerable thickness and crop out close to the existing road. These could provide a suitable local source of limestone.

PEAT

The deposits of as yet unworked and relatively easily accessible peat in Western Shetland are of considerable extent (pp. 278–280, Fig. 28). Peat is cut extensively by crofters as a domestic fuel, and in the area [308 525] a short distance west of Effirth, close to the head of Bixter Voe, it is cut on a small scale by the conventional methods for sale in Lerwick. In south-eastern Shetland peat briquettes were manufactured for a short time in a small factory on the shores of the Loch of Brindister [431 368], 3½ miles (5·5 km) SW of Lerwick. This venture, however, did not prove financially viable.

ORNAMENTAL AND SEMI-PRECIOUS STONES

Western Shetland is less well endowed with the raw materials for stones which can be used for lapidary and ornamental work than other parts of the Shetland Islands.

Agate. Agates up to 4 in (10 cm) in diameter are present in the vesicular tops of the basalt flows exposed on the south shore of Papa Stour between Hirdie Geo and Aesha Head [149 608] and between Scarvi Taing [175 593] and Kirk Sand [177 598]. These agates, however, commonly have a central core of barytes, which makes them unsuitable as gem stones. Many are cut by joints and break up easily.

Serpentine. A small outcrop of green and reddish mottled serpentine occurs close to the north-west corner of Maa Loch on the island of Vementry [297 604]. This serpentine has attractive green and deep red colour mottling and could prove an excellent material for making polished brooches and ornaments.

Scapolite. Scapolite forming a vein, up to 8 ft (2·4 m) thick, exposed in the south-east shore of Shelda Ness [303 405], 490 yd (440 m) NE of the southern tip of the peninsula, has an attractive silky sub-translucent lustre and ranges in colour from white to pale pink or reddish. It takes on a high polish, but is rather closely jointed and tends to break down into fairly small pieces. Other scapolite veins containing material suitable for polishing crop out on the east shore of Wester Wick [286 423].

REFERENCES

GROVES, A. W. 1952. Wartime Investigations into the Haematite and Manganese Ore Resources of Great Britain and Northern Ireland. *Ministry of Supply, Permanent Records of Research and Development.*

SUMM. PROG. 1934. *Mem. geol. Surv. Summ. Prog. for* 1933.

Appendix

LIST OF GEOLOGICAL SURVEY
PHOTOGRAPHS

(ONE-INCH GEOLOGICAL SHEET WESTERN SHETLAND)

Taken by Messrs. W. D. Fisher and A. Christie

Copies of these photographs are deposited for public reference in the library of the Institute of Geological Sciences, South Kensington, London SW7 2DE, and in the library of the Scottish office, West Mains Road, Edinburgh 9. Prints and lantern slides are supplied at a fixed tariff on application to the Director. All negatives are 5 in by 4 in. Colour slides of many of the listed photographs are also available.

GENERAL

Views northwards across St Magnus Bay from the south shore of of West Burra Firth	D 881, 882
View westward along the north shore of the Walls Peninsula to Papa Stour from the east end of West Burra Firth	D 883
View northwards from the Walls Peninsula across Vementry Island to the granophyre cliffs of Muckle Roe	D 903
General view of north coast of Vementry Island	D 907
View from Vaila west-south-westwards to Foula	D 932
View from Vaila northward across Vaila Sound towards Walls	D 933, 940
Coastal scenery along south shore of Vaila	D 936, 938
View looking north-eastward from Vaila across Green Head towards the hills of central Mainland	D 942
Views of the rugged metamorphic terrain along the north coast of the Walls Peninsula	D 876, 908, 950, 951, 987
Coastal scenery near Wester Wick, south end of Walls Peninsula	D 973
Views looking across Clousta Voe, showing ridges formed by the differential erosion of the Clousta Volcanic Rocks	D 992, 993

RECENT AND PLEISTOCENE

Possible meltwater channels, west slope of Hill of Melby	D 901
Glaciated pavement; north-west coast of Vementry Island	D 916
Belt of morainic drift, Papa Stour	D 931

MELBY FORMATION

Melby Fish Beds	D 887, 892, 896
Cross-bedded and convoluted sandstones in Melby Formation	D 888–891, 893–895
Brecciated rhyolite resting on sandstone of Melby Formation	D 885, 886

VOLCANIC ROCKS OF PAPA STOUR

WALLS FORMATION

SANDNESS FORMATION

MINOR INTRUSIONS

PLUTONIC COMPLEXES

U

Regular, closely spaced jointing in Sandsting Granite; coast at
    Culswick   D 975
Scapolite veins in granite-diorite complex; quarry at Mavis Grind   D 1344
Net-vein complex of basic diorite and granite, with hybridization
    and local flow-foliation; coast at Wilson's Noup, Northmaven   D 1345–1348

METAMORPHIC ROCKS

Extensive coastal exposures of gneiss with pegmatite and granite
    veining; north-west coast of Vementry   D 910, 912, 915
Early tight folds (D2) in limestone, hornblende-schist and quartz
    granulite; north coast of Walls Peninsula   D 952, 953, 983,
                                                                       985
Early linear structures (D2) in mica-schist and quartz-granulite;
    north coast of Walls Peninsula   D 977–979
Limestone and calc-silicate rock cut by complex of granite veins;
    north coast of Walls Peninsula   D 988
Interbanded gneiss and amphibolite with tight folds (D2), cut by
    small normal faults; west coast of Vementry   D 913
Tight conjugate folds (D4) in gneiss with granite and pegmatite
    veins; west coast of Vementry   D 911
Conjugate folds and kink bands in platy gneisses along north
    shore of Walls Peninsula   D 943, 946, 947,
                                                                       949, 981, 982

Amphibolite intensely net-veined by granite (agmatite); north
    coast of Walls Peninsula   D 880

INDEX

INDEX

Printed in Scotland by Her Majesty's Stationery Office at HMSO Press, Edinburgh
Dd 020030/3407 K16 11/75 (11907)